VOLUME ONE HUNDRED AND TWENTY THREE

PROGRESS IN
MOLECULAR BIOLOGY AND TRANSLATIONAL SCIENCE

Computational Neuroscience

VOLUME ONE HUNDRED AND TWENTY THREE

PROGRESS IN
MOLECULAR BIOLOGY AND TRANSLATIONAL SCIENCE

Computational Neuroscience

Edited by

KIM T. BLACKWELL

George Mason University,
Krasnow Institute for Advanced Study,
Fairfax, VA, USA

AMSTERDAM • BOSTON • HEIDELBERG • LONDON
NEW YORK • OXFORD • PARIS • SAN DIEGO
SAN FRANCISCO • SINGAPORE • SYDNEY • TOKYO
Academic Press is an imprint of Elsevier

Academic Press is an imprint of Elsevier
225 Wyman Street, Waltham, MA 02451, USA
525 B Street, Suite 1800, San Diego, CA 92101-4495, USA
Radarweg 29, PO Box 211, 1000 AE Amsterdam, The Netherlands
The Boulevard, Langford Lane, Kidlington, Oxford, OX5 1GB, UK
32 Jamestown Road, London NW1 7BY, UK

First edition 2014

Library of Congress Cataloging-in-Publication Data
A catalog record for this book is available from the Library of Congress

British Library Cataloguing in Publication Data
A catalogue record for this book is available from the British Library

ISBN: 978-0-12-397897-4
ISSN: 1877-1173

For information on all Academic Press publications
visit our website at store.elsevier.com

Working together
to grow libraries in
developing countries

www.elsevier.com • www.bookaid.org

CONTENTS

CONTRIBUTORS

Upinder S. Bhalla
National Centre for Biological Sciences, Bangalore, Karnataka, India

Kim T. Blackwell
Krasnow Institute for Advanced Study, George Mason University, Fairfax, VA, USA

Carmen C. Canavier
Department of Cell Biology and Anatomy, Neuroscience Center of Excellence, LSU Health Sciences Center, New Orleans, Louisiana, USA

Eamonn Dickson
Department of Physiology and Biophysics, University of Washington, Seattle, Washington, USA

Olivia Eriksson
Department of Numerical Analysis and Computer Science, Stockholm University, Stockholm, Sweden

Bjoern Falkenburger
Department of Neurology, RWTH Aachen University, Aachen, Germany

Erik Fransén
Department of Computational Biology, School of Computer Science and Communication, KTH Royal Institute of Technology, Stockholm, Sweden

Omar Gutierrez-Arenas
School of Computer Science and Communication, Royal Institute of Technology, Stockholm, Sweden

Bertil Hille
Department of Physiology and Biophysics, University of Washington, Seattle, Washington, USA

Shin Ishii
Graduate School of Informatics, Kyoto University, Uji, Kyoto, Japan

Ravi Iyengar
Department of Pharmacology and Systems Therapeutics, and Systems Biology Center New York, Mount Sinai School of Medicine, New York, USA

M. Saleet Jafri
School of Systems Biology and Department of Molecular Neuroscience, George Mason University, Fairfax, VA, USA

Tuula O. Jalonen
Department of Physiology and Neuroscience, St. George's University, School of Medicine, Grenada, West Indies

Alexandra Jauhiainen
Department of Medical Epidemiology and Biostatistics, Karolinska Institutet, Stockholm, Sweden

Peter Jung
Department of Physics and Astronomy and Quantitative Biology Institute, Ohio University, Athens, Ohio, USA

Alon Korngreen
The Mina and Everard Goodman Faculty of Life Sciences, and The Gonda Brain Center, Bar-Ilan University, Ramat-Gan, Israel

Jeanette H. Kotaleski
School of Computer Science and Communication, Royal Institute of Technology, and Department of Neuroscience, Karolinska Institutet, Stockholm, Sweden

Martin Kruse
Department of Physiology and Biophysics, University of Washington, Seattle, Washington, USA

Rashmi Kumar
School of Systems Biology, George Mason University, Fairfax, VA, USA

Angelika Lampert
Institute of Physiology and Pathophysiology, Friedrich-Alexander Universität Erlangen-Nuremberg, Erlangen, and Institute of Physiology, RWTH Aachen University, Aachen, Germany

Edwin S. Levitan
Department of Pharmacology and Chemical Biology, University of Pittsburgh, Pittsburgh, Pennsylvania, USA

Marja-Leena Linne
Computational Neuroscience Group, Department of Signal Processing, Tampere University of Technology, Tampere, Finland

Peter Lipton
Department of Neuroscience, University of Wisconsin, Madison, Wisconsin, USA

William W. Lytton
Department of Physiology & Pharmacology, SUNY Downstate, New York, USA

Khaled Machaca
Department of Physiology and Biophysics, Weill Cornell Medical College in Qatar, Doha, Qatar

Toma M. Marinov
UTSA Neurosciences Institute, University of Texas at San Antonio, San Antonio, Texas, USA

Ashutosh Mohan
Department of Physiology & Pharmacology, SUNY Downstate, New York, USA

Anu G. Nair
School of Computer Science and Communication, Royal Institute of Technology, Stockholm, Sweden

Honda Naoki
Imaging Platform for Spatio-Temporal Information, Graduate School of Medicine, Kyoto University, Kyoto, Kyoto, Japan

Samuel A. Neymotin
Department of Physiology & Pharmacology, SUNY Downstate, New York, USA

Padmini Rangamani
Department of Molecular and Cell Biology, University of California, Berkeley, California, and Department of Pharmacology and Systems Therapeutics, Mount Sinai School of Medicine, New York, USA

Jason S. Rothman
Department of Neuroscience, Physiology & Pharmacology, University College London, London, UK

Fidel Santamaria
UTSA Neurosciences Institute, University of Texas at San Antonio, San Antonio, Texas, USA

Paul D. Shepard
Department of Psychiatry, Maryland Psychiatric Research Center, University of Maryland School of Medicine, Baltimore, Maryland, USA

R. Angus Silver
Department of Neuroscience, Physiology & Pharmacology, University College London, London, UK

Zachary H. Taxin
Department of Physiology & Pharmacology, SUNY Downstate, New York, USA

Kristal R. Tucker
Department of Pharmacology and Chemical Biology, University of Pittsburgh, Pittsburgh, Pennsylvania, USA

Aman Ullah
Department of Physics and Astronomy and Quantitative Biology Institute, Ohio University, Athens, Ohio, USA

Ghanim Ullah
Department of Physics, University of South Florida, Tampa, Florida, USA

Granville Yuguang Xiong
Department of Pharmacology and Systems Therapeutics, Mount Sinai School of Medicine, New York, USA

Na Yu
Department of Cell Biology and Anatomy, Neuroscience Center of Excellence, LSU Health Sciences Center, New Orleans, Louisiana, USA

Ian G. Sula
School of Computer Science and Communications, Royal Institute of Technology, Stockholm, Sweden

Houda Saadi
Infancy University Institute of Capital Information, Graduate School of Medicine, Kyoto University, Sakyo, Kyoto, Japan

Samuel A. Bozzette
Department of Biostatistics & Epidemiology, Pfizer Y Corporate, New York, USA

Rafael Baquirann
Department of Medicine and Cell Biology, Division of Cardiology, and Department of Thrombosis and Vascular Biology, Mount Sinai School of Medicine, New York, USA

Jason S. Rothman
Department of Medicine and Statistics & Neurobiology, University College, London, UK

Ethel Steinmetz
UTAS Commonwealth Marine Laboratories Centre, Portsmouth, New Hampshire, USA

Paul T. Shippey
Department of Postdoctoral Research, Cancer Research Center, University of Maryland School of Medicine, Baltimore, MD, USA

R. Angus Silver
Department of Neuroscience, Physiology & Pharmacology, University College London, London, UK

Pauline H. Tesla
Division of Pharmacology & Neuroscience, NYU University, New York, USA

Edward R. Turkey
Department of Pharmacology and Chemical Biology, University of Pittsburgh, Pittsburgh, Pennsylvania, USA

Arnau Ullah
Department of Physics and Astronomy, and Translative Biology Institute, Ohio University, Athens, Ohio, USA

Chandra Ullah
Department of Physics, University of South Florida, Tampa, Florida, USA

Granville Yuguang Xiong
Department of Pharmacology and Systems Therapeutics, Mount Sinai School of Medicine, New York, USA

Na Yu
Department of Biology and Anatomy, Neuroscience Center of Excellence, LSU Health Sciences Center, New Orleans, Louisiana, USA

PREFACE

Neuronal function arises from the complex interplay among synaptic inputs, ionic channels, and the modulatory signaling pathways that are activated by G-protein-coupled receptors or calcium influx. Not only do these interactions span multiple time scales, but elongated dendrites with tiny dendritic spines introduce multiple spatial scales of interactions. Experiments using cutting-edge imaging and other novel techniques, rather than adding clarity, have instead added to the complexity. The nonlinear temporal and spatial interactions among diverse ionic channels and signaling pathways produce unexpected neuron behavior and hinder a deep understanding of how enzyme or ion channel mutations bring about abnormal behavior and disease.

Mathematical and computational modeling offers a powerful approach for examining the interaction between molecular pathways and ionic channels in producing neuron electrical activity. Modeling is an approach to integrate myriad data sources into a cohesive and quantitative formulation in order to evaluate hypotheses about neuron function. In particular, a validated model developed using *in vitro* data allows simulations of the response to *in vivo*-like spatiotemporal patterns of synaptic input. Such models can be utilized to perform *in silico* experiments, the results of which can guide experiments into more productive domains. Models further allow inspection of all molecular species and ion channels simultaneously at all spatial locations and time points, providing additional insight into mechanisms. Incorporating molecular signaling pathways and morphology into an electrical model allows a greater range of models to be developed, ones that can predict the response to pharmaceutical treatments, many of which target neuromodulator pathways. In this volume, we bring together different aspects of and approaches to molecular and multiscale modeling, with consequences for a diverse range of neurological diseases.

The book opens with the topic of ion channel modeling. Chapter 1 by Lampert and Korngreen describes an approach to combine electrophysiology and Markov kinetic modeling to understand how single channel mutations contribute to epilepsy and pain perception. The role of ion channels in pain is further explored in Chapter 2 by Fransén, which explains the interaction among diverse ion channels and synaptic channels in mediating peripheral pain. Due to complex interactions between channel activity, they demonstrate that novel and unexpected ion channel changes could explain neuron activity changes connected to pain sensitivity. Chapter 3 by Yu *et al.* continues the exposition on ionic channel interactions, by illuminating the ionic mechanisms controlling activity of dopamine neurons, which are the targets for antipsychotic and schizophrenia drugs.

The next section of the book presents models of calcium and other subcellular components of neurons. Chapter 4 by Ullah *et al.* shows how dynamics of calcium wave propagation are controlled by spatial distribution of calcium release channels on the endoplasmic reticulum. Though simulated in oocytes, similar principles likely operate in neurons. Chapter 5 by Jafri and Kumar investigates energy metabolism in mitochondria and shows how calcium and reactive oxygen species can modify energy metabolism, which can produce neuronal dysfunction. The next two chapters focus on two aspects of cell morphology. Chapter 6 by Naoki and Ishii demonstrates how transport of growth factors interacts with a bistable switch of biochemical reactions to allow selection of a single neurite as an axon during neuronal polarization. Chapter 7 by Rangamani *et al.* introduces a novel type of multiscale modeling of the interaction between the biochemical reaction networks governing the actin cytoskeleton and the biophysics of the membrane controlling the change in membrane geometry.

The next two chapters move outside of single neurons to discuss issues crucial for large-scale neuronal network and tissue modeling. Chapter 8 by Marinov and Santamaria explains the complexities of diffusion not only intracellularly, where spines can produce anomalous diffusion, but also extracellularly, where the tortuosity of the extracellular space creates a larger than expected path distance for diffusing molecules. Chapter 9 by Linne and Jalonen reviews models and the role of astrocytes in various diseases, which include astrocyte calcium dynamics and the interaction of astrocytes with neurons, especially at the tripartite synapse.

The last section of the book presents additional types of multiscale models. Chapter 10 by Hille *et al.* describes a remarkable integration of FRET imaging and electrophysiology with model development to produce a highly constrained and validated model of the control of a potassium channel by a GPCR signaling pathway. Chapter 11 by Taxin *et al.* brings together numerous concepts from previous chapters including astrocytes, mitochondria, calcium dynamics, and large-scale networks of biochemical reaction pathways to explore the consequences of neuronal ischemia. Chapter 12 by Nair *et al.* addresses issues crucial to all model development, namely, integration of data from diverse experimental sources, general issues of constraining parameters, and evaluating sensitivity to parameters. This chapter gives examples of large-scale models of signaling pathways in the basal ganglia, with the aim to understanding mechanisms of Parkinson's disease. Chapter 13 by Rothman and Silver presents a step-by-step approach for modeling synaptic responses, beginning with the standard equations for postsynaptic currents and adding in equations describing vesicle release controlled by depletion and facilitation resulting from vesicle cycling. The book closes with Chapter 14 on the issues facing an expansion of multiscale modeling to multiphysics modeling, including the issues of "synchronizing" the different spatial and temporal scales that are required for integrating biochemical activity with electrical activity and even changes to morphology.

A large diversity of software tools is represented here, including VCell, Neuron, Moose, XPPAUT, as well as the custom software required when embarking on new approaches. Similarly, a variety of numerical methods is presented, including Boolean networks, ordinary and partial differential equations and solvers, as well as stochastic methods such as Monte Carlo methods, some of which use the Gillespie method for exact stochastic stimulation. Several chapters discuss the selection of numerical techniques, in particular the question of stochastic versus deterministic simulation. Stochastic simulations are needed in small volumes where the number of interacting molecules is small and the approximation of concentration is not valid. The role of noise produced by stochasticity is also discussed.

I would like to express my gratitude to the contributing authors and coauthors of this book for their expertise, time, and contribution to making this volume a truly educational and informative work. I would also like to acknowledge the generous funding of my research through the NSF NIH CRCNS program (R01AA016022 & R01AA18066) and by ONR (MURI N00014-10-1-0198).

KIM "AVRAMA" BLACKWELL

CHAPTER ONE

Markov Modeling of Ion Channels: Implications for Understanding Disease

Angelika Lampert[*,†], Alon Korngreen[‡,§]

[*]Institute of Physiology and Pathophysiology, Friedrich-Alexander Universität Erlangen-Nuremberg, Erlangen, Germany
[†]Institute of Physiology, RWTH Aachen University, Aachen, Germany
[‡]The Mina and Everard Goodman Faculty of Life Sciences, Bar-Ilan University, Ramat-Gan, Israel
[§]The Gonda Brain Center, Bar-Ilan University, Ramat-Gan, Israel

Contents

Abstract

Ion channels are the bridge between the biochemical and electrical domains of our life. These membrane crossing proteins use the electric energy stored in transmembrane ion gradients, which are produced by biochemical activity to generate ionic currents. Each ion channel can be imagined as a small power plant similar to a hydroelectric power station, in which potential energy is converted into electric current. This current drives basically all physiological mechanisms of our body. It is clear that a functional blueprint of these amazing cellular power plants is essential for understanding the principle of all aspects of physiology, particularly neurophysiology. The golden path toward this blueprint starts with the biophysical investigation of ion channel activity and continues through detailed numerical modeling of these channels that will eventually lead to a full system-level description of cellular and organ physiology. Here, we discuss the first two stages of this process focusing on voltage-gated channels, particularly the voltage-gated sodium channel which is neurologically and pathologically important.

Progress in Molecular Biology and Translational Science, Volume 123
ISSN 1877-1173
http://dx.doi.org/10.1016/B978-0-12-397897-4.00009-7

We first detail the correlations between the known structure of the channel and its activity and describe some pathologies. We then provide a hands-on description of Markov modeling for voltage-gated channels. These two sections of the chapter highlight the dichotomy between the vast amounts of electrophysiological data available on voltage-gated channels and the relatively meager number of physiologically relevant models for these channels.

1. WHY DO WE NEED MODELING?

1.1. Ion channel diseases: Mutations focus attention on ion channels

Human cells possess an extensive variety of ion channels. Depending on its type, a single cell may express several hundred genes for different types of ion channels.[1] Recent years have seen a tremendous advance in our understanding of their individual roles as well as of their physiological interplay. Their specific fine-tuned function is so delicate that malfunction of a single-channel subtype, due to mutation or dysregulation, may result in serious and sometimes even life-threatening diseases.[2–4]

In this chapter, we focus on voltage-gated ion channels. Of the vast number of these channels, we shall focus on voltage-gated sodium channels (Nav), as these are responsible for the upstroke of the neuronal action potential and display complex gating mechanisms. They are also involved in a broad spectrum of diseases[5]; two of the most intensely investigated disorders based on Nav modulation are epilepsy and pain. In both cases, a single-point mutation leads to severe clinical symptoms.[6,7] The resulting biophysical changes are frequently not easily understood in the context of neuronal excitability and phenotype.

1.2. Structure and function of voltage-gated sodium channels

The α subunit of voltage-gated sodium channels is sufficient to form a functional channel. It contains four domains (DI–DIV) that are connected by intracellular linkers. Each domain consists of six transmembrane helices (Fig. 1.1A; Ref. 9). The S5 and S6 segments of each domain come together to form the pore (Fig. 1.1B and C), which is gated by voltage-dependent conformational changes. Positive charges in S4 sense changes in membrane potential and move within the voltage-sensing domain (VSD), formed by S1–S4. This movement is transferred to the pore by the S4–S5 linkers,

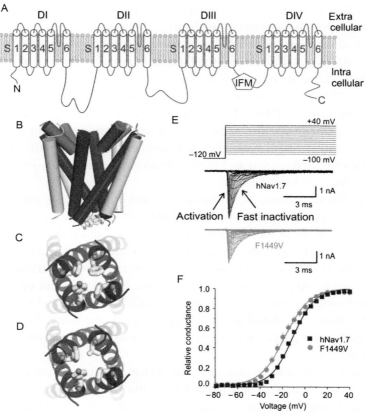

Figure 1.1 (A) 2D structure of mammalian voltage-gated sodium channels (α subunit) in the cell membrane. The four domains (DI–DIV) containing six segments each are connected by an intracellular linker. The IFM motif marks the inactivation particle. (B) Side view of the pore region of a homology model of Nav1.7. The pore is formed by the S5 and S6 segments of each domain (voltage-sensing domains are omitted for better visibility). (C) A ring of hydrophobic amino acids is situated at the intracellular face, which may act as the activation gate. (D) When mutated in the erythromelalgia mutation p.F1449V, the hydrophobic interaction may be interrupted and activation made easier. (E) Current recordings from transfected HEK cells expressing either hNav1.7 WT (black traces) or the erythromelalgia mutation p.F1449V (blue traces). Using the indicated voltage protocol, fast opening and closing voltage-gated sodium currents were recorded. (F) Conductance-voltage relation for hNav1.7 WT and its p.F1449V mutation show that the erythromelalgia mutation shifts the voltage-dependence of hNav1.7 to more hyperpolarized potentials. *(B)–(F) Adapted from Lampert* et al.[8] (See the color plate.)

and by their elevation toward the outside of the cell the S6 segments splay apart and open the permeation pathway.

Not all domains contain the same amount of gating charges (four in DI, five in DII, six in DIII, and eight in DIV) and, probably due to varying stiffness of the S6, their voltage sensitivity differs. In the skeletal muscle subtype Nav1.4, DII is quickest to move upon change in membrane potential, directly followed by DIII, whereas DI and especially DIV lag behind.[10] This complex gating behavior allows several nonconducting conformational states to manifest. Activation of more than one VSD increases the channel's probability of being open and may be reflected by several nonconducting states in Markov models.

The channel inactivates within milliseconds after opening: a hydrophobic motif on the DIII–DIV linker, the IFM motif/inactivation particle, binds to its receptor site at the cytosolic side of the pore and obliterates the permeation pathway. This inactivated state is nonconducting and needs to be reflected in stochastic models, as the channels must recover from inactivation at hyperpolarized potentials for a certain time before they can open again. Channels may not only be inactivated from the activated, conducting state, but also from the closed states, adding to the complexity that must be reflected by a computational model aiming to faithfully model the channel's gating behavior.

In addition to fast inactivation, sodium channels show a slow inactivated state which is occurring at a much longer time scale. Channels enter this nonconducting state within seconds at elevated potentials and recover from it within a similar time range. Slow inactivation is physiologically relevant when there are changes in resting membrane potential, for example, due to altered synaptic input or additional ion conductances in the cell membrane. It is relevant for dendritic integration[11] and affected by some disease causing mutations (such as small fiber neuropathy (SFN) or myopathy[12,13]). Some drugs target slow inactivation[14] and are used with some success for the treatment of epilepsy or diabetes-induced pain.

To date nine sodium channel genes have been identified in the human body (Scn1a to Scn6a, Scn8a to Scn11a), coding for the α subunits Nav1.1 to Nav1.9.[15] In the CNS, generally Nav1.1, 2, 3, and 6 are expressed, whereas the PNS additionally displays Nav1.7, 1.8, and 1.9. The latter two are slightly slower in their gating, as is subtype Nav1.5 (specific to the heart). Skeletal muscle mainly expresses Nav1.4. All sodium channel subtypes can be associated with up to two modulatory β subunits, for which four genes have so far been identified, Scn1b–Scn4b. The two disease entities that we focus on in this chapter can be caused by mutations in Nav1.7 (pain) and Nav1.1 (epilepsy).

Excitability of a neuron is shaped both by its ion channels and its morphology. Unfortunately, cell morphology often imposes severe limitations on recording from certain parts of the cell. It is technically possible to record from the cell soma, but it is incomparably harder to gain data from dendrites or axons. Therefore, most of our electrophysiological data are collected from the cell soma. We assume that excitability in the soma is comparable to that in other parts of the cell. It is feasible to create a Markov model of several conductances based on data from the cell soma or heterologously expressed channels (see below). *In silico* it is easy to place these conductances, for example, on the axon or nerve ending, and thereby assess the impact of gating changes, such as those induced by disease-causing mutations.

1.3. Epilepsy and pain can be induced by sodium channel mutations

Epilepsy and pain are both syndromes based on an increased neuronal excitability. The need for integrative methods that help us understand the role of channel alteration in causing this hyperexcitability is underlined by the >600 mutations found in Nav1.1 which are linked with mild to very severe forms of epilepsy. Many of these mutations are loss–of–function mutations (LOFs); Dravet Syndrome, for example, is a catastrophic early life epilepsy disorder, which is usually refractory to treatment and results in intellectual decline. This syndrome is caused mainly by LOFs. How can LOFs result in hyperexcitability[6]? Nav1.1 is expressed at high levels on inhibitory interneurons and loss of excitability here could result in decreased network inhibition.[16] On the other hand, interaction with other sodium channel subtypes, variation in expression and channel distribution may also play an important role. These are questions that are hard to address experimentally but can be investigated using computer models.

Four inherited pain syndromes are linked to mutations in Nav1.7, which is mainly expressed in sensory neurons of the dorsal root ganglia (DRGs). Complete loss of function of Nav1.7 leads to the absence of pain—affected patients mutilate themselves and have a reduced life expectance.[7] Gain–of–function mutations (GOFs), however, lead to the severe pain disorders erythromelagia (IEM), paroxysmal extreme pain disorder (PEPD) and SFN.

IEM patients suffer attacks of excruciating pain accompanied by reddening of the skin in their extremities triggered by warmth or mild exercise.

The patients describe the attacks, for example, ". . . as if hot lava is poured on the skin." Most of the patients do not profit from pharmaceutical treatment. Most IEM mutations render the channels more voltage sensitive, resulting in an opening of the channels at lower potentials (e.g., the mutation F1449V; Fig. 1.1F). This shift of the voltage dependence of activation eases action potential formation by normally subthreshold stimulation and is therefore likely to produce pain. In contrast, many other IEM mutations enhance slow inactivation, which may alleviate pain, as fewer channels are available.[17]

PEPD mutations, on the other hand, hamper fast inactivation, thereby producing persistent sodium currents. Most of the mutations display a slowed current decay indicating that the rate constant for the transition from the open to the inactivated state is slowed. This leads to prolonged opening of the channel, which favors the occurrence of resurgent currents[18]: an endogenous open channel blocker, most likely the c-terminal end of the β4 subunit,[19] occludes the open pore at positive potentials. As the blocking particle is positively charged, it is removed from the pore upon repolarization, and sodium current may flow until the inactivation particle finally binds. This resurgent current is most prominent in the cerebellum[20] where it supports high frequency firing. It is mostly carried by Nav1.6,[21] but other subtypes are also capable of producing resurgent currents.[22] It was recently shown that PEPD mutations also enhance the normally barely detectable resurgent currents of Nav1.7.[23] Slowing current decay increases the likelihood of binding of the endogenous blocker and thus potentially enhancing resurgent currents. This feature is only observed in PEPD mutations and not in mutated channels associated with IEM.[24] Thus, biophysically, PEPD is very different from IEM. The clinical presentation is also markedly different: attacks of PEPD triggered by cold or mechanical stimulation affect very young children within their first year of life. The pain is located in proximal body parts and is accompanied by a harlequin-type skin reddening.

To test for mutation-induced changes in excitability, PEPD or IEM mutations were transfected in rodent DRG neurons. Not surprisingly, both increased action potential firing and lowered their threshold (e.g., Ref. 25). What causes the remarkable difference in their clinical presentation? So far, we can only speculate and more detailed investigations, including cell morphology, temperature, age-dependent regulations, and possible changes in ion concentrations close to the cell membrane, may improve our understanding. It would be most helpful to have a valid computer model to help produce hypotheses that can be assessed experimentally (see Table 1.1 and Fig. 1.2).

Table 1.1 Electrophysiological recording of transfected rodent DRGs (rDRGs)

Advantages	Disadvantages
rDRGs are neurons with all necessary components (e.g., Kvs, Cavs, subunits, etc.,)	rDRGs are not human. Therefore all cellular components are based on rodent genetics
rDRGs are similar to human nociceptors	In whole-cell patch clamp •the intracellular ion concentrations are clamped •regulatory molecules are dialyzed out of the cell
The somata of rDRGs are easily accessible	The soma of the cell was investigated, but nociception occurs in the nerve endings, which are very hard to patch

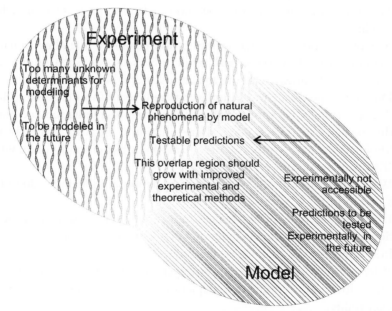

Figure 1.2 Scheme illustrating the applicability of *in vitro* experiments (left), computational models (right), and their potential interaction (middle).

The third pain disorder caused by mutations in Nav1.7, SFN, typically occurs later in life and leads to a degeneration of the peripheral nerve endings.[26] Patients report intense pain and symptoms of the autonomous nervous system and only occasionally reddening of the skin. SFN patients thus describe symptoms that can also be found in IEM or PEPD, although to a different extent and at a different age. Biophysically, the major effect of SFN mutations is impairment of slow inactivation, in contrast to the IEM mutations that often enhance this gating characteristic. For example, the I228M mutation does not display any major gating change other than an enhanced slow inactivation. Still, it increases excitability when transected into DRG neurons,[27] suggesting that impaired slow inactivation is capable of inducing disease.

Why do the mutations causing IEM mainly enhance slow inactivation, whereas those inducing SFN reduce it? Many possible *in vitro* experiments have already provided valuable insights, but this obvious difference remains hard to explain. *In silico* it is relatively easy to model the impact of mutant-induced slow inactivation in various disease conditions. Therefore, a reliable model of the sodium conductance would be most helpful.

The following questions may be addressed using an *in silico* modeling approach:

- A neuron expresses a large variety of ion channels, including different sodium channel subtypes and other conductances, such as Kv and Cav. How do disease-causing mutations affect their interplay?
- Different types of neurons display different sets of ion channels. How does the cellular background affect the impact of a mutation and could this explain variations in clinical presentation (e.g., why do PEPD and SFN patients have symptoms of the autonomous nervous system, while IEM patients do not)?
- What is the effect of cell morphology on neuronal excitability altered by a mutation?
- How can different gating changes induce the same clinical symptoms?
- Some patients present a genetic mosaic for their mutation—is the gene load important for cellular excitability?
- Why and how does temperature trigger attacks of hyperexcitability?
- How does ion accumulation/depletion close to the membrane affect excitability?

1.4. How can Markov models help understand the pathophysiology of channelopathies?

In physiological experiments, it is possible to determine the biophysical gating changes induced by single-point mutations, as outlined above. These parameters may be used to create a Markov model (see below), which not only helps us model the whole cellular excitability but also to understand

Table 1.2 Computational modeling of channelopathies

Advantages	Disadvantages
All conductances may be derived from human data	All conductances used for the model are results from experiments that are affected by experimental constraints
Conductances may be varied and adjusted, modeling different gene loads and neuronal cell types	There may be components necessary for cellular excitability that we do not yet know
Gating changes may be modeled in isolation	
Ion concentrations may be modeled and their change due to cellular activity may be monitored	
Cell morphology is taken into account	
Temperature can be implied	All predictions need to be tested experimentally to validate the model

occupancy of gating states (Table 1.2 and Fig. 1.2). These Markov models need to be validated by predicting gating changes that can be experimentally tested.

Markov models have only been rarely used for modeling channelopathies to date, probably because Markov models are assumed to make sense only when single-channel data are used. Nevertheless, there are already some examples, mostly models of Nav1.5 involved in arrhythmias, such as long QT syndrome.[28–31] A Markov model of an epilepsy mutation faithfully modeled persistent currents and predicted that inactivation of SCN1A open-state consists of two-steps, an initial gate closure followed by an additional mechanism to stabilize the inactivated state.[32] Interference with this mechanism may induce the epileptic symptoms of the Nav1.1 mutation investigated in this study.

2. MARKOV MODELS BUILT BASED ON WHOLE-CELL PATCH-CLAMP DATA

Having shown the need for valid computational models of voltage-gated ion conductances, we now focus on how these models are best generated. Ion channels are large transmembrane proteins constructed from a very large number of covalent bonds. From a purely chemical point of view, each bond contributes not only to the structure of the molecule but also to its dynamics. Each bond vibrates at a typical spring frequency that can be described using chemical kinetics. Thus, the "true" representation of the structural and dynamic properties of the ion channel can be defined as a set of equations roughly equaling the number of bonds in the protein. For a voltage-gated channel, we also need to add the sensitivities of some amino acid residues to the external electric field and the interactions of the protein with water, ions, and membrane lipids. All these equations must be implemented in a computer model and then time-intensive calculations must be performed to determine the activity of the channel. These molecular dynamics simulations, while of crucial importance to understanding the relationship between structure and function, are too slow and way too detailed to be helpful to the physiologist or even to the garden variety biophysicist.

We now show an approach that is a possible solution to this problem. All functional models are, by definition, phenomenological. Therefore, we will treat them according to standard formulations of chemical kinetics. First, let us consider the nature of the reaction. In mammals, ion channels (and all cellular reactions for that matter) operate at ~37 °C. Ion channels can move reversibly between the open and closed states. Thus, according to thermodynamical laws this reaction has low free energy and is close to equilibrium. Changes in extrinsic parameters, such as membrane potential or the concentration of a ligand, perturb the channel away from equilibrium. Similar to many systems at equilibrium, the channel responds to this perturbation by an apparently exponential relaxation. Formally, the simplest reaction of this kind can be written using the following chemical notation. This is the simplest "Markov model," just two links in the chain.

$$C \underset{k_{-1}}{\overset{k_1}{\rightleftharpoons}} O \tag{1.1}$$

$$
\begin{aligned}
I &= \bar{g}O(V - E) \\
\frac{dC}{dt} &= -k_1 C + k_{-1} O \\
\frac{dO}{dt} &= k_1 C - k_{-1} O
\end{aligned}
\tag{1.2}
$$

Where O denotes the open state, C the closed state, k is a kinetic rate constant having units of $1/s$, V is the membrane potential, I is the ionic current, E is the Nernst potential, and $\bar{g}$ is the maximal conductance. Here, we denote the forward reaction by a positive subscript and the reverse reaction by a negative subscript. Many Markov models of ion channels use a different notation with numbering related to the states and not to the transition. According to this alternative, we have two states and the rate constants will be k_{12} and k_{21}. We find both numbering systems equally confusing. Remembering that according to the principle of mass conservation $C + O = 1$, it is possible to write the differential equation for the open state as

$$\frac{dO}{dt} = k_1(1 - O) - k_{-1}O \tag{1.3}$$

From here, it can be simply shown that under steady-state conditions that $O_{inf} = k_1/(k_1 + k_{-1})$ and that the relaxation time constant of the reaction is $\tau = 1/(k_1 + k_{-1})$.

This general description of the system holds for every simple one-stage chemical or biological reaction that is close to equilibrium. Here, we focus on voltage-gated channels. Thus the rate constants should be a function of the membrane potential in the general exponential form:

$$k = Ae^{BV}$$

where V is the membrane potential in volts and the rate constants are in $1/s$. Thus, A has units of $1/s$ and B units of $1/V$. Let us start with two simple and opposite rate constants:

$$k_1 = 10e^{50V}; \quad k_{-1} = 10e^{-50V} \tag{1.4}$$

First, let us look at the steady-state condition:

$$O_{inf} = \frac{10e^{50V}}{10e^{50V} + 10e^{-50V}} = \frac{1}{1 + e^{-100V}}$$

Here, we can easily see the link between the rate constants and the classical Boltzmann curve that is so typical for the description of the voltage dependence of voltage-gated channels. For graphical convenience, we plotted k_1, k_{-1} (Fig. 1.3A), the open probability at steady state (O_{inf}) (Fig. 1.3B), and the time constant (Fig. 1.3C) all as a function of the membrane potential. Several insights can be extracted from Fig. 1.3 and from the above equation. First, it is instructive to note the direction and trend of the rate constants. The voltage dependence of the channel is a direct function of the voltage dependence of the rate constants. At hyperpolarized values of the membrane potential, $k_{-1} \gg k_1$, drawing the reaction toward the closed state. The opposite picture emerges at depolarized potentials where

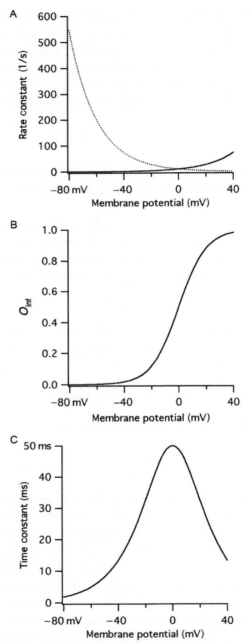

Figure 1.3 (A) The rate functions plotted as a function of the membrane potential of the kinetic scheme shown in Eq. (1.1). The forward rate constant is shown by the solid line and the backward rate constant by the dashed line. (B) Steady-state activation curve for this kinetic model. (C) Activation time constant for this model.

$k_{-1} \ll k_1$ and the reaction is drawn to the open state. When $k_1 = k_{-1}$ it is simple to see from Eq. (1.4) that $O_{inf} = 1/2$. This value of the membrane potential is often referred to by experimentalists as the potential of half activation or $V_{1/2}$. For the simple case, we are currently discussing this is also the potential at which the maximal value of τ can be observed (Fig. 1.3C). This correlation between $V_{1/2}$ and the maximal time constant holds true for many models and can serve as a simple sanity check of the integrity of one's model.

Now let us look at a slightly more complex model, one with two closed states and one open state. We write the differential equations explicitly here and will then display the same system as a matrix.

$$C_1 \underset{k_{-1}}{\overset{k_1}{\rightleftharpoons}} C_2 \underset{k_{-2}}{\overset{k_2}{\rightleftharpoons}} O \tag{1.5}$$

$$\frac{dC_1}{dt} = -k_1 C_1 + k_{-1} C_2$$
$$\frac{dC_2}{dt} = k_1 C_1 - k_{-1} C_2 - k_2 C_2 + k_{-2} O \tag{1.6}$$
$$\frac{dO}{dt} = k_2 C_2 - k_{-2} O$$

From this simple model it is possible to grasp the general rule describing the kinetics of a given transition. The rate of change is given as the sum of the rates entering that state minus the sum of the rates leaving it. What is also clear from this set of equations is that it is the product of a matrix of the rate constants with the vector of the states (Eq. 1.7a). This matrix is known as the Q matrix and is comprehensively described in depth in several publications.[33,34] It has been extensively used for simulating and analyzing single-channel activity. The reader will therefore find that most sources refer to the states in terms of probabilities. This is strictly correct, however, for our current discussion of whole-cell currents we can, *quasi* correctly, treat the states as a continuous level of occupancy.

When constructing the Q matrix two simple rules must be applied:
$q_{i,i} = -\sum(\text{transition rates leading away from the state } i)$
$q_{i,j} = (\text{transition rates from state } i \text{ to state } j)$
Thus the Q matrix for Eq. (1.5) appears as

$$Q = \begin{Bmatrix} -k_1 & k_1 & 0 \\ k_{-1} & -(k_2 + k_1) & k_2 \\ 0 & k_{-2} & -k_{-2} \end{Bmatrix} \tag{1.7}$$

With the equivalent of Eq. (1.6) in matrix form

$$
\left\{ \begin{array}{c} \mathrm{d}C_1/\mathrm{d}t \\ \mathrm{d}C_2/\mathrm{d}t \\ \mathrm{d}O/\mathrm{d}t \end{array} \right\} = \left\{ \begin{array}{ccc} -k_1 & k_1 & 0 \\ k_{-1} & -(k_2+k_{-1}) & k_2 \\ 0 & k_{-2} & -k_{-2} \end{array} \right\} \left\{ \begin{array}{c} C_1 \\ C_2 \\ O \end{array} \right\} \tag{1.7a}
$$

When the rate constants are not dependent on the membrane potential, it is possible to solve this matrix analytically to obtain the relaxation time constants of the channel.[33,34] When the rate constants are voltage dependent, it is usually not possible to provide an analytical solution. However, once the Q matrix is known, it is relatively simple to solve the system using Matlab or Mathematica. There are also several freely available software packages that are specifically designed for simulating Markov models of ion channels. We will discuss these packages later.

The third example we mention briefly can be expressed by the following kinetic scheme:

$$
C \underset{k_{-1}}{\overset{k_1}{\rightleftharpoons}} O \underset{k_2}{\overset{k_{-2}}{\rightleftharpoons}} I, \tag{1.8}
$$

where I represents an inactivated nonconducting state. We will not describe this scheme in depth, as it is less applicable to physiological cases. This model, once fully inactivated, must change to the open state in order to return to the closed state. Such channel activity has been observed but is rare and is usually considered only as part of a larger model. Another important point for the construction of a model with inactivation is the inversion of the transition rates between the open and inactivated state, as opposed to those between the closed and open states (cf. Eq. 1.5). This is required for the proper setting of the inactivated state. It is possible to use the following cyclic kinetic scheme to overcome the need to change to the open state on the way from inactivated to closed.

$$\tag{1.9}$$

In this model, given the correct setting of the rate constants, the channel moves from open to inactivated and from there to closed, without generating an unwanted current by reversing from inactivated back to the open state. The Q matrix for this model is

$$Q = \left\{ \begin{matrix} -(k_1 + k_3) & k_1 & k_3 \\ k_{-1} & -(k_{-2} + k_{-1}) & k_{-2} \\ k_{-3} & k_2 & -(k_2 + k_{-3}) \end{matrix} \right\} \quad (1.10)$$

Another important issue must be discussed when considering cyclic mechanisms. It is clear that the scheme depicted by Eq. (1.9) contains two opposing reaction cycles, one on the outside and one on the inside of the scheme. Again, considering this as a chemical reaction, one can see that even at equilibrium there is continuous cycling of material in opposite directions. Clearly, in order for the reaction to balance, the total rate of the external cycle should be equal to that of the internal one

$$k_1 k_{-2} k_{-3} = k_{-1} k_2 k_3 \quad (1.11)$$

Leading to

$$k_{-3} = \frac{k_{-1} k_2 k_3}{k_1 k_{-2}} \quad (1.12)$$

Thus, for cyclic mechanisms one of the rate constants is not independent but should be expressed as a function of the other constants reducing the system's degrees of freedom by one. This is called the principle of microscopic reversibility.[35] These four examples (Eqs. 1.1–1.12) can be viewed as basic building blocks that can be used to construct almost any functionally relevant Markov chain model of voltage-gated ion channels.

3. PRACTICAL CONSIDERATIONS FOR FITTING MODELS TO DATA

Having drawn the skeleton of the model, we must now select values for the rate constants. This is the difficult part. Considering again the kinetic scheme displayed in Eq. (1.5), we have four rate constants, each described by two parameters (Eq. 1.4) leading to eight free parameters. This number of free parameters explodes when more states are added to the model. The nonlinearity of the system impedes tuning and even eight free parameters are exceedingly hard to tune by hand.

Yet, before we address the tuning options, we survey the types of rate constants used in modeling voltage-gated ion channels. The rate constants we discussed previously (Eq. 1.4) can be derived from general kinetic theory.[36] However, several other equations have been used to describe rate constants in Markov models of voltage-gated channels. In an attempt to shift the activation curve along the membrane potential axis, rate constants have been often described by a modified version of Eq. (1.4):

$$k = Ae^{b(V-c)} \tag{1.13}$$

However, this formulation is misleading as can be demonstrated by simple algebra.

$$Ae^{b(V-c)} = Ae^{bV-bc} = \left(Ae^{-bc}\right)e^{bV} \tag{1.14}$$

The term in the parentheses is constant and, therefore, does not convey any shift of the activation curve along the membrane potential axis. It is easy to see that Eq. (1.13) is just an unnecessarily complex form of Eq. (1.4). The exponential rate constant can be extended using thermodynamic rationalization to include higher terms of voltage dependence.[37] These rate constants can account for voltage-dependent transitions that are rate limited while retaining a low number of states in the model:

$$k = Ae^{bV+cV^2} \tag{1.15}$$

Forsaking the attempt to use thermodynamically based rate constants, a popular way of calculating rate constants is via the Boltzmann equation.

$$k = \frac{a}{1 + e^{-b(c-V)}} \tag{1.16}$$

It is important to note that using Eq. (1.15) or Eq. (1.16) to describe the transition rates in Eq. (1.5) increases the number of free parameters for this model to 12. Here lies one of the major problems in constructing models for ion channels and particularly for voltage-gated ion channels. Even for a three-state model, the number of free parameters ranges from 8 to 12 and possibly even more. With such a high number of free parameters that require adjusting it is almost impossible to tune these models by hand.

One way to reduce this parameter space is to assume that some states are dependent on others. For example, writing the Markov chain model for the Hodgkin–Huxley potassium channel we obtain the following scheme:

$$C \underset{\beta}{\overset{4\alpha}{\rightleftharpoons}} C \underset{2\beta}{\overset{3\alpha}{\rightleftharpoons}} C \underset{3\beta}{\overset{2\alpha}{\rightleftharpoons}} C \underset{4\beta}{\overset{\alpha}{\rightleftharpoons}} O \tag{1.17}$$

Here, we have only two rate constants that are factored between transitions to reach the original Hodgkin–Huxley formalism after solving the Q matrix.

3.1. Fitting experimental data to Markov models

Another facet of ion channel modeling is fitting the model to a set of experimental results. This requires selecting the right data set, the right cost function and the appropriate fitting algorithm. In this review, we do not address the issue of single-channel recording and analysis, as most physiologically relevant kinetic analyses of voltage-gated ion channels in recent decades have been performed on ensemble currents obtained from whole-cell or excised patch recordings. Furthermore, almost all investigations so far have applied the analysis paradigm established by Hodgkin and Huxley.[38]

At the practical level this celebrated analysis methodology proceeds in several simple steps. First, one must construct steady-state activation curves of the investigated conductance (and inactivation curves if required). These curves are then used to estimate the number of activation gates in the Hodgkin–Huxley-like model that best fits the investigated conductance. Armed with the number of gates, one can now fit the current traces, one by one, to the model and secure the activation (and/or inactivation) time constants. This procedure has not changed considerably for some decades and we have often found that the original papers by Hodgkin and Huxley[38–41] are a better source of information than modern papers.

Still, this analysis procedure may introduce severe errors in the estimation of the kinetic constants of the investigated channel, especially if the activation and inactivation time constants are not sufficiently different.[42] The source of the problem is the separate (or disjoint, see Ref. 42) analysis of the steady-state and kinetic properties of the current. This severe error source can be overcome by simultaneous curve fitting of a Hodgkin–Huxley-like model to a data set including current traces recorded at several voltages.[42,43] Direct fitting of a model to a set of current traces was also recently introduced into the single-channel analysis software QUB,[44] allowing the fitting of Markov chain models to whole-cell currents. In contrast to the conventional method, this approach to data analysis and model fitting has been named global curve fitting or the full trace method.[42,43]

It is important to remember that fitting a full model to a set of data dramatically increases the number of free parameters. To complicate the problem, standard curve-fitting algorithms (such as, e.g., the simplex and gradient descent algorithms) tend to settle into a local minimum when dealing with many parameters.[45] Thus, to verify that the solution obtained is not a local minimum, it is important to select initial values that are sufficiently close to the global minimum. Alternatively, instead of guessing a set of initial values that may lead the minimization algorithm to a local minimum, it is possible to apply stochastic minimization algorithms to search a predefined parameter space.

Several popular stochastic minimization tools such as genetic algorithms[46] and simulated annealing[47] have been successfully applied to many problems of nonlinear curve fitting in biology. These algorithms usually perform tens to hundreds of thousands of iterations to find the global minimum. Their advantage over the more commonly used gradient descent algorithms is that genetic algorithms do not require an initial guess of the parameters.[42–44] With a large enough data set it is possible to arrive at the global minimum even from a random starting point.[48–51] This advantage of genetic algorithms is important as, given wrong starting parameters, gradient descent algorithms may arrive at a local minimum and mistakenly provide a model with "good" fitting parameters. However, the genetic algorithm method is not infallible, there may be a random combination of parameters that defy it. But our results show that a useable and functionally significant ion channel model describing whole-cell currents can be produced using a genetic algorithm with a data set containing a "complete" whole-cell activation of the channel as its input.[49,52]

We have encountered computational exhaustion due to the complexity of the models and the amount of electrophysiological data. To solve this computational bottleneck, we converted our optimization algorithm for work on a graphical processing unit using NVIDIA's CUDA. Parallelizing the process on a Fermi graphic computing engine from NVIDIA increased the speed $\sim 180 \times$ over an application running on an 80 node Linux cluster, considerably reducing simulation times. This application allows users to optimize models for ion channel kinetics on a single, inexpensive, desktop "super computer," greatly reducing the time and cost of building models relevant to neuronal physiology.[53,54]

4. CONCLUSION AND OUTLOOK

In this chapter, we deliberately split the discussion of ion channel function into two separate and only partially overlapping sections. We first elaborated on some of the advances in the study of the function and pathology of voltage-gated sodium channels. We then detailed the basic terminology and considerations behind building a successful Markov model for a voltage-gated channel. The dichotomy between these two parts of the chapter highlights the current state of the field of ion channel modeling that is graphically displayed in Fig. 1.2. There is a vast body of experiments from structure-function studies in many expression systems. However, the conversion of these studies to physiologically relevant kinetic models lags far behind. Several current trends may assist in closing this gap. First, computing power has increased in vast leaps during the last two decades and is now accessible to almost any researcher of ion channels. Secondly, while only 10 years ago fitting a model to a set of data required mastodonic parallel computers, through the low cost and availability of multicore desktop computers and the amazing advances in using graphic and DSP cards for parallel computing, the power of supercomputing will reach the investigator's desktop, if not today, then most definitely in the very near future. We predict that, given this surge in new and exciting hardware, we will soon see user friendly software that will enable any electrophysiologist to easily fit large quantities of voltage-clamp data to functional Markov models of ion channels in the comfort of his/her lab. This will be a major step toward better understanding the role of ion channels in cell function in health and disease.

ACKNOWLEDGMENTS

This work was supported by a joint grant from the German-Israeli Foundation to A. K. and A. L. (#1091-27.1/2010).

REFERENCES

1. Gabashvili I, Sokolowski BA, Morton C, Giersch AS. Ion Channel Gene Expression in the Inner Ear. *J Assoc Res Otolaryngol*. 2007;8:305–328.
2. Ashcroft F. *Ion Channels and Disease: Channelopathies*. Boston, MA: Academic; 2000.
3. Ashcroft FM. From molecule to malady. *Nature*. 2006;440:440–447.
4. Nilius B. A Special Issue on channelopathies. *Pflugers Arch*. 2010;460:221–222.
5. Eijkelkamp N, Linley JE, Baker MD, et al. Neurological perspectives on voltage-gated sodium channels. *Brain*. 2012;135:2585–2612.
6. Escayg A, Goldin AL. Sodium channel SCN1A and epilepsy: mutations and mechanisms. *Epilepsia*. 2010;51:1650–1658.
7. Lampert A, O'Reilly A, Reeh P, Leffler A. Sodium channelopathies and pain. *Pflugers Arch*. 2010;460:249–263.
8. Lampert A, O'Reilly AO, Dib-Hajj SD, Tyrrell L, Wallace BA, Waxman SG. A poreblocking hydrophobic motif at the cytoplasmic aperture of the closed-state Nav1.7 channel is disrupted by the erythromelalgia-associated F1449V mutation. *J Biol Chem*. 2008;283:24118–24127.
9. Catterall WA. Voltage-gated sodium channels at 60: structure, function and pathophysiology. *J Physiol*. 2012;590:2577–2589.
10. Chanda B, Bezanilla F. Tracking voltage-dependent conformational changes in skeletal muscle sodium channel during activation. *J Gen Physiol*. 2002;120:629–645.
11. Mickus T, Jung H, Spruston N. Properties of slow, cumulative sodium channel inactivation in rat hippocampal CA1 pyramidal neurons. *Biophys J*. 1999;76:846–860.
12. Lossin C, Nam TS, Shahangian S, et al. Altered fast and slow inactivation of the N440K Nav1.4 mutant in a periodic paralysis syndrome. *Neurology*. 2012;79:1033–1040.

13. Waxman SG. Painful Na-channelopathies: an expanding universe. *Trends Mol Med.* 2013;19:406–409.
14. Errington AC, Stoehr T, Heers C, Lees G. The investigational anticonvulsant lacosamide selectively enhances slow inactivation of voltage-gated sodium channels. *Mol Pharmacol.* 2007;73:157–169. http://dx.doi.org/10.1124/mol.107.039867.
15. Catterall WA, Goldin AL, Waxman SG. International Union of Pharmacology. XLVII. Nomenclature and structure-function relationships of voltage-gated sodium channels. *Pharmacol Rev.* 2005;57:397–409.
16. Ogiwara I, Miyamoto H, Morita N, et al. Nav1.1 localizes to axons of parvalbumin-positive inhibitory interneurons: a circuit basis for epileptic seizures in mice carrying an Scn1a gene mutation. *J Neurosci.* 2007;27:5903–5914.
17. Sheets PL, Jackson 2nd JO, Waxman SG, Dib-Hajj SD, Cummins TR. A Nav1.7 channel mutation associated with hereditary erythromelalgia contributes to neuronal hyper-excitability and displays reduced lidocaine sensitivity. *J Physiol.* 2007;581:1019–1031.
18. Bean BP. The molecular machinery of resurgent sodium current revealed. *Neuron.* 2005;45:185–187.
19. Grieco TM, Malhotra JD, Chen C, Isom LL, Raman IM. Open-channel block by the cytoplasmic tail of sodium channel beta4 as a mechanism for resurgent sodium current. *Neuron.* 2005;45:233–244.
20. Raman IM, Bean BP. Resurgent sodium current and action potential formation in dissociated cerebellar Purkinje neurons. *J Neurosci.* 1997;17:4517–4526.
21. Raman IM, Sprunger LK, Meisler MH, Bean BP. Altered subthreshold sodium currents and disrupted firing patterns in Purkinje neurons of Scn8a mutant mice. *Neuron.* 1997;19:881–891.
22. Rush AM, Dib-Hajj SD, Waxman SG. Electrophysiological properties of two axonal sodium channels, Nav1.2 and Nav1.6, expressed in mouse spinal sensory neurones. *J Physiol.* 2005;564:803–815.
23. Jarecki BW, Piekarz AD, Jackson 2nd JO, Cummins TR. Human voltage-gated sodium channel mutations that cause inherited neuronal and muscle channelopathies increase resurgent sodium currents. *J Clin Invest.* 2010;120:369–378.
24. Theile JW, Jarecki BW, Piekarz AD, Cummins TR. Nav1.7 mutations associated with paroxysmal extreme pain disorder, but not erythromelalgia, enhance Navbeta4 peptide-mediated resurgent sodium currents. *J Physiol.* 2011;589:597–608.
25. Dib-Hajj SD, Yang Y, Black JA, Waxman SG. The Na(V)1.7 sodium channel: from molecule to man. *Nat Rev Neurosci.* 2013;14:49–62.
26. Hoeijmakers JG, Faber CG, Lauria G, Merkies IS, Waxman SG. Small-fibre neuropathies—advances in diagnosis, pathophysiology and management. *Nat Rev Neurol.* 2012;8:369–379.
27. Estacion M, Han C, Choi JS, et al. Intra- and interfamily phenotypic diversity in pain syndromes associated with a gain-of-function variant of NaV1.7. *Mol Pain.* 2011;7:92.
28. Bankston JR, Sampson KJ, Kateriya S, et al. A novel LQT-3 mutation disrupts an inactivation gate complex with distinct rate-dependent phenotypic consequences. *Channels (Austin).* 2007;1:273–280.
29. Ruan Y, Denegri M, Liu N, et al. Trafficking defects and gating abnormalities of a novel SCN5A mutation question gene-specific therapy in long QT syndrome type 3. *Circ Res.* 2010;106:1374–1383.
30. Vecchietti S, Rivolta I, Severi S, Napolitano C, Priori SG, Cavalcanti S. Computer simulation of wild-type and mutant human cardiac Na+ current. *Med Biol Eng Comput.* 2006;44:35–44.
31. Vecchietti S, Grandi E, Severi S, et al. In silico assessment of Y1795C and Y1795H SCN5A mutations: implication for inherited arrhythmogenic syndromes. *Am J Physiol Heart Circ Physiol.* 2007;292:H56–H65.

32. Kahlig KM, Misra SN, George Jr AL. Impaired Inactivation Gate Stabilization Predicts Increased Persistent Current for an Epilepsy-Associated SCN1A Mutation. *J Neurosci.* 2006;26:10958–10966.
33. Colquhoun D, Hawkes AG. The principles of the stochastic interpretation if ion-channel mechanisms. In: Sakmann B, Neher E, eds. *Single-Channel Recording.* New York, NY: Plenum Press; 1995:397–482.
34. Colquhoun D, Sigworh JF. Fitting and staistical analysis of single channels records. In: Sakmann B, Neher E, eds. *Single-Channel Recording.* New York, NY: Plenum Press; 1995:483–589.
35. Colquhoun D, Dowsland KA, Beato M, Plested AJ. How to impose microscopic reversibility in complex reaction mechanisms. *Biophys J.* 2004;86:3510–3518.
36. Johnson FH, Eyring H, Stover GJ. *The Theory of Rate Processes in Biology and Medicine.* New York, NY: Wiley; 1974.
37. Destexhe A, Huguenard JR. Nonlinear thermodynamic models of voltage-dependent currents. *J Comput Neurosci.* 2000;9:259–270.
38. Hodgkin AL, Huxley AF. A quantitative description of membrane current and its application to conduction and excitation in nerve. *J Physiol.* 1952;117:500–544.
39. Hodgkin AL, Huxley AF. The components of membrane conductance in the giant axon of Loligo. *J Physiol.* 1952;116:473–496.
40. Hodgkin AL, Huxley AF. The dual effect of membrane potential on sodium conductance in the giant axon of Loligo. *J Physiol.* 1952;116:497–506.
41. Hodgkin AL, Huxley AF. Currents carried by sodium and potassium ions through the membrane of the giant axon of Loligo. *J Physiol.* 1952;116:449–472.
42. Willms AR, Baro DJ, Harris-Warrick RM, Guckenheimer J. An improved parameter estimation method for Hodgkin-Huxley models. *J Comput Neurosci.* 1999;6:145–168.
43. Willms AR. NEUROFIT: software for fitting Hodgkin-Huxley models to voltage-clamp data. *J Neurosci Methods.* 2002;121:139–150.
44. Milescu LS, Akk G, Sachs F. Maximum likelihood estimation of ion channel kinetics from macroscopic currents. *Biophys J.* 2005;88:2494–2515.
45. Press WH, Teukolsky SA, Vetterling WT, Flannery BP. *Numerical Recipes in C, the Art of Scientific Computing.* 2nd ed. Cambridge: Cambridge University Press; 1992.
46. Mitchell M. *An Introduction to Genetic Algorithms.* Cambridge, MA: MIT Press; 1996.
47. Kirkpatrick S, Gelatt CD, Vecchi MP. Optimization by simulated annealing. *Science.* 1983;220:671–680.
48. Gurkiewicz M, Korngreen A. Recording, analysis, and function of dendritic voltage-gated channels. *Pflugers Arch.* 2006;453:283–292.
49. Gurkiewicz M, Korngreen A. A numerical approach to ion channel modelling using whole-cell voltage-clamp recordings and a genetic algorithm. *PLoS Comput Biol.* 2007;3:e169.
50. Keren N, Peled N, Korngreen A. Constraining compartmental models using multiple voltage recordings and genetic algorithms. *J Neurophysiol.* 2005;94:3730–3742.
51. Keren N, Bar-Yehuda D, Korngreen A. Experimentally guided modelling of dendritic excitability in rat neocortical pyramidal neurones. *J Physiol.* 2009;587:1413–1437.
52. Gurkiewicz M, Korngreen A, Waxman SG, Lampert A. Kinetic modeling of Nav1.7 provides insight into erythromelalgia-associated F1449V mutation. *J Neurophysiol.* 2011;105:1546–1557.
53. Ben-Shalom R, Liberman G, Korngreen A. Accelerating compartmental modeling on a graphical processing unit. *Front Neuroinform.* 2013;7:4.
54. Ben-Shalom R, Aviv A, Razon B, Korngreen A. Optimizing ion channel models using a parallel genetic algorithm on graphical processors. *J Neurosci Methods.* 2012;206:183–194.

CHAPTER TWO

Ionic Mechanisms in Peripheral Pain

Erik Fransén
Department of Computational Biology, School of Computer Science and Communication, KTH Royal
Institute of Technology, Stockholm, Sweden

Contents

Abstract

Chronic pain constitutes an important and growing problem in society with large unmet needs with respect to treatment and clear implications for quality of life.

Computational modeling is used to complement experimental studies to elucidate mechanisms involved in pain states. Models representing the peripheral nerve ending often address questions related to sensitization or reduction in pain detection threshold. In models of the axon or the cell body of the unmyelinated C-fiber, a large body of work concerns the role of particular sodium channels and mutations of these. Furthermore, in central structures: spinal cord or higher structures, sensitization often refers not only to enhanced synaptic efficacy but also to elevated intrinsic neuronal excitability.

One of the recent developments in computational neuroscience is the emergence of computational neuropharmacology. In this area, computational modeling is used to study mechanisms of pathology with the objective of finding the means of restoring healthy function. This research has received increased attention from the pharmaceutical industry as ion channels have gained increased interest as drug targets. Computational modeling has several advantages, notably the ability to provide mechanistic links between molecular and cellular levels on the one hand and functions at the systems level on the other hand. These characteristics make computational modeling an additional tool to be used in the process of selecting pharmaceutical targets. Furthermore, large-scale simulations can provide a framework to systematically study the effects of several interacting disease parameters or effects from combinations of drugs.

In this chapter, we first briefly introduce pain for the unfamiliar reader. Second, we review how computational modeling has been used to address problems in pain. We finally describe how computational modeling combined with numerical search can be used in drug development. In these cases, modeling is used to search for the intervention, for example, ion channel blocker that changes the behavior of the diseased system and makes it function more like the healthy one.

Models are used in neuroscience to produce mechanistic explanations of experimentally observed phenomena. The validity of the model is assessed by replicating experimental findings. When the model provides predictions, these can be tested experimentally, further advancing knowledge and understanding of the problem. The models often include data from several sources and thereby provide a framework to relate separate findings and provide means to resolve seemingly contradictory findings. When combining data on a detailed scale to study a macroscopic function, for example, including ion channel data to study action potential propagation in an axon, the model provides a means to span levels from microscopic to mesoscopic scales. When a well-functioning model is obtained, it provides a mechanistic explanation for the problem. It can thereby be used to identify therapeutic targets as molecular components with a causal relationship to symptoms. It also addresses the mechanism of action (MOA) problem in drug discovery (see Section 3 for further discussion on this): the model provides a test bed to investigate macroscopic effects of a pharmacological substance on hypothesized molecular targets or can generate hypotheses regarding potential targets of a pharmacological substance given its effects on symptoms. The approach has further advantages with respect to experimental approaches in that it can be multifactorial. The modeling *in silico* approach thereby complements experimental *in vivo* and *in vitro* approaches.

1. BIOLOGICAL BACKGROUND

Pain is the sensation from thermal, mechanical, or chemical noxious stimuli. The origin of chemical stimuli can be external or endogenously produced after damage to the tissue or inflammation. Pain can be classified according to different schemes depending upon the location or system affected in the body, intensity, duration, etiology, etc.; however, a simplified subdivision into inflammatory, nociceptive (involving the detection organ), and neuropathic (involving the nervous system) pain has been suggested. Acute pain, when the stimulus is present, is relatively well treated pharmacologically by substances such as lidocaine. Pain may, however, outlast the stimuli or occur in the absence of obvious stimuli and is then referred to as chronic pain. Chronic pain is classified into several subtypes: (1) allodynia (when innocuous stimuli are perceived as painful), (2) hyperalgesia (an increased pain perception of noxious stimuli), and (3) spontaneous pain (pain in the absence of stimuli). For acute pain, there is evidence that discharge rates could be linearly correlated with the intensity of the noxious stimulation.[1–3] However, chronic pain states may be more complex, since pain sensation may outlast the period of nociceptor discharge and may persist when the rates have declined in higher order neurons.[4] In addition to rate, synchronicity between active nerve fibers has been suggested to play a role.[5]

The neural structures involved in pain can grossly be subdivided into those of the peripheral and central (spinal cord and brain stem as well as cortex and subcortical nuclei) nervous system. The peripheral part consists of nociceptors with cell bodies located in the dorsal root ganglia (body) or the trigeminal ganglia (face). Stimuli are detected at free nerve endings located in the skin or in internal organs. From the endings (termed peripheral branch or terminal), the axon runs to the spinal cord where it makes synaptic contacts to local interneurons, termed central terminals. Midway, the axon is connected to the cell body via a short branch, the stem axon. The axons conveying pain can be either thinly myelinated A–delta fibers, conveying the sharp immediate pain, or unmyelinated C-fibers, conveying the dull slow pain. It should be noted that whereas the action potential travels from the peripheral to the central terminal, chemical signals can travel either direction. This means that the descending neural activity from central regions also can affect the peripheral terminal by retrograde transport of substances which at the terminal take part in, for example, inflammatory processes. The central terminals terminate on interneurons in particular lamina of the dorsal horn. In the dorsal horn, one set of interneurons send spinal cord projections to neurons in the brain stem (rostral ventral medulla, periaqueductal gray, parabrachial nucleus) and thalamus. The thalamic neurons in turn send projections to sensory cortical areas, whereas the parabrachial nucleus sends projections to amygdala which in turn projects to anterior cingulate and insular cortex. In addition, orbital and prefrontal cortical areas are involved. Thus, there is no single pain center, but rather a distributed "pain matrix."

Peripheral pain, which is the topic of this chapter, constitutes changes to the C- or A-delta fiber. These changes may include sensitization of the sensory ending and may be the result of a complex interplay between the neuron and nonneuronal cell types of the skin such as keratinocytes. A large body of work in this area is investigating the intracellular and molecular mechanisms of nociceptive pain.[6] Changes may also occur on the axon,[7] axonal sensitization, and could be the result of changed ion channel densities dictated by the cell body. Finally, there can also be changes to the synaptic contact to the interneuron in the dorsal horn, contributing to the phenomena termed windup. Central sensitization includes NMDA receptor-mediated hypersensitivity, loss of tonic inhibition, changes in glia–neuronal interactions and changes in descending neuromodulation. It might be speculated that central sensitization, in part, is due to synaptic plasticity (reviewed in Ref. 8) as well as changes in intrinsic neuronal excitability[9,10] in pathways transmitting information to or within regions processing pain or in pathways modulating this transmission. Intriguingly, it seems that regardless of whether the initial cause to pain lies in the peripheral or in the central part, there will eventually be changes in both. Due to the multifaceted nature of pain, it has proved difficult to obtain an understanding of the underlying mechanisms. Computational modeling may, therefore, provide an additional path complementing *in vivo* and *in vitro* experiments.

2. MODELING OF PERIPHERAL PAIN

In pain research, the use of computational modeling is relatively limited. Following the seminal work of Hodgkin, Huxley, Katz, and Frankenhauser using computational modeling to study the propagation of action potentials in the giant squid axon, numerous studies have been conducted on the topic of propagation of action potentials in axons, both myelinated and unmyelinated. I have not included here computational studies of action potential propagation unless the study discussed the relevance to pain, as is the case, for example, in Amir and Devor[11] discussed more below. However, as propagation and particularly failure of propagation would also affect nociceptive processing, such studies may have relevance to the field of pain. Failures, particularly for high frequencies, are an integral part of axonal function, and it is conceivable that failures form part of a mechanism limiting nociceptive input.[12]

In an early study, Scriven[13] investigated the impact on action potential propagation by changing ion concentrations due to small intra- and extra-cellular volumes, further discussed below. Moreover, in a forward-looking review, Britton and Skevington[14] discussed methodological aspects of computational modeling applied to problems in pain and exemplified by discussing the role of the NMDA receptor in windup, discussed more below.

A large portion of pain research has been focusing on the aspect of sensitization, where the behavioral response of the animal is enhanced with respect to the stimuli. It has generally been assumed that this enhancement originates from changes in the detection threshold of the peripheral end located in the skin. However, it has not been possible to record from distal ends of C-fibers due to their small diameter, <0.5 μm and since it is hard to localize them in the skin. Studies using biochemical techniques have on the other hand produced vast amounts of information on the molecular interactions. Only a few modeling studies have, however, been conducted studying the intracellular signaling pathways. In one study, to elucidate the regulation of calcium, as a general mediator of neuronal excitability, the ATP-induced P2-receptor regulation of calcium, and thereby associated cellular excitability, was studied by Song and Varner.[15] The model contains 252 parameters which were estimated from nine different sets of data using optimization methods. They found relations for the cross talk between P2Y and P2X receptor activity explaining inhibition of P2X currents by P2Y activity.

In the absence of access to the end terminal, most electrophysiological studies of C-fibers have focused on recordings from cell bodies in the dorsal root ganglion. Questions addressing changes in detection threshold have thereby translated into studies of threshold for action potential initiation, changes in resting membrane potential, etc., and searches for the ionic mechanisms behind these changes. Excitability has also been studied in terms of changes in repetitive firing: spike frequency, frequency adaptation, and burst firing. Among the ion channels, sodium channels have a clear role in action potential generation and have, therefore, been extensively studied. There are also several human mutations of sodium channels linked to pain syndromes.

2.1. Role of Na currents

Sheets et al.[16] studied the effects of a hereditary mutation N395K in a sub-type of sodium channel, $Na_v1.7$, found in a pain patient (Fig. 2.1). The model represents a small diameter DRG soma and includes a range of Na and K currents ($Na_v1.7$, $Na_v1.8$, K_{dr}, K_A, and K_{leak}). They found that the shift in steady-state activation of the mutated Na channel is the major factor of the hyperexcitability induced by the mutation, but that changes in slow inactivation of the mutated sodium channel also can modulate the level of excitability.

Herzog et al.[17] studied the role of the slowly inactivating "persistent" sodium current $Na_v1.9$ (Fig. 2.2). Their model included a transient TTX-sensitive sodium current ($Na_v1.7$), the TTX-resistant $Na_v1.9$, a delayed rectifier potassium current, and a leak current. They found that $Na_v1.9$ contributes to the resting membrane potential as a consequence of its persistent characteristic. It also amplifies responses to subthreshold depolarizations, and it produces a propensity for rebound excitation follow-ing hyperpolarization. Maingret et al.[18] also used computational modeling to investigate the impact of the $Na_v1.9$ current (Fig. 2.3). They found a role in generation of plateau depolarizations and in generation of burst firing.

Contributions to the resting membrane potential can also come from the electrogenic ion pump Na/K-ATPase[19] and the Na-dependent K current,[20] both of which provide a hyperpolarizing current and become more activated upon repetitive spike activity. They have, therefore, been assumed to pro-vide negative feedback homeostatic control.

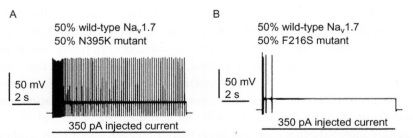

Figure 2.1 Enhanced spike generation by two human $Na_v1.7$ mutations, N395K (A) and F216S (B). Cells with control (healthy, wild type) $Na_v1.7$ channels, data not shown, respond to only one action potential produced at the onset of the pulse. *Source: Figure 4E,F in Ref. 16.*

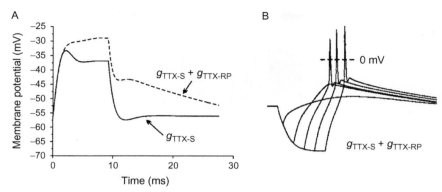

Figure 2.2 (A) Persistent Na current amplifies response to subthreshold stimulation. Soma membrane potential showing response to depolarizing current injection when only the transient TTX-sensitive (solid line, TTX-S, predominantly carried by $Na_v1.7$) or a combination of TTX-S and the persistent TTX-resistant (dashed line, TTX-RP, tentatively $Na_v1.9$) currents are present. (B) Persistent Na current contributes to post-inhibitory rebound spiking. Soma membrane potential responses to hyperpolarizing pulses of different durations when a combination of TTX-S and the persistent TTX-resistant (TTX-RP, tentatively $Na_v1.9$) currents is present. *(A) Source: Figure 6B in Ref. 17. (B) Source: Figure 10B1 in Ref. 17.*

2.2. Role of I_h

Kouranova et al.[21] studied the effect of the nonspecific cationic current I_h (the current from the HCN channel) on spiking in DRG somata (Fig. 2.4). Inclusion of I_h resulted in a depolarizing shift of the resting membrane potential, a reduction of the action potential width, and an increase of the medium afterhyperpolarization. It can also be noted that the presence of I_h might contribute to membrane potential oscillations discussed further in the following section.

2.3. Subthreshold spontaneous oscillations

Increases in excitability affecting action potential triggering may come from more than just "static" factors, such as increases in resting membrane potential. Neuronal membranes are also known to display limit cycle behavior, here in the form of subthreshold membrane potential oscillations (Fig. 2.5). These oscillations commonly appear from the interaction between two ion channels operating at a fast and slow timescale, respectively. These oscillations have been discussed in relation to occurrences of "spontaneous" ectopic (not triggered at the classical trigger point) spikes. Amir et al.[22] studied the interaction between the transient, TTX–sensitive Na current and a

Simulated nociceptor

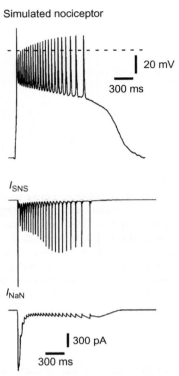

I_{SNS}

I_{NaN}

Figure 2.3 Sodium currents Na$_v$1.8 (SNS) and Na$_v$1.9 (NaN) contribute to depolarizing plateau and burst firing. Soma membrane potential (top), Na$_v$1.8 current (middle), Na$_v$1.9 current (bottom). Note how the persistent Nav1.9 remains open during the depolarization. *Source: Figure 2C in Ref. 18. © 2008 Rockefeller University Press. Originally published in Journal of General Physiology. 131: 211–225. http://dx.doi.org/10.1085/jgp. 200709935.*

leak potassium current in generation of membrane potential oscillations. They further found that the delayed rectifier potassium current reduces oscillation amplitude, by its effect on the input resistance, and that it increases the frequency of the oscillation. Moreover, including additional slower Na channel inactivation and persistent Na currents enabled improved agreement with experimental ectopic spike discharge patterns[23] (Fig. 2.6).

2.4. Repetitive spiking

In addition to mechanisms affecting the generation of a spike, several studies have addressed the capacity to generate spikes in bursts. The studies described earlier[17,18] both studied the role of the persistent Na$_v$1.9 current.

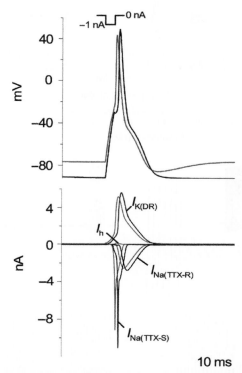

Figure 2.4 I_h contributes to membrane depolarization and afterhyperpolarization. Induction of an action potential by a depolarizing current (-1 nA, inward negative convention; 2 ms). Models containing I_h (black) or no I_h (gray) in addition to $Na_v1.7$ (Na TTX-S), $Na_v1.8$ (Na-TTX-R), and K_{dr} (K(DR)). As can be seen, I_h contributes to resting membrane depolarization and generation of a medium AHP. *Source: Figure 7C in Ref. 21.*

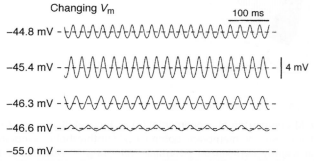

Figure 2.5 Subthreshold membrane potential oscillations can be produced by interplay between transient Na channels that show rapid inactivation and repriming kinetics and a passive leak K current. Membrane potential oscillations are triggered by depolarization and show maximal amplitude around -45 mV. Oscillation frequency 99 Hz is close to the experimental mean of 107 Hz. *Source: Figure 7A in Ref. 22.*

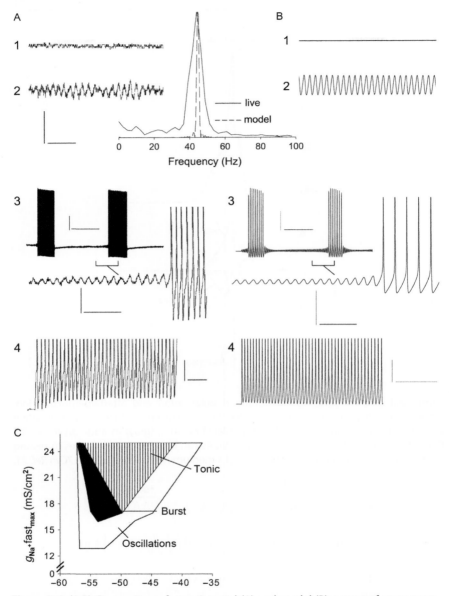

Figure 2.6 (A,B) Comparison of experimental (A) and model (B) output of soma membrane potential. (1) Resting membrane potential. (2) Subthreshold membrane potential oscillations at slightly depolarized potentials. (3) Firing in clusters/bursts at more depolarized potentials. (4) Tonic firing at even more depolarized potentials. Inset: Fourier analysis of subthreshold membrane potential oscillations. (C) Phase plane plot of conductance of the fast transient Na current (tentatively dominated by $Na_v 1.7$) versus membrane potential. The effect of reducing the conductance from its original value (top) reduces the potential window for tonic firing, increases the occurrences of bursting, and changes the occurrences of oscillations. *(A,B) Source: Figure 3 in Ref. 23. (C) Source: Figure 5A in Ref. 23.*

For sensory nonnociceptive fibers, the role of resurgent Na currents, for example, by $Na_v1.6$, has also been discussed. Furthermore, the study by Sheets et al.[16] on the $Na_v1.7$ mutation, discussed earlier, found that the mutation increased capacity of repetitive firing, decreasing the adaptation and increasing the number of action potentials generated. Changes in excitability may come not only from increases in spike generating factors but also in decreases of spike suppression factors. To study mechanisms in spike frequency adaptation, Scriven[13] included the effects of a limited intra-axonal space of a thin (unmyelinated) fiber and a limited extracellular space, the periaxonal space, to study action potential propagation (Fig. 2.7). By allowing Na^+, K^+, and Ca^{2+} ions to vary and thereby affect corresponding reversal potentials as well as electrogenic pump activity, it was found that the major contributor to spike frequency adaptation was hyperpolarization originating from, on the one hand, increased activity of the Na/K pump as a result of pump activation due to increased extracellular potassium or, on the

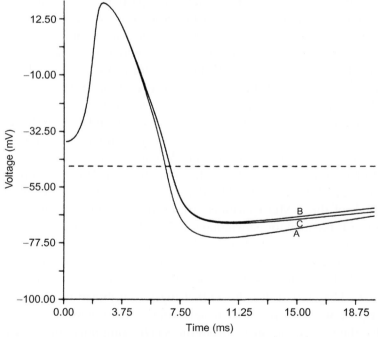

Figure 2.7 Membrane potential during an action potential for (A) absence of periaxonal K accumulation and absence of K-stimulated pump activity, (B) inclusion of periaxonal space K accumulation, (C) inclusion of periaxonal space K accumulation and K-stimulated pump activity. Dotted line represents resting membrane potential. *Source: Figure 1 in Ref. 13.*

other hand, from increased activation of a calcium–dependent potassium current activated by increased intra–axonal calcium accumulation. Our own studies[24,25] described below also included a periaxonal space and also found effects in excitability from changes in reversal potentials and pump activity.

Oscillations have, as discussed earlier, been implicated in the generation of spontaneous activity. When present at sufficient rates, this activity may be included in the term persistent activity. Elliott[26] studied generation of persistent firing either by changes of concentration of extracellular potassium or by changes of the steady-state activation curve of delayed rectifier potassium channels and cessation of this persistent firing by slow Na channel inactivation (Fig. 2.8).

2.5. Changes in the central terminal, windup

When peripheral C-fibers are stimulated repeatedly at low frequencies (0.2–2 Hz), the response of the spinal interneuron first stays at a lower rate for some 10 s, after which it increases to a higher level for a period which may outlast the stimulation by 10 s or more. This increase is termed windup, and it has been shown that the NMDA receptor plays a crucial role in this. We include modeling of this phenomenon even though it is considered a central sensitization as it may involve the synapse between the central terminal of the C-fiber and the interneuron and because it has been the subject of several modeling studies. In an early study by Britton et al.,[27] a model was constructed for the dorsal horn circuit including descending midbrain input. NMDA receptors were included in the input from small fibers (C-fibers) but not for large (A-beta) fibers. The depolarization-induced unblock of the Mg-block of the NMDA channel, arising from summed input from the C-fiber, was essential for generating windup of the interneuron. In a study by Farajidavar et al.,[28] effects from short-term synaptic plasticity (STP) as well as long-term spike-timing-dependent plasticity (STDP) changes on A-beta, A-delta, and C-fiber input to an interneuron were studied (Fig. 2.9). They found that the STP changes contributed to the band-pass property of windup (maximal neural responses for inputs at 0.3–3 Hz) and that the windup originated from the STDP component of the synaptic changes. Output from their model, shown in Fig. 2.9A, can be compared to experimental data shown in Fig. 2.9B. The link to importance of the NMDA receptor is thus implicitly included by the dependence of windup on STDP. Moreover, in a study by Aguiar et al.,[29] the integration of A–delta

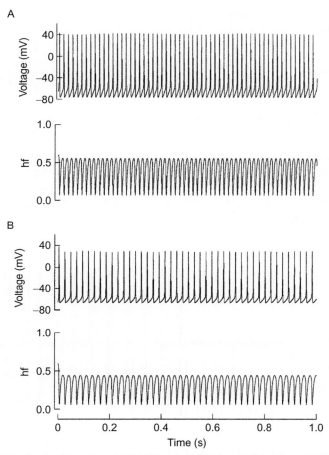

Figure 2.8 Persistent firing caused by (A) a depolarizing shift in the steady-state activation curve of the potassium channel and (B) a +10 mV shift in the potassium reversal potential. Soma membrane potential (top) and fast potassium channel inactivation particle (hf, bottom). *Source: Figure 1 in Ref. 26.*

and C-fiber input was studied in an interneuron containing a range of ionic conductances. In their study, windup was produced as interplay between the NMDA receptor and its magnesium block, and slow depolarizations produced by a calcium-dependent cationic conductance whose calcium came from the NMDA receptor as well as from a voltage-gated calcium channel. Thus, temporal summation was a combination of AMPA, NMDA, and cationic channel depolarization summing at fast, intermediate, and slow timescales.

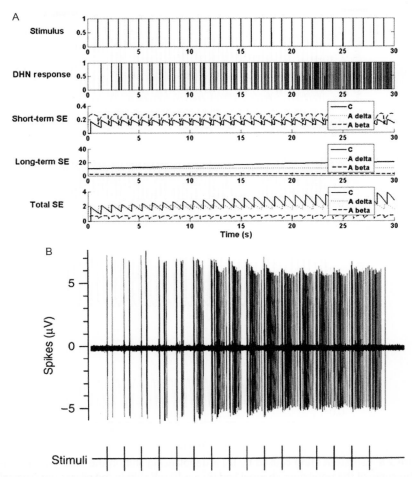

Figure 2.9 (A) Windup. Dorsal horn neuron (DHN) spiking and synaptic efficacy (SE) following 1 Hz stimulation. Note the increase in C-fiber synaptic efficacy (solid lines) and the gradual increase in DHN spiking response. Spiking response replicates well the experimental data shown in (B). (B) Enhanced spiking response, windup. Experimental extracellular recoding from a single spinal motor unit in the rat following 1 Hz stimulation. *(A) Source: Figure 6 in Ref. 28. (B) Source: Figure 9 in Ref. 28.*

2.6. Models of action potential propagation

The models discussed so far have all been one compartment models representing a small DRG somata or a space-clamped axon. Spatial models enabling the study of propagating action potentials have, as mentioned in the beginning, been used extensively in the study of axon function, but to a small extent in problems in pain, despite ample evidence that changes in propagation are present in pain patients.[30]

Amir and Devor[11] studied mechanisms by which the action potential, traveling along the axon of an A–delta fiber from peripheral to central terminals, may invade the stem axon and the soma. This may constitute a mechanism to couple the activity conveyed by the fiber to the soma and thereby affect its metabolic activity, and thereby link electrical activity to synthesis of, for example, ion channels. This is of particular interest to studies of pain since changes in ion channel expression or activity is one of the key hypotheses for the excitability changes observed in patients. They simulated a piece of the peripheral branch and a piece of the central branch, both connected to the stem axon, in turn connected to a soma.

We have also used a spatial representation of a nociceptive C–fiber in our work. In Tigerholm *et al.*,[25] we developed a model representing the distal part of the C–fiber including the thin end (branch) and thicker central part (parent) comprising 2430 isopotential compartments. Each compartment was equipped with a set of ion channels ($Na_v1.7$, $Na_v1.8$, $Na_v1.9$, K_{dr}, K_A, K_M, K_{Na}, h) and the Na/K-ATPase ion pump. Similar to Scriven,[13] the extracellular space adjacent to the axon, the periaxonal space, was represented so that changes in intra- and extracellular concentrations of Na^+ and K^+ ions could be represented and used to compute, for example, reversal potentials. The model replicates basic physiological properties of mechano–insensitive nociceptive fibers such as conduction velocity as well as changes of conduction velocity, both activity–dependent slowing (ADS) developing on the minute timescale and the faster acting recovery cycles (RC) including the velocity speedup denoted supernormal phase. ADS and RC as produced by the model, Figs. 2.10A and 2.11A, respectively, can be compared to human data, Figs. 2.10B and 2.11B, respectively. The model could also replicate velocity changes from pharmacological manipulation, block of the h–channel and the TTX–sensitive $Na_v1.7$ channel, as well as changing the temperature and the extracellular sodium ion concentration. The rationale for investigating the mechanisms behind ADS was that ADS is reduced in pain patients[30] and may thereby represent one of the excitability factors in pain.

We have subsequently[24] used our C–fiber model to investigate potential changes in ion channel expression underlying the changes in C–fiber properties observed in chronic pain patients[30] and in animal experiments using application of nerve growth factor.[33] The data show a change in proportion of mechano–insensitive presumed nociceptive to mechano–sensitive

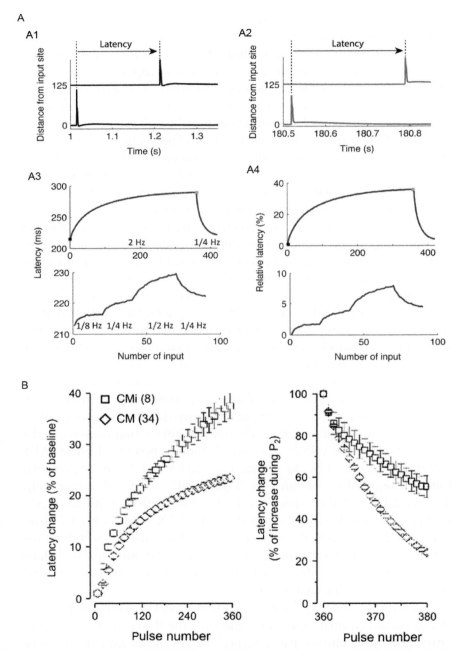

Figure 2.10 (A) Activity-dependent slowing of conduction velocity. Stimulus evoked action potentials for (A1) the first and (A2) 360th stimulation. Time frame shown is

presumed sensory fibers from 2:1 to 1:2, suggesting that a proportion of the nociceptive fibers have acquired properties rendering them mechano-sensitive, and by this contributing to enhanced pain. In our work, we used the model to analyze which changes to ion channel or ion pump conductances that would make a mechano-insensitive fiber function more like a mechano-sensitive (Fig. 2.12). Relative to nociceptive mechano-insensitive fibers, mechano-sensitive fibers show a higher initial conduction velocity, less activity-dependent conduction velocity slowing, and less prominent post-spike supernormality and a lower propensity for conduction failure. We found involvement of $Na_v1.7$ and $Na_v1.8$, consistent with several lines of experimental work. However, whereas the density of $Na_v1.7$ was increased (as expected), density of $Na_v1.8$ was decreased. Further, the results also showed support for an increase of the Na/K pump activity. Interestingly, the largest contribution making a nociceptive fiber behave as mechano-sensitive fibers came from a reduction of the delayed rectifier K current, commonly associated with action potential repolarization.

3. *IN SILICO* PHARMACOLOGY

3.1. Pharmaceutical target approach

In drug discovery addressing metabolic, cardiac, and nervous system diseases, ion channels are gaining interest as targets.[34] Ion channel-specific blockers and enhancers are, therefore, developed. The pharmaceutical industry is often designing new pharmacological compounds by targeting single receptors, enzymes, and ion channels, a strategy denoted "target" approach. However, production and testing of new drug candidates is costly. Repeated cycles of this process to explore alternative candidates or optimize

---

expanded from simulations shown in (A3) and (A4) around the time of the first and last AP, respectively. (A3) Latency during repetitive stimulation (upper panel: 360 pulses at 2 Hz, 60 pulses at 0.25 Hz; lower panel: 20 pulses at 0.125 Hz, 20 pulses at 0.25 Hz, 30 pulses at 0.5 Hz, 20 pulses at 0.25 Hz). (A4) Relative latency changes for both the high-frequency (upper panel) and low-frequency (lower panel) protocols. The latency during repetitive stimulation was normalized to the initial latency. Compare upper panel to CMi-data from human shown in (B). (B) Activity-dependent slowing in human. The figures show the change in latency (time between stimulation and recording electrode) relative to the latency of the first pulse over 360 stimulations at 2 Hz. CM mechano-sensitive fiber, CMi mechano-insensitive nociceptive fiber. *(A) Source: Figure 3 in Ref. 25. (B) Source: Figure 7B in Ref. 31.*

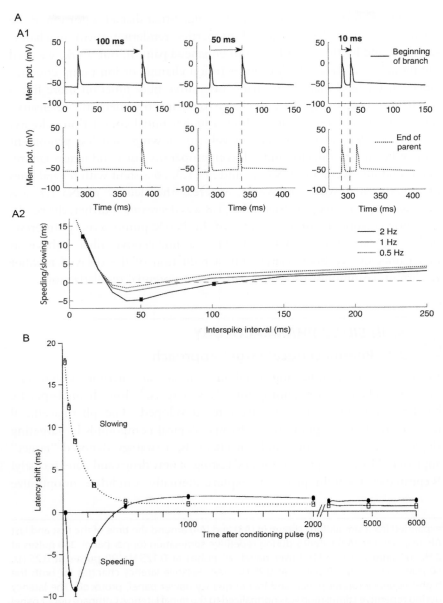

Figure 2.11 (A) Recovery cycle velocity changes evoked by double pulse stimulation. The figure shows the slowing/speeding during the recovery cycle protocol. The repetition frequency is 0.5, 1, or 2 Hz and the interstimulus interval (ISI) vary between 10 and 250 ms. (A1) The membrane potential for different ISIs. The upper graphs represent the membrane potential at the beginning of the branch axon and the lower graphs show

function are time consuming. Moreover, with a target approach, it may be difficult to effectively treat the clinical symptoms. An initial positive outcome on a cell line where the new compound shows high affinity for the targeted channel may later *in vivo* tests turn out negative with regard to treating the symptoms. In neuroscience, this may be particularly common due to the complexity of interactions at each level (cell, tissue, and organ). However, using computational neuroscience, we can perform extensive searches among alternatives of targets and can do systems-level function testing by using a combination of biophysical modeling and computational search strategies. By using a biophysical model, we are able to span several levels from the ion channel to the single neuron and networks of neurons. Thus, the tests are done on a systems level where effects on physiological function can be assessed, and characterization of the target is achieved at the molecular (ion channel) level. The approach, therefore, has the advantage of producing biophysical target characteristics usable for design of drug candidates. A computational approach thereby complements experimental approaches. Importantly, the model derived, due to its mechanistic nature, will also constitute a hypothesis of MOA, a problem central in drug development.

3.2. Advantages of modeling approaches

In the following, we discuss a few issues where a modeling approach can provide advantages. Items labeled [A]–[F] will be key points that can be addressed by a modeling approach and we will return to these below. In clinical pain studies, findings of human single point mutations in patients have provided valuable insights, for example, mutations of the sodium channel 1.7. Intriguingly, patients with a loss–of–function mutation have a complete loss of sense of pain, whereas those with a gain-of-function mutation suffer

---

the membrane potential at the end of the parent axon. (A2) The slowing/speeding for different ISIs. Note the interval between 25 and 75 ms where the second AP propagates faster than the first, the so-called supernormal phase. Compare to mechano-insensitive fiber–data from human shown in (B). (B) Velocity recovery cycles in human C-fibers. The figures show the difference in latency between the first and second pulse in a double pulse protocol applied at 0.25 Hz when the time between the two pulses was varied between 20 and 6000 ms. A positive difference indicates the second pulse propagated slower than the first. Mechano-sensitive fibers (open boxes) show how slowing increases for decreasing interstimulus interval (ISI). Mechano-insensitive nociceptive fibers show slowing for short and long ISIs but speeding for ISIs between 20 and 400 ms, the so-called supernormal phase. *(A) Source:Figure 6 in Ref. 25. (B) Source: Figure 4 in Ref. 32.*

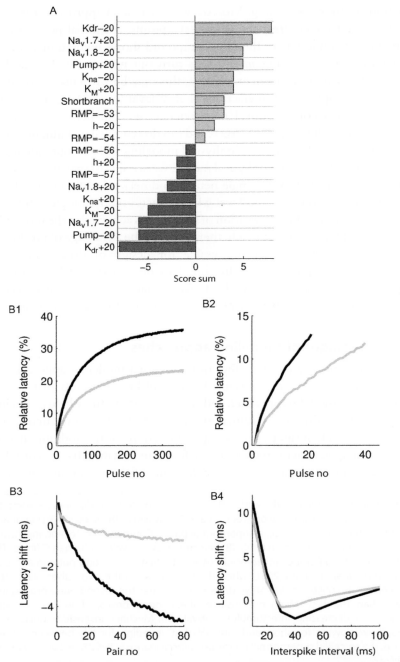

Figure 2.12 (A) Results from weighting together results from all stimulation protocols by use of a scoring system. Large bars should be interpreted as strong candidates for

from Paroxysmal extreme pain disorder (PEPD).[35] Genetic mutation studies, however, have some limitations. First, it is unknown whether also additional factors are involved, for example, compensatory upregulation of other Na channels, role of channel-binding partners, modifier genes, epigenetic factors, etc. Second, it is reasonable to assume that the more general sources of pain, trauma, inflammation, and metabolic disorders include changes in more than one ion channel. Taken together, this would call for a multifactorial approach to address pain symptoms [A].

In *in vivo* animal experiments (often under general anesthesia), the following advantages can be seen:

+ experimental models of pain induction can be applied and new stimulation/diagnostic protocols can be developed
+ activity-dependent changes in fiber conduction velocity (CV and ADS) can be measured and compared to healthy human subjects and patients

However, disadvantages are also present:

− it is hard to control experimental factors: agent concentration, temperature, and physiological status of the animal
− as in human experiments, there is no access to C-fiber intracellular membrane potential [B]

The use of *in vitro* preparations (cell cultures, blocks of neural tissue) can provide additional, complementary information, particularly regarding the problem *in vivo* of control of experimental factors at cellular level. Advantages include:

+ experimental factors: agent concentration, temperature, etc., can be well controlled
+ *in vitro* experiments on somata of DRG neurons provides threshold and shape of action potential
+ *in vitro* experiments on somata of DRG neurons give characteristics of individual ion channel subtypes

---

explaining what change makes a CMi fiber more (light gray) or less (dark gray) CM-like. The results suggest that, compared to CMi fibers, CM fibers have less K_{dr} channels, more $Na_V1.7$ channels, etc. (B) Simulations with one set of parameters changed to enable the nociceptive mechano-insensitive (CMi) model (black, default values) to display sensory mechano-sensitive (CM) fiber-like function (gray). Compared to the default CMi model (black), the CM-like model (gray) has 50% lower K_{dr} channel density and 10% higher pump density and a shorter branch length (1 rather than 2 cm). Figures show how (B1) ADS, (B2) conduction failures, (B3) supernormal phase type-1, and (B4) supernormal phase type-2 were affected by these changes. *(A) Source: Figure 5A in Ref. 24. (B) Source: Figure 6 in Ref. 24.*

However, disadvantages include:
— the resting membrane potential *in vitro* is different from *in vivo* [C]
— there is no fiber conduction (no traveling action potential) [D]
— there are no activity-dependent propagation changes (ADS, RC) [E]
— the link between models of pain at the cellular level, effects at the systems level, and symptoms of patients, that is, part of the problem of specifying MOA, is complicated [F]

A computational modeling approach provides some of the information miss-ing in the methodologies listed earlier and thus constitutes a complementary approach. Using a computational approach, we have addressed the problems [A]–[F] listed earlier. More specifically:
+ the model includes a range of ion channels expressed in C-fibers as well as ion transporters and thus allows for a multifactorial approach [A].
+ the model fills the gap between *in vitro* DRG somata recordings and *in vivo* microneurography by providing intracellular membrane potential measurements [B], producing C-fiber action potential conduction [D], and producing activity-dependent velocity changes [E]
+ the model is not based on *in vitro* resting membrane potentials [C]
+ the model addresses part of the problem of MOA [F] as the criteria for constructing the model are taken from observations of changes observed in patients

3.3. Constructing models from data

Optimization and parameter search are extensively used in computational biology and systems biology to construct models from data (e.g., Refs. 36,37). In computational neuroscience, parameter optimization is less com-mon, but has, for example, been used to find parameters for, for example, the Hodgkin–Huxley equations, given voltage clamp and current clamp data.[38,39] The goal in these cases is to produce a model that accurately describes experimental data of a neuron. It can be applied to find the value of a parameter either because the value has not been measured experimen-tally, as for instance the density/conductance of an ion channel in a thin den-drite or axon, and is, therefore, unknown or because the value is different under different conditions, as, for example, the concentration of a neuro-modulator, and the parameter, therefore, is "free."

The aim of using optimization in most cases is to obtain a model that accurately describes data. In our approach, we instead want to find the parameter changes that reverse a pathological function into normal function.

For example, we want to find optimal concentrations of a drug that affect the characteristics of a diseased model neuron in order to change its pathological response characteristics back to healthy function.[40,41] Thus, starting from a model producing output that replicates the activity of the diseased cell, the search method searches for parameter values that change ion channel characteristics in a way that the behavior of the model neuron replicates the behavior of the healthy cell (Fig. 2.13). The procedure is iterative. Output from the pathological model simulations constitutes the input to the optimizer. The optimizer searchers for cases of reduced error (decreased difference between pathological and healthy model function) by adjusting parameter values of the drug concentrations which in turn change ion channel properties. An advantage of the way we formulate the research problem is that improvements are searched for rather than attempting to find the global minimum representing the "true" model.[40,41] Improvements are manifold (see Fig. 2.14), but the global optimum is single and it is in general impossible to show it was found.

3.4. Quality of solution and model

Parameter search by definition searches for the minimum of the error. In our studies, solutions are often concentrations of drug substances. Due to biological variation (limited precision in specification of dose for an individual, variation in dose between administration time points, variation in disease status of the patient, variation in physiology between humans, etc.), it is important that adequate treatment is obtained for a wide range of concentrations. Thus, concentration values similar to the selected solution should also show improved cellular function. More specifically, we want the neighborhood of a solution to be as large as possible. We are thus more interested in the "width" of the base of the error function, rather than the smallness (depth) of the minimum point. Similar use of classification of simulation output and estimation of class volume has been used for model evaluation in computational biomechanics.[42] Thus, we believe that for target identification, robustness of the solution constitutes a much more important issue than, for example, finding the local minimum of the error (here making the difference between the pathological function and normal function as small as possible).[40] In addition, to assess model quality, parameter sensitivity may be calculated. First-order sensitivity index[43] is one such measure estimating the contribution of each parameter to the output of the model. One of the advantages of this method is that it provides the contribution of each of a

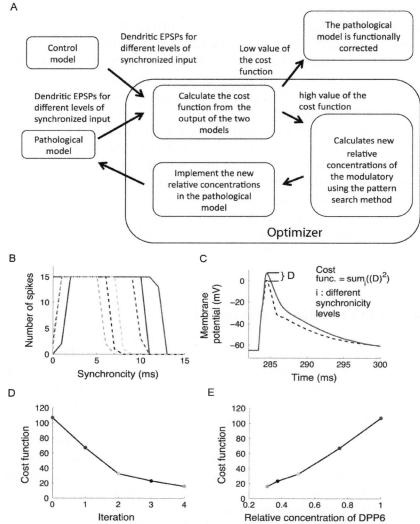

Figure 2.13 Optimization method. (A) The output of the model, the compound EPSP from the pathological cell, was compared to the control model to generate the cost function value. If the value was high, new relative concentrations of the modulators were calculated by the optimizer. The effects on channels by the new modulator concentrations were subsequently implemented in the model, a new simulation was run, and a new set of EPSPs were generated. A new cost function value was computed, and when this fell below the stop criteria, the model was labeled functionally corrected. (B) Output from the pathological model (red) is used as the starting point of the optimization. The goal is defined by the control model output (blue). For illustration, output from each iteration cycle during the optimization is also shown (dashed lines). Final

set of parameters to the output at each time point. It is thereby possible to assess contributions of a parameter to different phases of the simulation, for example, during the applied external stimulus versus during the subsequent systems response. The index is, however, a local property, calculated around a single point (the proposed model), and thus does not provide information about, for example, all models possible within, for example, the volume spanned by the physiological interval of each parameter and so even less for all possible physiological models.

3.5. An example of applying computational search and addressing robustness

In a recent project, we investigated how we could make a neuron with pathological properties behave more normal (Tigerholm et al., 2011). We constructed pathological cell models representing three channelopathies, in this work linked to epilepsy, producing enhanced excitability. In the first case, we had enhancement of the transient sodium channel; in the second case, a suppression of an A-type potassium channel; and in a third case, we modeled the characteristics of a mutated potassium channel from a patient. In each case, our question was whether we could change the property of a specific potassium channel so that the neuron would behave like a healthy neuron. The channel was selected due to its implication in epileptogenesis and due to multiple data on modulation of its characteristics by intracellular messengers and binding partners.

We further selected which cellular function characteristic of the pathology that we should reverse. In this case, we chose to measure production of action potentials, specifically how many nerve impulses which are produced from a set of synaptic input with a relatively high synchronicity. The rationale for looking at production of nerve impulses is that epilepsy is characterized by excess production of nerve impulses. The rationale for looking at synchronicity is that epilepsy is characterized by enhanced synchronicity.

---

solution found is indicated in green. Each color in (B), (D), and (E) corresponds to the same iteration cycle. Panels (C–E) illustrate the optimization cycle. (C) Model output, dendrite membrane potential, produced in a control case (blue) and a pathological case (red). The difference in peak EPSP amplitude at the input site for the first and second input cycles was used to calculate the cost function value. (D) The cost function value at each iteration cycle during the optimization. (E) The relative concentration of the modulator (here the relative abundance of the auxillary subunit DPP6 coassembeled with the ion channel) for each iteration cycle of the optimization. *Source: Figure 3 in Ref. 40.*

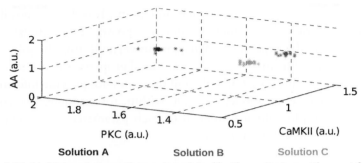

Figure 2.14 Multiple solutions. Data points visited by the optimizer with a cost function value below 2 are shown for the three solutions A (black), B (dark gray), and C (light gray). Axes values for AA (arachidonic acid), PKC (protein kinase C), and CaMKII (calmodulin-dependent protein kinase II) are in arbitrary concentration units. *Source: Figure 8a in Ref. 40.*

Many computational studies including modulation of ion channels are based on very simplified actions to the channel, such as a shift in the voltage dependence or a change in conductance. The advantage of using experimental characteristics of changes to the channel is that we will know that the properties of the channel can be realized experimentally. The ion channel that we chose to affect, K_A (in pyramidal neurons assembled from $K_v4.2$ and other A-type potassium channel genes), has a particular property in healthy controls. This channel is assembled together with two additional proteins, KChIPs (attached on the inside) and DPPXs (attached on the outside). There are four of each bound to the functional channel. These additional proteins come in several splice variants. Depending on the variant, the channel displays different properties. By including the different properties reported in experiments, we can have a range of functional properties of this channel. The advantage of using these proteins is that they are highly specific to the potassium channel.

We now let the numerical optimizer search for combinations of the additional proteins so that the resulting properties of the potassium channel enable the diseased neurons to behave more like the healthy neuron. When the diseased neuron responds similarly to the healthy control we say it is functionally corrected. In this case, the elevated response to the very highest degrees of synchronous input characteristic of the pathologies should be reduced to the lower levels displayed by the healthy model.

Furthermore, as an alternative to using these auxiliary proteins, we also used kinases and lipid messengers known to change the function of the A-type potassium channel. The advantage of this test is that the range of effects on the channel is larger and hence a better correction is possible. The disadvantage is that these substances are not specific to this channel. However, we still know the necessary binding sites do exist on the channel. To be useful in drug discovery, one would have to make a synthetic compound that selectively targets only the binding sites of this particular channel.

For the case of kinases and lipid messengers, we obtained several different solutions (Fig. 2.14). We then subsequently investigated which solution, among the three with smallest error found, would be best in terms of robustness. As described earlier, robustness here refers to the capacity of a solution to fulfill the criteria setup for approved function. We found that the robustness was different among the solutions where the solution with the largest robustness was not the one with the smallest error.

4. CONCLUSION

In pain research, models have been used to study mechanistic relationships between cellular components: ion channels, pumps, intracellular signaling pathways, and cellular function and in particular cellular hyperexcitability such as reduced action potential threshold or increased action potential firing. When the effects of pharmacological substances on ion channel properties are included, the pharmacological effects on cellular function or disease-related dysfunction can be studied. Furthermore, the use of parameter search (parameter optimization) enables the systematic study of relationships between pharmacological effects on targets and disease symptoms expressed on the cell or systems level.

REFERENCES

1. Dubner R, Beitel R. Peripheral neural correlates of escape behavior in rhesus monkey to noxious heat applied to the face. In: Bonica J, Albe-Fessard D, eds. *Advances in Pain Research and Therapy*. Vol. 1. New York: Raven Press; 1976:155–160.
2. Fors U, Ahlquist M, Skagerwall R, Edwall L, Haegerstam G. Relation between intradental nerve activity and estimated pain in man—a mathematical analysis. *Pain*. 1984;18:397–408.
3. Zimmerman M. Encoding in dorsal horn interneurons receiving noxious and non noxious afferents. *J Physiol Paris*. 1977;73:221–232.
4. Lenz F, Kwan H, Dostrovsky J, Tasker R. Characteristics of the bursting pattern of action potentials that occurs in the thalamus of patients with central pain. *Brain Res*. 1989;496:357–360.
5. Sandkuhler J. Neurobiology of spinal nociception: new concepts. *Prog Brain Res*. 1996;110:207–224.
6. Basbaum A, Bautista D, Sherrer G, Julius D. Cellular and molecular mechanisms of pain. *Cell*. 2009;139:267–284.

7. Krishnan AV, Lin CS, Park SB, Kiernan MC. Axonal ion channels from bench to bedside: a translational neuroscience perspective. *Prog Neurobiol.* 2009;89:288–313.
8. Ji R-R, Kohno T, Moore K, Woolf C. Central sensitization and LTP: do pain and memory share similar mechanisms? *TINS.* 2003;26:696–705.
9. Daoudal G, Debanne D. Long-term plasticity of intrinsic excitability: learning rules and mechanisms. *Learn Mem.* 2003;10:456–465.
10. Zhang W, Linden D. The other side of the engram: experience-driven changes in neuronal intrinsic excitability. *Nat Rev Neurosci.* 2003;4:885–900.
11. Amir R, Devor M. Electrical excitability of the soma of sensory neurons is required for spike invasion of the soma, but not for through-conduction. *Biophys J.* 2003;84:2181–2191.
12. Sun W, Miao B, Wang X, et al. Reduced conduction failure of the main axon of polymodal nociceptive C-fibres contributes to painful diabetic neuropathy in rats. *Brain.* 2012;135:359–375.
13. Scriven DR. Modeling repetitive firing and bursting in a small unmyelinated nerve fiber. *Biophys J.* 1981;35:715–730.
14. Britton N, Skevington S. On the mathematical modelling of pain. *Neurochem Res.* 1996;21:1133–1140.
15. Song S, Varner J. Modeling and analysis of the molecular basis of pain in sensory neurons. *PLoS One.* 2009;4:e6758.
16. Sheets PL, Jackson 2nd JO, Waxman SG, Dib-Hajj SD, Cummins TR. A Nav1.7 channel mutation associated with hereditary erythromelalgia contributes to neuronal hyperexcitability and displays reduced lidocaine sensitivity. *J Physiol.* 2007;581:1019–1031.
17. Herzog RI, Cummins TR, Waxman SG. Persistent TTX-resistant Na+ current affects resting potential and response to depolarization in simulated spinal sensory neurons. *J Neurophysiol.* 2001;86:1351–1364.
18. Maingret F, Coste B, Padilla F, et al. Inflammatory mediators increase Nav1.9 current and excitability in nociceptors through a coincident detection mechanism. *J Gen Physiol.* 2008;13:211–225.
19. Hamada K, Matsuura H, Sanada M, et al. Properties of the Na+/K+ pump current in small neurons from adult rat dorsal root ganglia. *Br J Pharmacol.* 2003;138:1517–1527.
20. Bischoff U, Vogel W, Safronov BV. Na+activated K+ channels in small dorsal root ganglion neurones of rat. *J Physiol.* 1998;510:743–754.
21. Kouranova EV, Strassle BW, Ring RH, Bowlby MR, Vasilyev DV. Hyperpolarization-activated cyclic nucleotide-gated channel mRNA and protein expression in large versus small diameter dorsal root ganglion neurons: correlation with hyperpolarization-activated current gating. *Neuroscience.* 2008;153:1008–1019.
22. Amir R, Liu C-N, Kocsis J, Devor M. Oscillatory mechanism in primary sensory neurons. *Brain.* 2002;125:421–435.
23. Kovalsky Y, Amir R, Devor M. Simulation in sensory neurons reveals a key role for delayed Na+ current in subthreshold oscillations and ectopic discharge: implications for neuropathic pain. *J Neurophysiol.* 2009;102:1430–1442.
24. Petersson ME, Obreja O, Lampert A, Carr RW, Schmelz M, Fransén E. *Differential axonal conduction patterns of mechano-sensitive and mechano-insensitive nociceptors—a combined experimental and modelling study.* 2014[submitted].
25. Tigerholm J, Petersson ME, Obreja O, et al. Modelling activity-dependent changes of axonal spike conduction in primary afferent C-nociceptors. *J Neurophysiol.* 2013; [Epub ahead of print].
26. Elliott J. Slow Na+ channel inactivation and bursting discharge in a simple model axon: implications for neuropathic pain. *Brain Res.* 1997;754:221–226.
27. Britton N, Chaplain M, Skevington S. The role of N-methyl-D-aspartate (NMDA) receptors in wind-up. *IMA J Math Appl Med Biol.* 1996;13:193–205.

28. Farajidavar A, Saeb S, Behbehani K. Incorporating synaptic time-dependent plasticity and dynamic synapse into a computational model of wind-up. *Neural Netw.* 2008;21:241–249.
29. Aguiar P, Sousa M, Lima D. NMDA channels together with L-type calcium currents and calcium-activated nonspecific cationic currents are sufficient to generate windup in WDR neurons. *J Neurophysiol.* 2010;104:1155–1166.
30. Ørstavik K, Weidner C, Schmidt R, et al. Pathological C-fibres in patients with a chronic painful condition. *Brain.* 2003;126:567–578.
31. Obreja O, Ringkamp M, Namer B, et al. Patterns of activity-dependent conduction velocity changes differentiate classes of unmyelinated mechano-insensitive afferents including cold nociceptors, in pig and in human. *Pain.* 2010;148:59–69.
32. Weidner C, Schmidt R, Schmelz M, Hilliges M, Handwerker HO, Torebjörk HE. Time course of post-excitatory effects separates afferent human C fiber classes. *J Physiol.* 2000;527(Pt 1):185–191.
33. Obreja O, Ringkamp M, Turnquist B, et al. Nerve growth factor selectively decreases activity-dependent conduction slowing in mechano-insensitive C-nociceptors. *Pain.* 2011;152:2138–2146.
34. Li S, Gosling M, Poll CT, Westwick J, Cox B. Therapeutic scope of modulation of nonvoltage-gated cation channels. *Drug Discov Today.* 2005;10:129–137.
35. Dib-Hajj SD, Cummins TR, Black JA, Waxman SG. From genes to pain: Na v 1.7 and human pain disorders. *Trends Neurosci.* 2007;30:555–563.
36. Rajasethupathy P, Vayttaden SJ, Bhalla US. Systems modeling: a pathway to drug discovery. *Curr Opin Chem Biol.* 2005;9:400–406.
37. Snoep JL. The silicon cell initiative: working towards a detailed kinetic description at the cellular level. *Curr Opin Biotechnol.* 2005;16:336–343.
38. Druckmann S, Banitt Y, Gidon A, Schurmann F, Markram H, Segev I. A novel multiple objective optimization framework for constraining conductance-based neuron models by experimental data. *Front Neurosci.* 2007;1:7–18.
39. Vanier MC, Bower JM. A comparative survey of automated parameter-search methods for compartmental neural models. *J Comput Neurosci.* 1999;7:149–171.
40. Tigerholm J, Fransén E. Reversing nerve cell pathology by optimizing modulatory action on target ion channels. *Biophys J.* 2011;101:1871–1879.
41. Tigerholm J, Börjesson SI, Lundberg L, Elinder F, Fransén E. Dampening of hyperexcitability in CA1 pyramidal neurons by polyunsaturated fatty acids acting on voltage-gated ion channels. *PLoS One.* 2012;7:e44388.
42. Yakovenko S, Gritsenko V, Prochazka A. Contribution of stretch reflexes to locomotor control: a modeling study. *Biol Cybern.* 2004;90:146–155.
43. Saltelli A, Chan K, Scott EM. *Sensitivity Analysis.* New York, NY: Wiley; 2008.

CHAPTER THREE

Implications of Cellular Models of Dopamine Neurons for Schizophrenia

Na Yu[*], **Kristal R. Tucker**[†], **Edwin S. Levitan**[†], **Paul D. Shepard**[‡], **Carmen C. Canavier**[*]

[*]Department of Cell Biology and Anatomy, Neuroscience Center of Excellence, LSU Health Sciences Center, New Orleans, Louisiana, USA
[†]Department of Pharmacology and Chemical Biology, University of Pittsburgh, Pittsburgh, Pennsylvania, USA
[‡]Department of Psychiatry, Maryland Psychiatric Research Center, University of Maryland School of Medicine, Baltimore, Maryland, USA

Contents

Abstract

Midbrain dopamine neurons are pacemakers *in vitro*, but *in vivo* they fire less regularly and occasionally in bursts that can lead to a temporary cessation in firing produced by depolarization block. The therapeutic efficacy of antipsychotic drugs used to treat the positive symptoms of schizophrenia has been attributed to their ability to induce depolarization block within a subpopulation of dopamine neurons. We summarize the results of experiments characterizing the physiological mechanisms underlying the ability of these neurons to enter depolarization block *in vitro*, and our computational simulations of those experiments. We suggest that the inactivation of voltage-dependent Na^+ channels, and, in particular, the slower component of this inactivation, is critical in controlling entry into depolarization block. In addition, an ether-a-go-related gene (ERG) K^+

53

current also appears to be involved by delaying entry into and speeding recovery from depolarization block. Since many antipsychotic drugs share the ability to block this current, ERG channels may contribute to the therapeutic effects of these drugs.

Dopamine is a neuromodulatory transmitter released by several groups of neurons in the brain; here, we focus on those in the ventral tegmental area (VTA) and substantia nigra pars compacta (SNc) located in the midbrain. Although relatively few in number, these cells make extensive connections with other brain regions and subserve a variety of important brain functions.[1] Dopamine neurons in the SNc or A9 project principally to the dorsal striatum, whereas those in the medially adjacent VTA or A10 innervate the ventral striatum as well as cortical and subcortical regions of the limbic system.[2] As suggested by this pattern of connectivity, midbrain dopamine neurons are necessary for normal motor function[3] and a wide array of behavioral processes including reward signaling, affect, addiction, and stress.[4–6] Clinically, VTA and SNc dopamine neurons have been implicated in both neurodegenerative and psychiatric disorders, such as Parkinson's disease and schizophrenia. The selective death of SNc dopamine neurons is a hallmark feature of Parkinson's disease,[3] and deranged reward signaling mediated by dopamine neurons in the VTA appears critical for the initiation of addiction and relapse.[4] Most relevant for the purposes of this chapter, dopamine receptor blockers are effective therapeutics for psychosis and are often used to treat the positive symptoms of schizophrenia including hallucinations and delusions.[7,8]

1. DOPAMINE NEURON ELECTROPHYSIOLOGY

In vivo, midbrain dopamine neurons fire irregular single spikes at 3–8 Hz[9] interspersed with bursts of spikes with higher intraburst frequencies.[10] In freely moving rats, reward–related stimuli increase both the incidence and intensity of bursting.[11] In awake monkeys, bursts signal behaviorally relevant information about appetitive rewards.[5] Deprived of their afferent inputs in a slice preparation, these neurons are spontaneous pacemakers that fire single spikes between 1 and 4 Hz.[12] Depolarizing current steps elicit steady firing rates up to 10 Hz, but additional depolarization causes a cessation of spiking, which is often referred to as "depolarization block."[13] Other depolarizing events *in vitro* can also lead to a train of high frequency action potentials followed by depolarization block, including spontaneous plateau potentials that occur when small conductance (SK) calcium–activated potassium channels are blocked[14] and in response to simulated activation of an NMDA receptor conductance.[15]

2. DEPOLARIZATION BLOCK HYPOTHESIS OF ANTIPSYCHOTIC DRUG ACTION

All antipsychotic drugs currently used to treat schizophrenia block CNS dopamine receptors to varying degrees.[7,8] Although receptor blockade is a rapid process, antipsychotic drugs had been reported to take up to 3 weeks to achieve their maximal therapeutic effects.[16] This led to speculation that a mechanism with a longer time course than acute receptor block is required to account for the delayed therapeutic efficacy of these drugs. An important clue was provided by studies comparing the acute and chronic effects of antipsychotic drugs on the activity of dopamine neurons in anesthetized rats. Although acute administration of antipsychotic drugs increases the proportion of spontaneously active neurons, 3 weeks of repeated treatment with these drugs has the opposite effect, decreasing the number of spontaneously firing dopamine cells. Several lines of evidence point to the involvement of depolarization block in this phenomenon. For example, although local application of glutamate fails to restore normal levels of population activity, compounds with inhibitory effects including dopamine, and GABA agonists are effective in increasing the number of spontaneously firing neurons.[17–18] Intracellular recordings obtained from dopamine neurons in rats chronically treated with antipsychotic drugs are consistent with these results and are characterized by small amplitude spikes and large depolarizing shifts in membrane potential indicative of depolarization-induced inactivation of spike generation (Fig. 3.1). Notably, the therapeutic efficacy of antipsychotic drugs is correlated with their ability to induce selective depolarization block in the VTA, whereas the ability to produce extrapyramidal motor side effects is correlated with their ability to induce depolarization block in the SNc.[19–21] Therefore, the depolarization block hypothesis of antipsychotic drug action posits that the induction of depolarization block mediates the therapeutic effects of antipsychotic drugs so that the time course of induction of depolarization block determines the time course of the therapeutic effects.[22]

The mechanism(s) responsible for the induction of depolarization block in dopamine neurons are incompletely understood but appear to involve long–loop homeostatic mechanisms. These mechanisms are hypothesized to require several weeks to fully develop and to compensate for a decrease

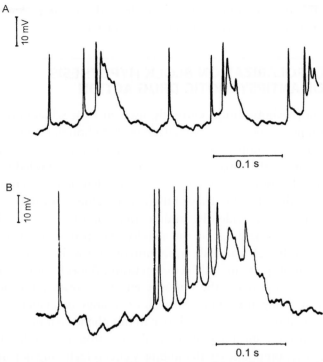

Figure 3.1 Intracellular recordings of spontaneously firing DA neurons in haloperidol-treated rats. Although most of the DA neurons in treated rats are not spontaneously active, a few firing neurons may nonetheless be impaled with a recording electrode. These neurons typically demonstrate rapid, high-amplitude burst activity unlike that commonly observed in DA neurons in control animals. (A) Rapid, high-amplitude bursting in a DA cell presumably on the threshold of depolarization block. (B) A DA cell in a less depolarized state. Final spikes in burst are inactivated. *Reproduced with permission from Figure 2 of Grace and Bunney[17] with the original figure caption.*

in dopaminergic activation of downstream targets by increasing the net excitation of the dopaminergic population. Lesions of the striatum and the nucleus accumbens, areas which form reciprocal connections with the SNc and VTA, respectively, prevent the development of depolarization block in the SNc and VTA, respectively.[21,23] This effect of striatal input is likely mediated by inhibition of local inhibitory neurons, which increases the net level of excitation of the SNc and VTA dopaminergic neurons via disinhibition. Another possible slow homeostatic mechanism may be to increase the level of net excitation to dopamine neurons from the cortex or the subthalamic nucleus.[22] Ultimately, the progressive increases in net excitation lead to depolarization block, which relieves the excessive activation of dopamine systems thought to be responsible for psychosis in schizophrenia, but compromises the ability of the dopaminergic system to subserve its normal signaling functions.

A challenge to the depolarization block hypothesis of antipsychotic drug action arose when it was shown that significant therapeutic benefits in patients are actually achieved within the first days of antipsychotic administration—well before the presumed onset of depolarization block.[24] It is well established that response to antipsychotic drugs is profoundly different in individuals with and without schizophrenia. This led to speculation that the effects of antipsychotic drugs may be different in animal models of schizophrenia than in control animals. In a recent study, Valenti *et al.*[25] assessed the effects of first and second generation antipsychotic drugs on the activity of dopamine neurons in a well-characterized neurodevelopmental model of schizophrenia. In these rats, which exhibit increased baseline excitation in the VTA, acute administration of antipsychotics increased the incidence of depolarization block. However, several weeks of repeated administration were required to achieve the maximal decrease in the population activity. This suggests that in idiopathic schizophrenia, depolarization block may be involved in both the acute and delayed response to antipsychotics drugs. The implied therapeutic significance of depolarization block in the treatment of schizophrenia provides strong motivation to understand the cellular and ionic mechanisms contributing to its development.

3. COMPUTATIONAL MODEL OF PACEMAKING AND DEPOLARIZATION BLOCK

In order to study mechanisms of depolarization block in dopamine neurons, we developed a computational model[26] of a spiking pacemaker that enters depolarization block before achieving firing rates much faster than 10 Hz. Our model (Fig. 3.2) is based on data obtained from the SNc subpopulation of dopamine neurons. There may be important differences in the intrinsic properties of subpopulations of dopamine neurons, but it is clear that both SNc and VTA dopamine neurons share a propensity to exhibit depolarization block. The model consists of a fast spiking sodium current I_{Na}, a delayed rectifier $I_{K,DR}$, a transient outward potassium current $I_{K,A}$, an SK calcium-activated potassium channel $I_{K,SK}$, an L-type calcium current $I_{Ca,L}$, a leak current I_L, and an applied external stimulus current I_{STIM}

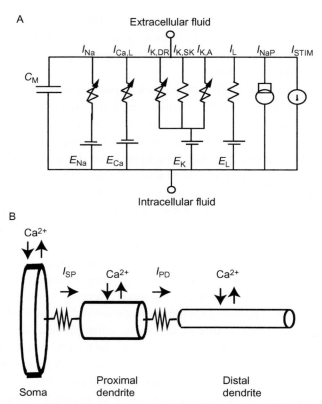

Figure 3.2 Model of dopamine neuron as a spiking pacemaker. (A) Conductance-based equivalent circuit for each compartment (I_{STIM} is confined to the soma). The arrows indicate time- and voltage-dependent conductances. (B) Schematic representation of the multicompartmental model and the calcium balance.

(Fig. 3.2A). The model has three spatial compartments representing the soma, proximal, and distal dendrites (Fig. 3.2B), and a material balance on calcium is performed in each compartment. A material balance is also performed on sodium, and an electrogenic sodium pump current, I_{NaP}, contributes a nearly constant hyperpolarizing current to the model.

Dopamine neurons exhibit an oscillation in intracellular calcium concentration during spontaneous repetitive firing of action potentials,[27] but in our model, this oscillation in calcium is driven by the repetitive spiking rather than the other way around as suggested in some previous models.[28] Dopamine neurons have no stable resting potential, but gradually depolarize in between action potentials. During pacemaking, the following sequence of events repeats itself (Fig. 3.3). L-type Ca^{2+} channels in midbrain dopamine neurons are active at relatively hyperpolarized potentials,[29] and as the model depolarizes, calcium entry via this channel increases the free Ca^{2+} concentration in the cytosol. When an action potential is triggered, the potassium currents are activated. They repolarize the action potential, then cause a prolonged afterhyperpolarizing potential that turns off the L-type calcium channel current (blue trace in Fig. 3.3C). Since the SK potassium channel responsible for the prolonged afterhyperpolarization is voltage independent,

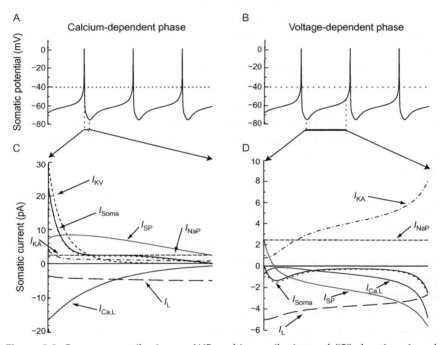

Figure 3.3 Currents contributing to AHP and interspike interval (ISI) duration. A and B show the somatic membrane potential during spontaneous spiking activity, simulated from schematic DA neuron model. Also indicated is the portion of the ISI analyzed below. C and D show the most important currents that contribute to the total somatic current during the time intervals of the ISI indicated in panels A and B. Labeling: I_{KA}, hyperpolarization-activated K^+ current; I_{NaP}, electrogenic Na^+ pump; $I_{Ca,L}$, Ca^{2+} L-type channel current, I_{SP} dendritic current; I_L, leak current; I_{Soma}, net contribution from I_{KA}, I_{NaP}, $I_{Ca,L}$, I_{SP}, and I_L to total somatic current is indicated by solid curve. The total somatic current was obtained by multiplying the somatic current densities in the model by the somatic surface area. *Reproduced from Figure 3 of Kuznetsova et al.[26] with kind permission from Springer Science and Business Media.* (See the color plate.)

the time course of this portion of the cycle (Fig. 3.3A) is dependent on the time course of the removal of Ca^{2+}. In our model, the SK current is primarily activated in the dendrites due to their larger surface area to volume ratio, which causes larger excursion in free cytosolic Ca^{2+} concentration in the smaller diameter dendrites. The effect of this current in the soma therefore manifests itself only indirectly as the axial flow of current between the soma and proximal dendrites (I_{SP}, red trace in Fig. 3.3C). In addition to closing the L-type Ca^{2+} channels, the afterhyperpolarization removes sodium channel inactivation so that there is sufficient sodium channel availability for the next slow depolarization to trigger an action potential. During the hyperpolarization following the action potential, calcium concentration decreases due to the very low level of activation of $I_{Ca,L}$, turning off the SK channel current.

During the Ca^{2+}-independent portion of the interspike interval (Fig. 3.3B), the membrane potential changes very slowly as the L-type calcium channel opens. Activation of the L-type channel is not slow, but the net current is small[30] during the interspike interval. This current (blue trace in Fig. 3.3C) turns on gradually with increasing depolarization. In fact, in the model, the membrane potential changes so slowly over the "voltage-dependent" phase of the interspike interval that L-type Ca^{2+} channel activation remains near its steady-state value as a function of membrane potential. The slow depolarization continues as the L-type Ca^{2+} channels open again, and the cycle repeats.

If a depolarizing current step of sufficient magnitude is applied to the model soma (Fig. 3.4), depolarization block ensues. During spontaneous pacemaking, sufficient inactivation of the sodium channel is removed during the afterhyperpolarizing phase of each action potential so that there is sufficient sodium channel availability to trigger the next full-height action potential. In the presence of a sufficiently strong depolarizing current pulse (Fig. 3.4A), there is insufficient afterhyperpolarization following the action potentials to remove sodium channel inactivation (Fig. 3.4B) resulting in depolarization (DP) block.

4. AVAILABILITY OF SODIUM CURRENT CONTROLS ENTRY INTO DP BLOCK

Kuznetsova et al.[26] predicted that increasing the availability of sodium channels by increasing the sodium channel conductance density would delay entry into depolarization block by allowing spiking to continue at values of

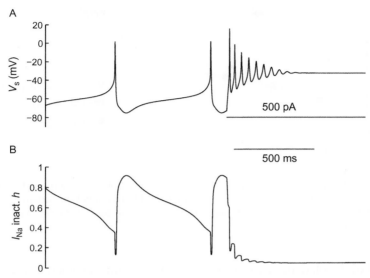

Figure 3.4 Simulated induction of depolarization block by simulated somatic current injection. (A) Somatic membrane potential trace from schematic DA neuron model under control conditions before simulated applied current and during application of 500 pA pulse. In the latter case, a cessation of the spiking response is observed. (B) Inactivation variable (*h*) of fast Na$^+$ current as a function of time for the simulation in A. At the larger value of applied current, the inactivation of Na$^+$ current is not removed leading to depolarization block. *Adapted with permission from Figure 6 of Kuznetsova* et al.[26].

injected current that caused entry into depolarization block under control conditions. Figure 3.5 shows that for the control value of the sodium channel conductance density, the model produces sustained spiking at up to 10 Hz at 400 pA of depolarizing current injected at the soma; if the level of injected current is increased, the neuron ceases to spike, indicated by the end of the red curve at those values. Increasing the sodium channel conductance density allows sustained spiking at faster frequencies and higher levels of depolarizing current than was possible with the control value.

In order to test this prediction in dopamine neurons, we used the dynamic clamp, which allows the time course of virtual conductances to be calculated in real time based on the time course of the measured value of the somatic membrane potential. A virtual conductance is realized by injecting the current produced by the product of the instantaneous value of the conductance and the driving force into the soma. This technique enabled Tucker et al.[31] to show that augmenting the sodium conductance indeed allowed dopamine neurons to continue spiking in the presence of

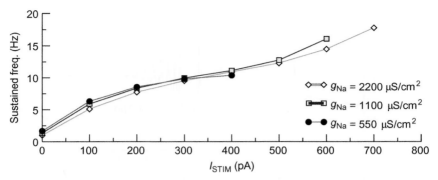

Figure 3.5 Spiking current conductance densities control the maximum frequency in the schematic model. Frequency current plots at different levels of sodium conductance density (g_{Na}). *Adapted with permission from Figure 8(a) of Kuznetsova et al.[26].* (See the color plate.)

a level of depolarizing current that caused entry into depolarization block under control conditions (Fig. 3.6). Current pulses that supported continuous spiking under control conditions led to depolarization block in the presence of submaximal concentrations of the sodium channel blocker, TTX. These effects were reversed by reintroducing a sodium conductance via dynamic clamp.[31] Understanding the precise mechanism of depolarization block in dopamine neurons may lead to improved therapeutics.

5. THE ETHER-A-GO-GO-RELATED GENE POTASSIUM CHANNEL AND SCHIZOPHRENIA

Although all antipsychotic drugs act as dopamine receptor blockers, many share another common effect, namely, blockade of ether-a-go-go-related gene (ERG) Kv11.1 potassium channels (Table 3.1).[32] The first evidence that midbrain dopamine neurons might express these channels was provided by a study showing a slow, calcium-independent component of an afterhyperpolarization that was partially blocked by the antipsychotic drug haloperidol.[33] The main effect of ERG blockade in that study was a reduction in the duration of a pause in spontaneous firing that occurs following trains of spikes elicited by a square pulse of depolarizing current. This poststimulus pause in firing may be analogous to the characteristic pause in spontaneous activity that follows a burst of spikes.[34] Blocking the ERG current with the sulfonanilide antiarrhythmic drug E-4031 prolonged

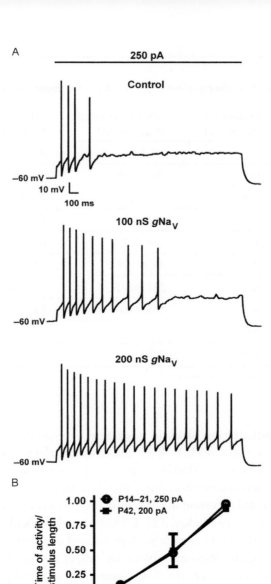

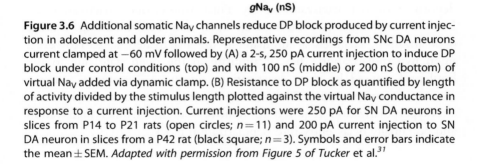

Figure 3.6 Additional somatic Na$_V$ channels reduce DP block produced by current injection in adolescent and older animals. Representative recordings from SNc DA neurons current clamped at −60 mV followed by (A) a 2-s, 250 pA current injection to induce DP block under control conditions (top) and with 100 nS (middle) or 200 nS (bottom) of virtual Na$_V$ added via dynamic clamp. (B) Resistance to DP block as quantified by length of activity divided by the stimulus length plotted against the virtual Na$_V$ conductance in response to a current injection. Current injections were 250 pA for SN DA neurons in slices from P14 to P21 rats (open circles; $n = 11$) and 200 pA current injection to SN DA neuron in slices from a P42 rat (black square; $n = 3$). Symbols and error bars indicate the mean ± SEM. *Adapted with permission from Figure 5 of Tucker et al.*[31]

Table 3.1 Comparative affinity of antipsychotic drugs for dopamine D_2 receptors and hERG K^+ channels

Drug	D_2 receptor[*] (K_i nmol/L)	hERG K^+ channel (IC_{50}, nmol/L)	Relative potency (hERG IC_{50}/D_2 K_i)
Pimozide	0.7	18–54.6	26–78
Haloperidol	0.7	26.8–1000	38–1429
Sertindole	1.2	2.7–14.7	2–12
Chlorpromazine	1.3	21,600	16,615
Risperidone	1.7	148–167	87–98
Thioridizine	2.3	33.2–191	14–83
Olanzapine	6.4	231–6013	36–940
Ziprasidone	8.4	125–169	15–20
Clozapine	82	320–250	4–30
Quetiapine	155	5765	37

Adapted with permission from Shepard et al.[32]

episodes of depolarization block in dopamine neurons *in vitro*.[35] Ji et al.[36] provided evidence that both E-4031 and another selective ERG channel blocker, the peptide toxin, rBeKm-1, had similar effects on dopamine neurons activity. As it is extremely unlikely that these structurally disparate drugs would have a second common site of action, it was concluded that the effects of both compounds are mediated by ERG channel blockade.

Activation of the ERG current with prolonged depolarization may contribute to delaying depolarization block by simply making the net current less depolarizing. However, the ERG current is also uniquely suited to remove sodium channel inactivation after a spike because of its kinetic scheme, which is shown in Fig. 3.7. This channel activates slowly with-depolarization but inactivates rapidly. Furthermore, once inactivated, the ERG channel must pass through an open state before closing. During depolarization, channels accumulate in the inactivated state. However, upon hyperpolarization, they accumulate in the open state due to the bottleneck caused by slow deactivation, causing a "resurgent" current after an action potential. This resurgent current can contribute to the afterhyperpolarization required to remove sodium channel inactivation and potentially delay entry into depolarization block.

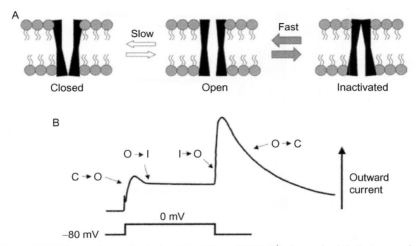

Figure 3.7 Voltage-dependent characteristics of ERG K$^+$ channels. (A) Cartoon schematic illustrating the three conductance states of ERG K$^+$ channels. (B) Macroscopic current corresponding to the conductance states in illustrated (A). At hyperpolarized membrane potentials, ERG channels are closed (C). Depolarizing the membrane potential from -80 to 0 mV (lower trace) slowly opens the channels (C$\rightarrow$O) resulting in an initial outward current as K$^+$ ions diffuse out of the cell. However, ERG channels rapidly inactivate (O$\rightarrow$I) by entering a second nonconductive state that limits the outward current produced by the initial depolarization. Partial repolarization of the membrane potential induces a rapid transition from an inactive to an open conformation (I$\rightarrow$O). As the rate of deinactivation (I$\rightarrow$O) exceeds the rate at which the channels can close, a large "resurgent" current is generated as the membrane potential repolarizes. *Reprinted with permission from Figure 1 of Shepard* et al.[32]

6. ERG CONDUCTANCE BOTH DELAYS ENTRY INTO AND SPEEDS RECOVERY FROM DEPOLARIZATION BLOCK

In order to determine what effect ERG channel blockade might have on depolarization block, Ji et al.[36] induced depolarization block in a slice preparation in two ways. First, the dynamic clamp was utilized to inject a square pulse of virtual NMDA conductance that increased the spike frequency but led to depolarization block at the end of the train of spikes. Second, application of a negative modulator of the SK current led to a spontaneous depolarization that manifested as a high frequency train of spikes followed by depolarization block. These protocols were used to mimic in an *in vitro* preparation the natural bursting that occurs *in vivo* (Fig. 3.1).

Figure 3.8A shows a control trace from a spontaneously pacemaking dopamine neuron in a slice preparation and the effect of applying a square

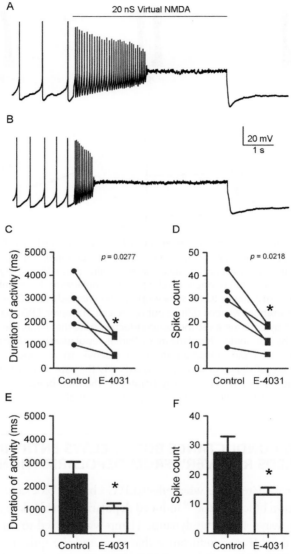

Figure 3.8 The prototypical ERG channel blocker E-4031 promotes depolarization block by virtual NMDA receptor stimulation. (A) and (B) are representative whole-cell dynamic clamp recordings of DA neurons stimulated with 20 nS of virtual NMDA current for 5 s to induce depolarization block before (A) and after (B) superfusion with 10 μM E-4031. The horizontal line above the traces indicates virtual NMDA conductance application. Summary of the effects of 10 μM E-4031 on the duration (C and E) and spike count (D and F) from dynamic clamp simulations of NMDA-induced bursting activity. *$P < 0.05$. *Adapted with permission from Figure 4 of Ji et al.*[36]

pulse of virtual NMDA conductance (black bar above the membrane potential trace). This virtual excitatory conductance activates instantaneously with depolarization and has a synaptic reversal of potential of 0 mV. The resultant injected current at first increases the spike rate, then leads to depolarization block. After ERG block by E-4031 (Fig. 3.8B), spiking is maintained for a much shorter duration and far fewer spikes are evoked. These results support our proposition that, under control conditions, the ERG current delays entry into depolarization block.

As described in the Section 3 earlier, the SK channel current plays a critical role in regular pacemaking. In the model, the prolonged afterhyperpolarization produced by the SK channel closes the L-type Ca^{2+} channels, ensuring that each slow depolarization leads to only a single spike. If the SK channel is blocked by apamin[14,37] or if its affinity for calcium is decreased by the negative modulator NS-8593, a runaway depolarization unopposed by the SK channel can occur, resulting in a depolarized plateau (Fig. 3.9A). If, under these conditions, the ERG channels are blocked by E-4031, the duration of the plateau is markedly lengthened (Fig. 3.9B and C). We conclude that under control conditions, the ERG current contributes to spontaneous repolarization of the plateau. We suspect that this repolarization occurs via regenerative deactivation of the L-type Ca^{2+} channel that supports the plateau.[14,38] In some neurons, ERG channel blockade caused dopamine neurons to remain depolarized indefinitely (Fig. 3.9D). In these neurons, hyperpolarizing pulses could briefly reestablish spiking prior to reentry into depolarization block.

These results show that blockade of the ERG current affects the propensity of dopamine neurons to enter and remain in state of a depolarization block. Since the therapeutic action of antipsychotics has been associated with their propensity to induce depolarization block, the possibility exists that part of this therapeutic action may be mediated by their ability to block the ERG current and thereby promote depolarization block.[32]

7. COMPUTATIONAL MODEL OF BURSTING AND DP BLOCK

Our computational model of pacemaking, taken from Kuznetsova et al.[26] and described above (Figs. 3.3 and 3.4), was not sufficient to capture burst firing induced by a virtual pulse of NMDA, the spontaneous plateaus induced by negative modulation of the SK current, or repolarization of those plateaus. Since the model did not contain an ERG current, it was also not capable of modeling the effects of ERG block on these phenomena. The

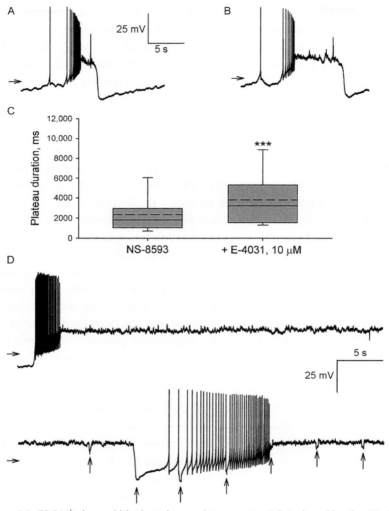

Figure 3.9 ERG K$^+$ channel block prolongs plateau potentials induced by the SK channel negative modulator, NS8593. (A) Representative tracing from a DA neuron in the presence of NS8593 (6 μM). Note the "burst" of action potentials elicited during the initial phase of a depolarizing plateau potential. (B) Addition of E-4031 (10 μM) prolongs the duration of the plateau. (C) Box (25th and 74th percentiles) and whisker (10th and 90th percentiles) plot illustrating the effects of E-4031 (10 μM) on the duration of spontaneous plateau potentials elicited by negative modulation of SK channels. Solid and dashed lines inside the box represent the median and mean, respectively. ***$P < 0.003$ versus NS8593. (D) Perfusion with E-4031 (10 μM) results in the loss of spontaneous activity through depolarization block following removal of a negative bias current. Brief hyperpolarizing current pulses (0.03 nA, 200 ms; vertical arrows) are capable of repolarizing the neuron which leads to recovery of spontaneous spiking followed by the rapid return of depolarization block. *Adapted with permission from Figure 3 of Ji* et al.[36]

computational model presented in Ji *et al.*[36] (Fig. 3.10) made the following major modifications to the model of Kuznetsova *et al.*[26]: the addition of the slow ERG current ($I_{K,ERG}$) and the addition of a second, slower component of sodium channel inactivation. The ERG current was modeled using the kinetic scheme shown in Fig. 3.7, fit to the data presented in Figure 1A and B of Ref. 39. The three-compartment schematic structure of the model was preserved (Fig. 3.10B). Here, we explain why these modifications were necessary and present an improved parameter set (see Appendix for details) for the modified model that better captures the effects observed in brain slices. The A-type potassium current and the sodium pump current were not included in the model for bursting. However, the nonspecific hyperpolarization–activated cation current (I_H) based on the description of Migliore *et al.*[40] and an M-type K^+ current based on the description of

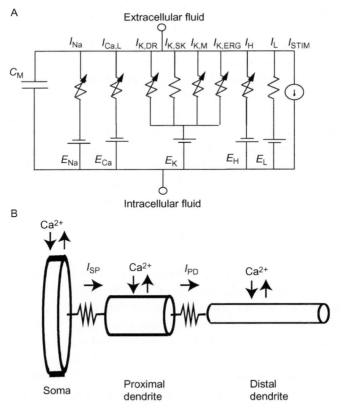

Figure 3.10 Model of dopamine neuron extended to simulate bursting and pacing. Details given in Appendix.

Drion et al.[41] were included. These modifications were not critical for either pacing or bursting activity.

For the passive properties of the model, the axial resistivity was set to 40 Ω cm and the capacitance to 1 $\mu F/cm^2$. First, the model was silenced by simulated injection of sufficient hyperpolarizing current into the soma in order to hold the model neuron at -65 mV. Then the time constant and input resistance were measured using small amplitude (40 pA) hyperpolarizing pulses. The membrane time constant of the model was 31 ms compared to reported values of 30–40 ms in dopamine neurons[42] and the input resistance was 476 MΩ under control conditions, within the reported range of 100–500 MΩ.[12,36,42] ERG block, simulated by setting the conductance for I_{ERG} to zero, resulted in a 502-MΩ input resistance. These results were similar to those observed in brain slices following ERG block.[36]

In order to simulate the experimental results presented in Figs. 3.6, 3.8, and 3.9, it was necessary to incorporate a second, slow component of sodium channel inactivation that was recently described in midbrain dopamine neurons.[43] Figure 3.11A1 shows the voltage dependence of the two sodium inactivation variables (h and hs) in the model. The description of the Hodgkin–Huxley-type inactivation variable h was based on experimental data from nucleated patches.[44] The time course of hs is slower than that of h, and its steady-state voltage dependence has not been characterized in dopamine neurons. Two components of inactivation are evident during recovery from inactivation, but entry into the slow inactivation state cannot be observed during a single pulse protocol because it is obscured by entry into the fast inactivated state. However, during multiple pulses that are spaced sufficiently close together, there is incomplete recovery from the slowly inactivating state, so that the sodium current evoked by multiple pulses declines as the train progresses. This protocol was applied in the model (Fig. 3.11A2) using 3 ms voltage clamp steps separated by 100 ms and fitted to the data of Ding et al.[43]

In order to simulate the experimental rescue from depolarization block by additional sodium conductance (Fig. 3.6), the model neuron was held hyperpolarized at -65 mV and a 2-s current depolarizing step was applied to drive the model cell into depolarization block (Fig. 3.11B1). An additional amount of sodium conductance was injected into the soma only, in order to prevent depolarization block (Fig. 3.11B2). For consistency with the dynamic clamp protocol used in Tucker et al.,[31] the injected virtual conductance did not have the hs component modeled in this simulated sodium conductance. The half inactivation voltage for hs is -51 mV,

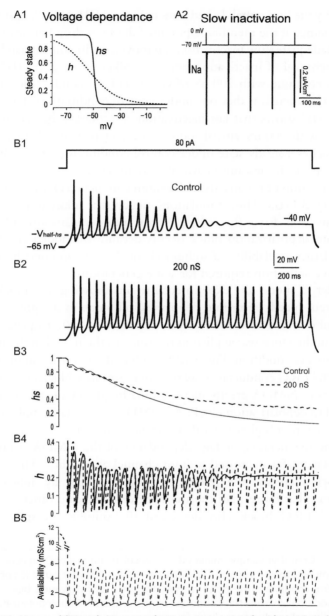

Figure 3.11 How additional Na$_V$ conductance rescues depolarization block. (A1) Steady-state voltage dependence for fast inactivation (*h*) and slow inactivation (*hs*). (A2) The peak of the sodium current elicited by a10-Hz train of 3 ms voltage steps from −70 to 0 mV decays during the train due to accumulation of slow inactivation *hs*. This

(Continued)

indicated by the horizontal dashed lines in Fig. 3.11B1 and B2. Spiking cannot be sustained if the membrane potential does not drop below this value, because inadequate slow inactivation is removed during the interspike interval. The removal of inactivation between spikes (see Fig. 3.11B3) allows spiking to continue with 200 nS of virtual sodium conductance added in Fig. 3.11B2, whereas it dies out under control conditions in Fig. 3.11B1. Figure 3.11B4 shows that fast inactivation continues to be removed after each spike with 200 nS virtual sodium conductance added (Fig. 3.11B2) but approaches a steady state under control conditions (Fig. 3.11B1). The key difference is the amount of available sodium conductance prior to each spike (the fraction of noninactivated sodium conductance), indicated by the peaks in Fig. 3.11B5. These simulations suggest that failure to recover from slow inactivation drives the model neuron into depolarization block, and rescue from depolarization block occurs because the added conductance results in larger availability of sodium channels that can contribute to the regenerative activation required for spike generation.

In order to model the experimental results showing that blocking the ERG current speeds entry into depolarization block in the presence of an applied pulse of virtual NMDA conductance (Fig. 3.8), a square pulse of NMDA conductance was applied in the soma to the model in the free running pacemaker condition (Fig. 3.12A) both under control conditions (top) and with the ERG conductance set to zero to simulate blockade of this current (bottom). NMDA greatly increased the frequency of spiking and ultimately led to depolarization block. An NMDA conductance pulse increases the frequency to a higher level than can be achieved with a current pulse because the regenerative voltage dependence of the NMDA conductance can compensate for decreased sodium channel availability. The simulated ERG block increased the frequency of pacemaking by approximately 100%, somewhat more than the mean percent increase observed experimentally.[36] ERG block also decreased the number of spikes in the

---

Figure 3.11—Cont'd simulation mimics the protocol in Figure 8 of Ding et al.[43] (B) This simulation mimics the protocol in Fig. 3.6 in which a 2-s current pulse is applied. (B1) Under control conditions, the pulse causes entry into depolarization block. (B2) The simulated addition of virtual Na_V conductance prevents entry into depolarization block. (B3) The additional conductance allows for more recovery from slow inactivation. (B4) At similar levels of hs, the recovery under control from h is similar to that with added Na_V, but deteriorates as the difference in hs is established. (B5) The major difference is that the noninactivated (available) conductance after each spike is increased with additional conductance.

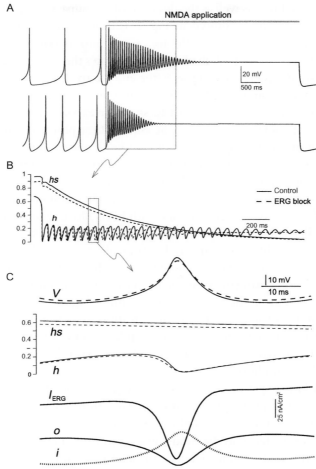

Figure 3.12 How the ERG current delays entry into depolarization block. (A, top) A square pulse of simulated virtual NMDA conductance applied to a pacemaking model neuron induces a burst of spikes followed by depolarization block. (A, bottom) Simulated ERG current block increases the pacemaker rate and accelerates entry into depolarization block, shortening both the bursting duration and the number of spikes in the burst. (B) ERG block clearly decreases recovery from slow sodium channel inactivation (*hs*) during the square pulse in conductance with a more ambiguous effect on fast sodium inactivation (*h*). The addition of ERG channels enhances *hs*, thus the fraction of available sodium channels (*h**hs). (C) During a single spike, the potential during the interspike interval is elevated in the absence of the ERG current (dashed trace) compared to control (solid trace), resulting in more slow inactivation (lower levels of *hs*). The change in the level of *h* due to a spike is greater under control conditions due to the extra afterhyperpolarization due to the ERG current. The time course of the ERG current and its open state (*o*), inactivation state (*i*) is shown for the control case only.

NMDA-evoked burst as well as the duration of the burst. The gradual accumulation of inactivation, shown in Fig. 3.12B as a decrease in the non-inactivated fraction, is responsible for entry into depolarization block (cf. Fig. 3.11). ERG block (dashed trace) accelerates this decrease compared to control (solid trace), resulting in faster entry into depolarization block. The contribution from the faster component of inactivation is not immediately clear from Fig. 3.12B, so Fig. 3.12C shows an expanded view of a single action potential under each condition at a similar time point. The membrane potential during the interspike interval is more depolarized with the ERG block (dashed line) compared to control (solid line), explaining the faster decrease in noninactivated channels since the time constant for slow inactivation is faster at more depolarized potentials. In addition, there is less removal of fast inactivation after a spike with ERG block than under control conditions. Thus, fewer spikes are elicited in an NMDA-evoked burst when the ERG current is blocked compared to control. Due to the kinetic scheme that characterizes the ERG current (Fig. 3.7), the open fraction of ERG channels decreases during the rising phase of the action potential and exhibits a small increase during the afterhyperpolarization (Fig. 3.12C). This results in more removal of sodium channel inactivation, which in turn leads to a larger spike height, a greater number of spikes, and a longer duration of bursting spikes when NMDA is applied for the control case compared to the ERG block condition.

To model the experimental results showing that blocking the ERG current delays or prevents recovery from depolarization block during plateaus induced by block (or negative modulation) of the SK current (Fig. 3.9), the SK channel conductance in the model was set to zero. This disrupted pacemaking activity and produced bursts of action potentials terminating in depolarization block and a depolarized plateau (Fig. 3.13A1). This plateau repolarized spontaneously and abruptly as in the experimental data (compare to Fig. 3.9A). The plateau resulted from the failure of the action potentials to turn off the L-type calcium current shown in Fig. 3.13B1 (compare to the calcium-dependent phase in the pacemaking model in Fig. 3.3A and C in which the L-type Ca^{2+} current essentially turns off during the AHP following an action potential). The Ca^{2+}-activated SK potassium current is largely responsible for this portion of the AHP, and in its absence, bursts of spikes can replace pacemaking in which a single spike rides each depolarizing wave. The excessive unopposed depolarization produced by the regenerative, autocatalytic L-type Ca^{2+} current results in depolarization block, which is sustained until the slowly activating ERG current (Fig. 3.13C1) accumulates

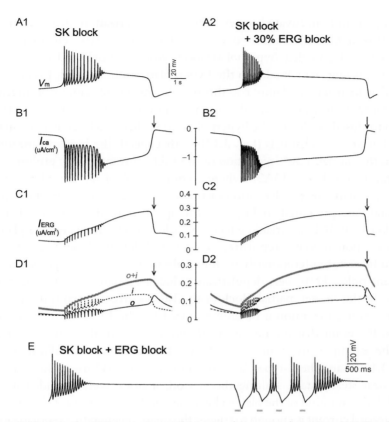

Figure 3.13 How the ERG current contributes to plateau repolarization. (A1) Simulated block of the SK current produces a burst of spikes followed by DP block. (A2) A 30% reduction in the ERG conductance again shortens the burst of spikes and also prolongs the depolarized plateau. (B) The time course of the L-type Ca^{2+} current, with the arrow indicating the regenerative decrease in this current that drives the sharp plateau repolarization in A. (C) The time course of the ERG current, with the arrow showing the point at which this current tips the balance in favor of the outward currents to initiate plateau repolarization. (D) The ERG channel fraction in the open state (*o*) and inactivation state (*i*) as well as the slow pool of *ss* (bold gray curve) show the slow time course of the combined pool as well as the bump in the open state during repolarization. (E) A complete block of the ERG channels induces permanent depolarization block at −40 mV. Brief (200 ms) hyperpolarizing somatic current pulses (gray bars) temporarily relieve depolarization block.

sufficiently during the depolarization to change the total net current from inward to outward at the arrow in panels B1,C1, and D1. The actual slow variable is not the fraction of activated channels, but rather the sum of the open and inactivated channels due to the bottleneck, or slow transitions between the open and closed states compared to the fast transitions between the open and inactivated states. The hyperpolarization that terminates the plateau is sudden for two reasons. First, the L-type channels close regeneratively, in that hyperpolarization closes channels and reduces the inward current, which adds to the hyperpolarization. Second, ERG channels that are inactivated must pass through the open state prior to closing, and the transition from inactivated to open is fast relative to the transition from open to closed, so during the hyperpolarization there is a brief increase in the number of open channels (Fig. 3.13D1) that speeds the hyperpolarization.

In the model, a 30% reduction of the ERG current greatly prolongs the plateau phase (Fig. 3.13A2), whereas complete blockade completely prevents spontaneous repolarization of the plateau (Fig. 3.13E). In the experimental data presented in Fig. 3.9, the application of the ERG channel blocker E-4031 had variable effects. In at least some instances, the plateau did not spontaneously repolarize, as in Fig. 3.9D. In these cases, spiking could be restored temporarily by the application of a brief hyperpolarizing current pulse. However, depolarization block reestablished itself in these cases. In the model, partial block of the ERG current extends the plateau by increasing the amount of time required for the ERG current to build up to the point that the net outward current exceeds the inward current. In the example given in Fig. 3.13A2, the ERG current illustrated in Fig. 3.13C2 and the L-type Ca^{2+} current in Fig. 3.13B2 are in close balance and near steady state. During this period, if noise were included in the model, the duration of the plateau becomes random because a slight increase in outward current is enough to trigger the steep, regenerative repolarization of the plateau. Although the driving force for the L-type Ca^{2+} current is increasing during this repolarization, the current drops sharply due to the steepness of the voltage dependence in this potential range. Similarly, the fraction of open ERG channels increases during the repolarization, partially offsetting the decrease in driving force for this potassium channel. For the case of complete ERG block in the model (Fig. 3.13E), brief hyperpolarizing pulses relieve depolarization block by turning off the L-type Ca^{2+} current. As this current gradually depolarizes the membrane, depolarization block is reestablished. The experimentally observed variability in the effect of the ERG channel block on plateau duration, and the fact that ERG block does not cause permanent depolarization block in every case, indicates that there is heterogeneity between neurons in the contribution of the ERG conductance to repolarization of the plateau. It is likely that some other slow current contributes to plateau repolarization in addition to the ERG current, and that this current is responsible for repolarization of the plateau in the neurons that did not remain in depolarization block after the ERG block.

8. CONCLUSIONS

Here, we have used a model dopamine neuron to explore the roles of slow inactivation of the sodium channel and of the ERG potassium channel in depolarization block. During prolonged depolarization, sodium channels accumulate in the inactivated state due to a slow component of inactivation, and by increasing the sodium conductance, depolarization block can be reversed. Blocking the ERG current accelerates entry into depolarization block by promoting entry into and hindering recovery from Na^+ channel inactivation. Furthermore, blocking the ERG current delays or even prevents recovery from spontaneous episodes of depolarization block when the SK channel is negatively modulated. Since antipsychotic drugs block ERG as a side effect, it is conceivable that some of the therapeutic benefits of these drugs are derived by increasing the tendency of dopamine neurons to go into depolarization block.

If increasing the tendency for midbrain ventral tegmental dopamine neurons to enter depolarization block has therapeutic value in schizophrenia, then drugs tailored to directly increase depolarization block in these neurons may have potential to provide improved relief from positive symptoms. Possible strategies include selectively reducing the afterhyperpolarization following an action potential, increasing the fraction of sodium channels expressing slow inactivation, increasing the L–type calcium channel current responsible for plateau potentials, and/or decreasing the slow potassium currents activated by depolarization. Conversely, if depolarization block in nigral dopamine neurons causes extrapyramidal side effects, then drugs that counter depolarization block in these neurons by the opposite strategies would minimize side effects. Clearly, it will be important to understand the heterogeneous aspects of intrinsic membrane properties of dopaminergic populations in the midbrain to design better therapeutics.

ACKNOWLEDGMENT

This work was supported by NIH Grant 5R01NS061097-05 to C. C. C and P30-GM103340.

APPENDIX: FULL MODEL EQUATIONS AND PARAMETERS

The voltage-gated currents were modeled as follows:

$$I_{Na} = g_{Na}m^3hhs(v-60), \quad I_{Ca,L} = g_{Ca,L}l(v-63), \quad I_{K,DR} = g_{K,DR}n^3(v+85),$$

and $I_H = g_H m_H(v+34.25)$, where the gating variables m, h, hs, l, n and m_H

obey equations of the form $dx/dt = (x_{inf} - x)/\tau_x$, with $x_{inf} = 1/[1 + \exp (-(v - x_{half})/x_k)]$. Parameters are given in Table A1.

Table A1 Time and voltage dependence of model gating variables

x	x_{half} (mV)	x_k (mV)	Time constants: τ_x (ms)
m	−30.0907	9.7264	$0.01 + 1/(a+b)$, $a = -(15.6504 + 0.4043v)/[\exp(-19.565 - 0.50542v) - 1]$ $b = 3.0212 \exp(-7.4630 \times 10^{-3}v)$
h	−54.0289	−10.7665	$0.4 + 1/(a+b)$, $a = 5.0754 \times 10^{-4} \exp(-6.3213 \times 10^{-2}v)$ $b = 9.7529 \exp(0.13442v)$
hs	−51	−1	$20 + 580/(1 + \exp(v+3))$
n	−25	12	$1 + 20 \exp(-\{\ln[1 + 0.05(v+53)]/0.05\}^2/350)$
l	−43	9	$18 \exp[-(v+60.0)^2/625.0] + 0.30$
m_H	−80	8	$625 \exp[0.075(v+112)]/\{1 + \exp[0.083(v+112)]\}$

The M and ERG potassium currents are described as follow:

$$I_{K,M} = g_{K,M}q(v+85)$$
$$dq/dt = a_q(1-q) - b_q q$$

where $\alpha_q = 0.02/(1 + \exp(-(v+20)/5))$ and $\beta_q = 0.01\exp((v+43)/18)$;

$$I_{K,ERG} = g_{K,RRG}o(v+85)$$
$$do/dt = \alpha_a(1 - o - i) + \beta_i i - (\alpha_i + \beta_a)o$$
$$di/dt = \alpha_i o - \beta_i i,$$

where o and i are the fraction of open and inactivated channels, respectively, with transition rates given by $\alpha_a = 0.0061 \exp(0.1085v)$, $\beta_a = 2.1469 \times 10^{-5} \exp(-0.0570v)$, $\alpha_i = 227.775 \exp(0.1123v)$, and $\beta_i = 21 \exp(0.0712v)$.

The "leak current" was actually modeled as a combination of an inward rectifying potassium current $I_{K,IR}$, a sodium leak channel ($I_{L,Na}$), and a small calcium leak ($I_{L,Ca}$),required for the calcium balance.

$$I_L = I_{K,IR} + I_{L,Na} + I_{L,Ca}, I_{K,IR} = g_{K,IR}(v+85)/(1 + \exp((v+45)/20)), I_{L,Na}$$
$$= g_{L,Na}(v-60), I_{L,Ca} = g_{L,Ca}(v-63)$$

The soma, proximal dendrites, and distal dendrites are modeled as cylindrical with diameters denoted as d_s, d_p, and d_d, respectively ($d_d = 1.5$ μm, $d_p = 3$ μm, and $d_s = 15$ μm). L_s, L_p, and L_d denote their lengths ($L_d = 350$ μm, $L_p = 150$ μm, and $L_s = 25$ μm). The total resistance between the soma and each proximal dendrite is $R_{sp} = 0.5 L_s R_a / (0.25 \pi d_s^2 L_s) + 0.5 L_p R_a / (0.25 \pi d_p^2 L_p)$, and the total conductance is $G_{sp} = 1/R_{sp}$. The intensive conductances are $g_{sp} = 4 G_{sp} / \pi d_s L_s$ and $g_{ps} = G_{sp} / \pi d_p L_p$, where the factor of 4 reflects that there are four proximal dendrites but only one soma. The axial current in intensive units flowing into the soma from all four proximal dendrites is $I_{sp} = g_{sp}(v_s - v_p)$. The axial current for an individual proximal dendrite is $I_{ps} = g_{ps}{}^*(v_p - v_s)$. Similarly, the total resistance between the proximal dendrites and each distal dendrite is $R_{pd} = 0.5 L_d R_a / (0.25 \pi d_d^2 L_d) + 0.5 L_p R_a / (0.25 \pi d_p^2 L_p)$, and the total conductance is $G_{pd} = 1/R_{pd}$. The intensive conductances are $g_{pd} = 2 G_{pd} / \pi d_p L_p$ and $g_{dp} = G_{pd} / \pi d_d L_d$, where the factor of 2 reflects that there are two distal dendrites for each proximal one. The axial current in intensive units flowing into the proximal dendrite from both of its distal dendrites is $I_{pd} = g_{pd}(v_p - v_d)$. The axial current for an individual distal dendrite is $I_{dp} = g_{dp}{}^*(v_d - v_p)$.

The differential equations for transmembrane potential v_i are as follows:

$$C dV_s/dt = -I_{Na,s} - I_{K,DR,s} - I_{K,M,s} - I_{K,ERG,s} - I_{K,SK,s} - I_{Ca,L,s} - I_{L,Na,s}$$
$$- I_{L,Ca,s} - I_{GIRK,s} - I_{H,s} + I_{stim} - I_{sp} - I_{NMDA}$$

$$C dV_p/dt = -I_{Na,p} - I_{K,DR,p} - I_{K,M,p} - I_{K,ERG,p} - I_{K,SK,p} - I_{Ca,L,p} - I_{L,Na,p}$$
$$- I_{L,Ca,p} - I_{GIRK,p} - I_{H,p} - I_{ps} - I_{pd}$$

$$C dV_d/dt = -I_{Na,d} - I_{K,DR,d} - I_{K,M,d} - I_{K,ERG,d} - I_{K,SK,d} - I_{Ca,L,d} - I_{L,Na,d}$$
$$- I_{L,Ca,d} - I_{GIRK,d} - I_{H,d} - I_{dp}$$

The calcium balance is given by

$$\frac{d[Ca]}{dt} = \frac{-2 f_{Ca}\left(I_{L,Ca} + I_{Ca,p} + I_{Ca,L}\right)}{F d}$$

$$I_{Ca,p} = \frac{I_{Ca,p\_,max}}{1 + 0.00055/[Ca]}$$

where [Ca] (in millimolar) is the Ca^{2+} concentration in each compartment, $f_{Ca} = 0.015$ is the fraction of free calcium, d is the diameter of the compartment and F is Faraday's constant. Extrusion of Ca^{2+} from the compartments is modeled as a nonelectrogenic pump with $I_{Ca,p\_max} = 6$ μA/cm^2.

The calcium-dependent conductance was modeled as

$$I_{K,SK} = \frac{g_{K,SK}(v+85)}{1+(0.00019/[Ca])^4}$$

And the virtual NMDA conductance was modeled as

$$I_{NMDA} = \frac{-g_{NMDA} * v}{1+(Mg/3.57)*\exp(-0.062v)}$$

where $Mg = 1.5$ mM.

Maximal conductances in $\mu S/cm^2$ were $g_{Na}=2300$, $g_K=1200$, $g_{L,ca}=19.8$, $g_{girk}=32.5$, $g_{Na,leak}=2.1$, $g_{K,ERG}=115$, $g_{SK}=300$, $g_{KM}=40$, $g_{Ca,leak}=0.05$, $g_H=20$.

REFERENCES

1. Girault JA, Greengard P. The neurobiology of dopamine signaling. *Arch Neurol.* 2004;61:641–644.
2. Bloom FE, Roth RH, Cooper JR. *The Biochemical Basis of Neuropharmacology.* Oxford: Oxford University Press; 1991.
3. Bernheimer H, Birkmayer W, Hornykiewicz O, Jellinger K, Seitelberger F. Brain dopamine and the syndromes of Parkinson and Huntingdon: clinical, morphological, and neurochemical correlations. *J Neurosci.* 1973;20:415–455.
4. Kalivas PW, Volkow ND. The neural basis of addiction: a pathology of motivation and choice. *Am J Psychiatry.* 2005;162:1403–1413.
5. Schultz W. Predictive reward signal of dopamine neurons. *J Neurophysiol.* 1998;80:1–27.
6. Tanaka S. Dopaminergic control of working memory and its relevance to schizophrenia: a circuit dynamics perspective. *Neuroscience.* 2006;139:153–171.
7. Carlsson A, Lindqvist M. Effect of chlorpromazine or haloperidol on formation of 3-methoxytyramine and normetanephrine in mouse brain. *Acta Pharmacol Toxicol.* 1963;20:140–144.
8. Kapur S, Remington G. Dopamine D2 receptors and their role in atypical antipsychotic action: still necessary and maybe even sufficient. *Biol Psychiatry.* 2001;50:873–883.
9. Grace AA, Bunney BS. The control of firing pattern in nigral dopamine neurons: single spike firing. *J Neurosci.* 1984;4:2866–2876.
10. Grace AA, Bunney BS. The control of firing pattern in nigral dopamine neurons: burst firing. *J Neurosci.* 1984;4:2877–2890.
11. Hyland BI, Reynolds JNJ, Hay J, Perk CG, Miller R. Firing modes of midbrain dopamine cells in the freely moving rat. *Neuroscience.* 2002;114:475–492.
12. Grace AA, Ohn S-P. Morphology and electrophysiological properties of immunocytochemically identified rat dopamine neurons recorded in vitro. *J Neurosci.* 1989;9:3463–3481.
13. Richards CD, Shiroyama T, Kitai ST. Electrophysiological and immunocytochemical characteristics of GABA and dopamine neurons in the substantia nigra of the rat. *Neuroscience.* 1997;80:545–557.
14. Ping HX, Shepard PD. Apamin-sensitive Ca^{2+}-activated K^+ channels regulate pacemaker activity in nigral dopamine neurons. *NeuroReport.* 1996;7:809–814.
15. Deister CA, Teagarden MA, Wilson CJ, Paladini CA. An intrinsic neural oscillator underlies dopaminergic neuron bursting. *J Neurosci.* 2009;29:15888–15897.

16. Johnstone EC, Crow TJ, Frith CD, Carney MW, Price JS. Mechanism of the antipsychotic effect in the treatment of acute schizophrenia. *Lancet.* 1978;1:848–851.
17. Grace AA, Bunney BS. Induction of depolarization block in midbrain dopamine neuron by repeated administration of haloperidol: analysis using in vivo intracellular recording. *J Pharmacol Exp Ther.* 1986;238:1092–1100.
18. Pucak ML, Grace AA. Effects of haloperidol on the activity and membrane physiology of substantia nigra dopamine neurons recorded in vitro. *Brain Res.* 1996;713:44–52.
19. Chiodo LA, Bunney BS. Typical and atypical neuroleptics: differential effects of chronic administration on the activity of A9 and A10 midbrain dopaminergic neurons. *J Neurosci.* 1983;3(8):1607–1619.
20. White FJ, Wang RY. Differential effects of classical and atypical antipsychotic drugs on A9 and A10 dopamine neurons. *Science.* 1983;221:1054–1057.
21. White FJ, Wang RY. Comparison of the effects of chronic haloperidol treatment on A9 and A10 dopamine neurons in the rat. *Life Sci.* 1983;32:983–993.
22. Grace AA, Bunney BS, Moore H, Todd CL. Dopamine-cell depolarization block as a model for the therapeutic actions of antipsychotic drugs. *Trends Neurosci.* 1997;20:31–37.
23. Bunney BS, Grace AA. Acute and chronic haloperidol treatment: comparison of effects on nigral dopaminergic cell activity. *Life Sci.* 1978;23:1715–1727.
24. Kapur S, Seeman P. Antipsychotic agents differ in how fast they come off the dopamine D2 receptors. Implications for atypical antipsychotic action. *J Psychiatry Neurosci.* 2000;25:161–166.
25. Valenti O, Cifelli P, Gill KM, Grace AA. Antipsychotic drugs rapidly induce dopamine neuron depolarization block in a developmental rat model of schizophrenia. *J Neurosci.* 2011;31(34):12330–12338.
26. Kuznetsova AY, Huertas MA, Kuznetsov AS, Paladini CA, Canavier CC. Regulation of the firing frequency in a computational model of a midbrain dopaminergic neuron. *J Comput Neurosci.* 2010;28:389–403.
27. Wilson CJ, Callaway JC. Coupled oscillator model of the dopaminergic neuron on the substantia nigra. *J Neurophysiol.* 2000;83:3084–3100.
28. Kuznetsov AS, Kopell NJ, Wilson CJ. Transient high-frequency firing in a coupled-oscillator model of the mesencephalic dopaminergic neuron. *J Neurophysiol.* 2006;95:932–937.
29. Durante P, Cardenas C, Whittaker J, Kitai S, Scroggs R. Low-threshold L-type calcium channels in rat dopamine neurons. *J Neurophysiol.* 2004;91:1450–1454.
30. Puopolo M, Raviola E, Bean BP. Roles of subthreshold calcium current and sodium current in spontaneous firing of mouse midbrain dopamine neurons. *J Neurosci.* 2007;27:645–656.
31. Tucker K, Huertas M, Horn J, Canavier CC, Levitan ES. Pacemaker rate and depolarization block in nigral dopamine neurons: a sodium channel balancing act. *J Neurosci.* 2012;32:14519–14531.
32. Shepard PD, Canavier CC, Levitan ES. Ether-a-go-go related gene (ERG) potassium channels. What's all the buzz about? *Schizophr Bull.* 2007;33(6):1263–1269.
33. Nedergaard S. A Ca^{2+}-independent slow afterhyperpolarization in substantia nigra compacta neurons. *Neuroscience.* 2004;125:841–852.
34. Shepard PD, Bunney BS. Repetitive firing properties of putative dopamine-containing neurons in vitro: regulation by an apamin-sensitive Ca^{2+}-activated K^+ conductance. *Exp Brain Res.* 1991;86:141–150.
35. Canavier CC, Oprisan SA, Callaway J, Ji H, Shepard PD. Computational model predicts a role for ERG current in repolarizing plateau potentials in dopamine neurons: implications for modulation of neuronal activity. *J Neurophysiol.* 2007;98:3006–3022.

36. Ji H, Tucker K, Putzier I, et al. Functional characterization of ether-a-gogo-related potassium channels in midbrain dopamine neurons: implications for a role in depolarization block. *Eur J Neurosci.* 2012;36(7):2906–2916.

37. Johnson SW, Wu YN. Multiple mechanisms underlie burst firing in rat midbrain dopamine neurons in vitro. *Brain Res.* 2004;1019(1–2):293–296.

38. Nedergaard S, Flatman JA, Engberg I. Nifedipine- and omega-conotoxin-sensitive Ca^{2+} conductances in guinea-pig substantia nigra pars compacta neurones. *J Physiol.* 1993;466:727–747.

39. Ficker E, Jarolimek W, Kiehn J, Baumann A, Brown AM. Molecular determinants of dofetilide block of HERG K^+ channels. *Circ Res.* 1998;82:386–395.

40. Migliore M, Cannia C, Canavier CC. A modeling study suggesting a possible pharmacological target to mitigate the effects of ethanol on reward-related dopaminergic signaling. *J Neurophysiol.* 2008;99(5):2703–2707.

41. Drion G, Bonjean M, Waroux O, et al. M-type channels selectively control bursting in rat dopaminergic neurons. *Eur J Neurosci.* 2010;31:827–835.

42. Johnson SW, North RA. Two types of neurone in the rat ventral tegmental area and their synaptic inputs. *J Physiol.* 1992;450:455–468.

43. Ding S, Wei W, Zhou FM. Molecular and functional differences in voltage-activated sodium currents between GABA projection neurons and dopamine neurons in the substantia nigra. *J Neurophysiol.* 2011;106:3019–3034.

44. Seutin V, Engel D. Differences in Na^+ conductance density and Na^+ channel functional properties between dopamine and GABA neurons of the rat substantia nigra. *J Neurophysiol.* 2010;103:3099–3114.

CHAPTER FOUR

The Role of IP$_3$ Receptor Channel Clustering in Ca^{2+} Wave Propagation During Oocyte Maturation

Aman Ullah[*], Peter Jung[*], Ghanim Ullah[†], Khaled Machaca[‡]

[*]Department of Physics and Astronomy and Quantitative Biology Institute, Ohio University, Athens, Ohio, USA
[†]Department of Physics, University of South Florida, Tampa, Florida, USA
[‡]Department of Physiology and Biophysics, Weill Cornell Medical College in Qatar, Doha, Qatar

Contents

Abstract

During oocyte maturation, the calcium-signaling machinery undergoes a dramatic remodeling resulting in distinctly different calcium-release patterns on all organizational scales from puffs to waves. The dynamics of the Ca^{2+} release wave in mature as compared to immature oocytes are defined by a slower propagation speed and longer duration of the high Ca^{2+} plateau. In this chapter, we use computational modeling to identify the changes in the signaling machinery, which contribute most significantly to the alterations observed in Ca^{2+} wave propagation during *Xenopus* oocyte maturation. In addition to loss of store-operated calcium entry and internalization of plasma membrane pumps, we propose that spatial reorganization of the IP$_3$ receptors in the plane of the ER membrane is a key factor for the observed signaling changes in Ca^{2+} wave propagation.

Progress in Molecular Biology and Translational Science, Volume 123
ISSN 1877-1173
http://dx.doi.org/10.1016/B978-0-12-397897-4.00006-1

1. INTRODUCTION

Calcium (Ca^{2+}) is the universal signal for egg activation at fertilization.[1] The dynamics of the calcium-release waves vary between species. In some species, sperm–egg fusion triggers multiple calcium oscillations, while in others a single wave is observed.[1] Furthermore, calcium-signaling patterns can show dramatic modifications during development. In immature *Xenopus* oocytes, calcium waves are saltatory. That is, calcium release from stores through small clusters of channels is brief and jumps from cluster to cluster, while it is longer in the egg and propagates more continuously.[3–6] In response to subthreshold signals that mobilize Ca^{2+} from intracellular stores, immature *Xenopus* oocytes exhibit transient calcium waves, which rise and decay rapidly on the order of seconds and travel across only a fraction of the cell.[4] Eggs, on the other hand, support calcium waves that sweep the entire cell and raise the cytosolic calcium levels for minutes.[3–6] The sweeping calcium wave in the egg travels about a factor 2–3 slower than the transient wave in the immature oocyte. Furthermore, sustained stimulation of calcium release in eggs upon uncaging IP_3 leads to an elevation of intracellular calcium for minutes, while it decays in seconds in the immature oocyte.

Multiple changes in the calcium-signaling machinery during oocyte maturation have been reported: plasma membrane pumps, responsible for calcium extrusion from the cytosol, are internalized and hence are not functional in the egg[4]; store-operated calcium entry (SOCE) is only present in the immature oocyte[7,8]; calcium-induced regeneration of IP_3 through PLC_δ,[2] and increased sensitivity of IP_3-dependent Ca^{2+} release has been reported in egg.[4] Furthermore, the ER remodels during *Xenopus* oocyte maturation, resulting in large patches that concentrate IP_3 receptors.[2,3] Indeed, we have recently shown that ER remodeling contributes significantly to sensitizing IP_3-dependent Ca^{2+} release during oocyte maturation.[22] While all these changes have been established, their relative contributions to the mode of propagation of the Ca^{2+} release wave during oocyte maturation remain unclear. Given the complexity and close interaction of Ca^{2+} signaling pathways, it is difficult to address the contribution of individual pathways in isolation in the physiologic context.

In this study, we use mathematical modeling to investigate the causes underlying the observed differences in the mode of Ca^{2+} wave propagation during oocytes maturation. The model starts on the organizational scale of calcium release through clusters of IP_3Rs and takes into account calcium extrusion through PMCAs (plasma membrane calcium ATPases), Ca^{2+} sequestration through SERCA (sarco–endoplasmic reticulum calcium ATPases), Ca^{2+} influx through SOCE, and calcium–induced IP$_3$ production through PLC$_\delta$ to predict cellular calcium dynamics. It allows us to test the effects of changes in PMCA, SOCE, and spatial distributions of the release channels, that is, mimicking ER remodeling as reported in by Terasaki et al.[9] and Sun et al.,[22] for calcium signals on the cellular scale. The goal is to use the experimental observations (changes in wave speed, signal duration, sensitivity to stimulation by IP$_3$, and spatial reorganization) in conjunction with the computational model to extract the most critical determinants of calcium-signaling differentiation during oocyte maturation. Our study extends well beyond the study in Ref. 2 as we start our modeling on the scale of elemental calcium-release events from clusters of IP$_3$Rs (with a model that has been tested on that scale[10]), include changes in spatial organization of these clusters as part of the equation, and utilize not only calcium wave propagation but also decay of cytosolic calcium in the wake of the wave and upon global stimulation.

2. METHODS

The calcium concentration in the cytoplasm is determined by the fluxes across the plasma membrane and intracellular calcium stores. Calcium ions (Ca^{2+}) enter the cytoplasm from the extracellular matrix through SOCE and leave the cytoplasm through PMCAs. Similarly, Ca^{2+} is released from intracellular stores such as endoplasmic reticulum (ER) into the cytosol through IP$_3$-activated calcium-release channels (IP$_3$Rs) and is pumped back into the ER by ATP-dependent SERCA pumps. Thus, representing the above-mentioned fluxes by J_{SOCE}, J_{PMCA}, J_{IP_3Rs}, J_{pump}, respectively, the equations governing the intracellular calcium dynamics are given by

$$\frac{dc}{dt} = D_{cyt}\nabla^2 c + \rho_{IP_3}\left(\overrightarrow{x}\right) - J_{SERCA} + J_{leak} + \delta(J_{SOC} - J_{PMCA}), \quad (4.1A)$$

$$\frac{dc_{ER}}{dt} = D_{ER}\nabla^2 c_{ER} + \gamma\left(J_{SERCA} - J_{leak} - \rho_{IP_3}\left(\overrightarrow{x}\right)\right), \quad (4.1B)$$

$$\frac{dp}{dt} = D_{IP_3}\nabla^2 p + \beta(p^* - p) + J_{PLC} + J_{stim}, \quad (4.1C)$$

where c and c_{ER} represent the cytoplasmic and luminal calcium concentrations, respectively, and p the IP$_3$ concentration. The first terms on the right-hand side of Eq. (4.1A–4.1C) represent the spatial diffusion of cytosolic calcium, luminal calcium in the ER, and IP$_3$, with diffusion coefficients D_{cyt}, D_{ER}, and D_{IP_3}, respectively. All intracellular Ca^{2+} buffers are incorporated through the use of effective calcium diffusion coefficients, obtained from confocal line scans.[11] The factor γ in Eq. (4.1B) describes the ratio of cytoplasm to ER volume and is used to account for different volumes of the two compartments. Sources of cytosolic Ca^{2+} are described by the source function

$$\rho_{IP_3}\left(\overrightarrow{x}\right) = \sum_n J_{IP_3}(n)\delta\left(\overrightarrow{x} - \overrightarrow{x}_n\right), \qquad (4.2)$$

where $\overrightarrow{x}_n$ denotes the positions of clusters of IP$_3$Rs, $\delta\left(\overrightarrow{x}\right)$ Dirac's delta functions, and $J_{IP_3}(n)$ the fluxes through the respective clusters, that is,

$$J_{IP_3}(n) = N_{open}(n)v_{channel}(c_{ER} - c). \qquad (4.3)$$

Here, $N_{open}(n)$ denotes the number of open channels, and $v_{channel}$ the maximum calcium flux density generated by an open channel. The number of open channels varies stochastically, as each channel can open and close stochastically. Since we are interested in signaling patterns on the cellular scale, it is not necessary to spatially and temporally resolve release of calcium from single channels. In previous work,[10] we have established that a stochastic version of this cluster-release model applied to a single cluster of IP$_3$Rs successfully reproduces the statistical distribution of durations and amplitudes of elementary calcium-release events (calcium puffs) in immature oocytes and eggs.

To simulate the number of open channels for each cluster, we adopt a kinetic model for the IP$_3$ receptor, developed by Sneyd and Dufour (SD).[12] According to the SD model, IP$_3$Rs are composed of four identical and independent subunits (see Fig. 4.1). Each subunit has four binding sites, one for IP$_3$, two for Ca^{2+} activation, and one for Ca^{2+} inactivation. A single subunit can exist in six states: the resting state (R), the inactivated states (I$_1$ and I$_2$), the open state (O), shut state (S), and activated state (A). State R represents the state where neither IP$_3$ nor Ca^{2+} is bound. The subunit can switch from state R to I$_1$ by binding Ca^{2+} to its inactivation site. To account for sequential binding of IP$_3$ and Ca^{2+}, the subunit has to transit first from state R to O by binding IP$_3$ and subsequently to A by binding Ca^{2+} to its activation site. The subunit can switch to the shut state S if IP$_3$ is bound. It can also transition to an inactivated state I$_2$ from state

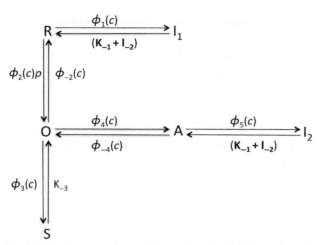

Figure 4.1 Simplified schematic of one of four subunits of the Sneyd–Dufour model for IP$_3$ receptor.[12] R denotes the unbound receptor, O the IP$_3$-bound state, A the active state, I$_1$ and I$_2$ the inactivated states, and S the shut state. The subunit can switch from state R to the active state A by binding IP$_3$ with the rate $\Phi_2(c)$ and Ca^{2+} with the binding rate $\Phi_4(c)$. Once in the active state, the receptor can inactivate to the state I$_2$ by binding Ca^{2+} with the binding rate $\Phi_5(c)$. The receptor can also inactivate directly from the unbound state R or through the shut state after IP$_3$ has bound. The symbols p and c denote cytosolic IP$_3$ and Ca^{2+} concentration, respectively. For details of the calcium-dependent rate functions Φ_{1-5} which are slightly different from those in Ref. 13, see Ref. 10.

A by binding Ca^{2+} to its inactivating site. An IP$_3$R conducts Ca^{2+} when all four subunits are in active state.

Leakage of Ca^{2+} from the ER to cytoplasm through nonspecific processes is given by

$$J_{\text{leak}} = \nu_{\text{leak}}\left(c_{\text{ER}} - c\right), \qquad (4.4)$$

where ν_{leak} denotes a constant (see Table 4.1). Cytosolic calcium is removed by SERCA into the ER and by PMCA into the extracellular domain. These pumps are distributed uniformly on the ER membrane. The fluxes are modeled as[13]

$$J_{\text{SERCA}} = \nu_{\text{SERCA}} \frac{c^2}{k_{\text{SERCA}}^2 + c^2}, \qquad (4.5)$$

$$J_{\text{PMCA}} = \nu_{\text{PMCA}} \frac{c^2}{k_{\text{PMCA}}^2 + c^2}. \qquad (4.6)$$

Values for the dissociation constants k_{SERCA} and k_{PMCA} are supplied in Table 4.1.

SOCE channels located in the plasma membrane open rapidly in response to intracellular store depletion.[14] This flux is described phenomenologically by

$$J_{\text{SOC}} = \alpha_1 + \alpha_2 p, \qquad (4.7)$$

where the second term introduces sensitivity to ER depletion.[13] Phospholipase C (PLC) in the plasma membrane stimulates production of IP$_3$ upon increasing cytosolic Ca^{2+} levels, that is,

$$J_{\text{PLC}_\delta} = \nu_{\text{PLC}} \frac{c^2}{k_{\text{PLC}}^2 + c^2}. \qquad (4.8)$$

PLC is believed to be an important signaling effector in egg as it increases sensitivity to IP$_3$. A copropagating IP$_3$ wave has been postulated in Ref. 2 as

Table 4.1 Model parameters used for the simulation

ν_{channel}	$2049.0\ \text{s}^{-1}$
ν_{leak}	$0.0002\ \text{s}^{-1}$
ν_{SERCA}	$27.0\ \mu\text{M}^2/\text{s}$
ν_{PMCA}	$40\ \mu\text{M}/\text{s}$
PLC_δ	$0.5\ \text{nM}/\text{s}$
k_{SERCA}	$0.18\ \mu\text{M}$
k_{PMCA}	$0.425\ \mu\text{M}$
k_{PLC}	$0.5\ \mu\text{M}$
α_1	$0.6\ \text{nM}/\text{s}$
α_2	$0.004\ \text{s}^{-1}$
γ	5.405
β	$0.03\ \text{s}^{-1}$
p^*	$70\ \text{nM}$
D_{cyt}	$30\ \mu\text{m}^2/\text{s}$
D_{ER}	$10\ \mu\text{m}^2/\text{s}$
D_{IP_3}	$280\ \mu\text{m}^2/\text{s}$

an integral part of the sweeping calcium wave. In eggs, SOCE is blocked due to MPF activation[7,15] and PMCA is internalized.[4] Hence, the parameter δ in Eqs. (4.1A–4.1C) vanishes. The spatiotemporal dynamics of IP$_3$ is modeled by Eq. (4.1C). The first term on the right-hand side describes cytosolic diffusion, while the second and third terms describe linear degradation of IP$_3$ with a time constant $1/\beta$ and generation of IP$_3$ by calcium through PLC$_\delta$, respectively. The use of a more detailed and Ca^{2+}-dependent mechanism of IP$_3$ degradation[2] does not alter our results (not shown).

Equations (4.1A–4.1C) are solved numerically on a two-dimensional membrane. Clusters of IP$_3$Rs are placed regularly on the membrane with various arrangements in terms of cluster size and distance (see results). We used a first-order, fully explicit solver in combination with a four-point discretization of the Laplacian to solve the coupled PDEs in Eqs. (4.1A–4.1C). No-flux boundary conditions, an integrating time step of 200 μs, and a discretization step of 0.5 μm were used for all simulations in this chapter. Smaller time steps and discretization steps have been used for control but have not changed the results. The δ-function for the cluster point sources are replaced by appropriate discrete representations (see, e.g., in Ref. 16).

Modeling calcium signaling in immature oocytes, we use small clusters of 20 IP$_3$Rs per cluster and include SOCE and PMCA. All channels comprising a single cluster are placed on one bin of the discretized ER membrane with size 0.5 μm × 0.5 μm. This makes the distribution consistent with the channel distribution reported in Ref. 23.

3. RESULTS

One of the most important features of Ca^{2+} signals in eggs is the sustained Ca^{2+} rise in response to Ca^{2+} store mobilization. Figure 4.2 shows the time course of calcium elevation in an egg loaded with caged-IP$_3$ and oregon-green-BAPTA-1 in response to a 1-s UV uncaging pulse as described in Refs. 3,4. This leads to a sustained Ca^{2+} rise in the region of uncaging in eggs compared to a transient Ca^{2+} signal in oocytes (Fig. 4.2).

Another indicator of a change in the calcium-signaling machinery during oocyte maturation is the propagation of the fertilization wave, induced by sperm or by locally uncaging caged-IP$_3$. In Ref. 3, global calcium waves were generated through uncaging of IP$_3$ for 20 s in the immature oocyte and for 1 s in the egg. The wave in the immature oocyte propagates at 22 μm/s, that is, at about twice the rate than in the egg. If the immature oocyte is stimulated with IP$_3$ at the same rate as the egg, the resulting wave

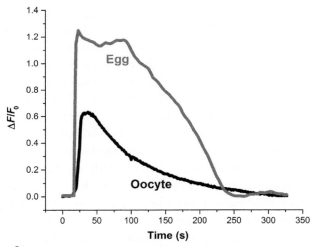

Figure 4.2 Ca^{2+} transients in oocyte and egg with caged-IP$_3$ (10 μM) and Oregon-green-BAPTA-1 (40 μM). IP$_3$ was uncaged globally for 1 s and the course of the ensuing Ca^{2+} transient in the region of interest recorded. (For color version of this figure, the reader is referred to the online version of this chapter.)

is not self-sustained and dies off rapidly, indicating that the calcium–signaling machinery in the egg is more excitable than in the immature oocyte.

To model the response of oocytes to a global stimulus, we homogeneously distribute 25,599 clusters of IP$_3$Rs with 20 channels per cluster at a distance of 2.5 μm over an area of 400×400 μm^2. These numbers are approximately consistent with reports in Ref. 22 and 23. PMCA and SOCE are present, while PLC$_\delta$ is not. A global stimulus of 10 μM/s is applied for 1 s causing a global release of Ca^{2+} into the cell. As shown in Figure 4.3, the calcium concentration decays rapidly to baseline in a similar manner as seen in the experiment (compare with Figure 4.2).

In order to simulate a calcium wave in the oocyte triggered by local uncaging of IP$_3$, we again consider a sheet of 400×400 μm^2 with clusters of IP$_3$Rs each at a distance of 2.5 μm. We stimulate one edge of the square sheet with a flux of $J_{stim} = 100$ μM/s for 2 s. The average speed of the resulting wave of 17.0 μm/s was determined by the elapsed time during which the wave traversed half of the sheet. This speed of the wave is consistent with values reported in Ref. 3. The snapshots of the propagating wave in Fig. 4.4 also reveal that the wave is—like in the experiments described above—not self-sustained, that is, the wave front only travels through a fraction of the cell before it disappears. While a more intense and longer lasting stimulation of the wave will result in a wave that reaches the other end of the

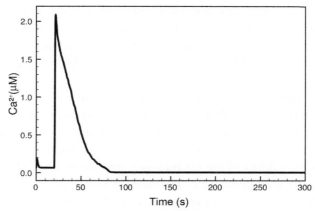

Figure 4.3 Ca^{2+} signaling response to global uncaging of IP$_3$ in oocytes *in silico*. At time $t = 20$ s, a uniform stimulus was applied for the duration of 1 s at a rate of 10 µM/s.

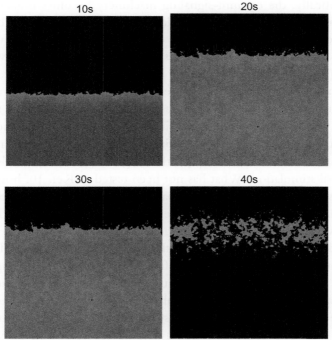

Figure 4.4 Snapshots of Ca^{2+} wave in oocytes (PMCA and SOCE present, PLC$_\delta$ not present, cluster distance = 2.5 µm, 20 IP$_3$Rs per cluster) at the indicated times after injecting 100 µM of IP$_3$ within 2 s at the bottom. The sheet has a size of 400 µm × 400 µm. Lighter gray scales indicate larger Ca^{2+} concentrations.

$400 \, \mu m$ sheet, the wave is still not self-sustained, as it would not propagate fully through a larger sheet. In conclusion, our model with SOC and PMCA functional and PLC_δ absent, with clusters of 20 channels $2.5 \, \mu m$ apart, mimics very well the calcium-signaling behavior observed in the (immature) oocyte.

In the following, we explore the changes in the calcium-signaling machinery during oocyte maturation that may lead to the observed differences in signaling behavior at global stimulation and at local stimulation.

Since the observed calcium wave in eggs is self-sustained, we have to increase the excitability of the calcium-signaling machinery. There are several options to accomplish this. The first option is to activate PLC_δ (as suggested in Refs. 2,17). In the presence of PLC_δ, IP_3 will be generated upon release of Ca^{2+}, which rapidly diffuses and triggers calcium release at neighboring clusters, amplifying the calcium wave. Activating PLC_δ sufficiently to make the wave self-sustained has the consequence that the global stimulation leads to a calcium signal which will never fall back to baseline. Mathematically, the calcium-signaling machinery becomes bistable and is trapped in a stable state with high-calcium levels. Although such behavior is consistent with previous modeling efforts in Refs. 2,18, in view of the observation that global calcium levels come down to baseline levels after minutes (see Fig. 4.2), such behavior is physiologically unrealistic. Hence, activation of PLC_δ cannot be the single cause for the transition from an abortive calcium wave in the oocyte to a sustained calcium wave in the egg. Another way to increase the excitability of the signaling machinery is to increase the IP_3 receptor-binding affinity to IP_3. Such differentiation has been suggested in Ref. 10 to explain the long-lasting calcium levels in eggs after global stimulation. What has not been tested in Ref. 10, however, is whether such a modification of the channel is consistent with the formation of a sustained calcium wave in eggs after local stimulation and with the speed of such wave, being slower than in the immature oocyte.

In order to test this hypothesis for the egg, we simulate our model in the absence of PMCA and SOCE and a fivefold increased IP_3-binding ($\phi_2(c)$) (see Ref. 10) affinity to IP_3, everything else left the same as in the model for the oocyte. Instead of global stimulation, we uncage IP_3 along one edge of the sheet. We indeed observe a self-sustained calcium wave (for snapshots, see Fig. 4.5). The speed of $14.28 \, \mu m/s$, however, is well above those reported in Refs. 3,2.

Yet, another way to increase the excitability of the calcium-signaling machinery is to increase the binding affinity of the IP_3 receptor for Ca^{2+} (ϕ_2

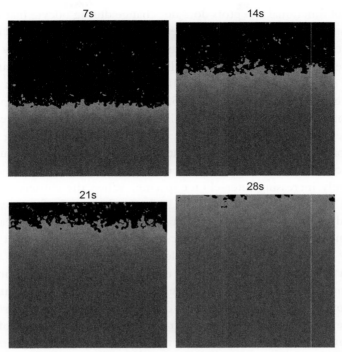

Figure 4.5 Snapshots of Ca^{2+} wave in egg (PMCA and SOCE absent, IP₃ affinity increased by a factor of 5, PLC_δ activated, cluster distance = 2.5 μm, 20 IP₃Rs per cluster) at the indicated times after injecting 20 μM of IP₃ in 2 s at the bottom. The sheet has a size of 400 μm × 400 μm. Lighter gray scales indicate larger Ca^{2+} concentrations.

in Fig. 4.1). This, however, leads to increased spontaneous activity in the absence of stimulation by IP₃ which is not observed in the egg, and it does not generate the long-lasting calcium elevations when the cell is globally stimulated (see Ref. 10). At this point, we found that increased IP₃ affinity of the IP₃ receptor is the only way we can generate a fully self-sustained calcium wave after local stimulation and a long-lasting but finite calcium elevation after global stimulation with IP₃ (see Fig. 4.8). Our simulations, however, are at odds with the observation that the Ca^{2+} wave speed in eggs is smaller by approximately a factor of 2 in comparison to the immature oocyte.

One aspect of oocyte maturation we have not yet considered is remodeling of the ER during development. It has been reported in Refs. 3,4,19–22 that clusters of IP₃Rs in mature oocytes are significantly larger than in immature oocytes. We are starting with the assumption that the number of functional receptors does not change during oocyte maturation. In our model, we have generated larger clusters by combining small clusters with 20 channels (as in the oocyte model), leaving channel density within the cluster and the total amount of IP$_3$Rs invariant. We studied arrangements of clusters with 20, 80, 320, 500, 720, and 980 channels each at their respective distances of 2.5, 5, 10, 12.5, 15, and 17.5 μm while keeping the total number of channels constant. Clusters of about 1000 channels have been used successfully to mimic the feature of calcium signaling in ER patches of mature egg.[22] We find that the speed of the fully self-sustained wave decreases substantially with increasing cluster size (see Fig. 4.6) or equivalently increasing inter-cluster spacing. When the clusters are comprised of 980 channels separated by 17.5 μm, the speed of the wave has decreased to 8.0 μm/s. When the cluster distance reaches a critical value (approximately 20 μm in our specific situation), the wave cannot sustain itself any more as the clusters decouple.

We have shown in Fig. 4.6 that we can reduce the speed of calcium wave to 8.0 μm/s, that is, almost half of the speed of the wave in the immature oocyte, by keeping the total number of channels constant and rearranging the clusters of IP$_3$ receptors into larger clusters. Hence, clustering of IP$_3$Rs is a determinant of the speed of the calcium wave. ER remodeling, however, does not only generates large clusters it also maintains membrane reticular

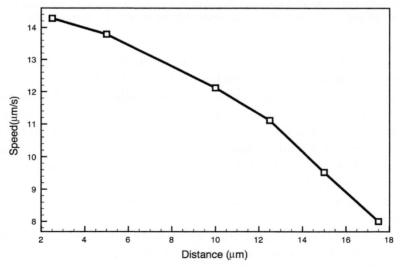

Figure 4.6 Speed of the intracellular calcium wave in egg (PMCA and SOCE absent, PLC$_\delta$ activated, IP$_3$ affinity increased by a factor of 5) as a function of the cluster distance.

areas with small clusters.[3,4,22] The total mass of IP$_3$Rs may be larger in the egg. To arrive at an understanding of the role of small and large clusters of IP$_3$Rs for the feature of the calcium wave in the egg, we perform simulations of such arrangements.

An example of such simulation is shown in Fig. 4.7, where we place 361 clusters with 980 channels each at a distance of 21 μm on a 400×400 μm^2 membrane sheet and small clusters of 20 channels at a distance of 3 μm to mimic a realistic distribution of IP$_3$Rs. In this spatial arrangement of clusters, the speed of the calcium wave has dropped to 10.0 μm/s. This speed is only slighter larger than the speed of the wave in the presence of the large clusters, suggesting that the larger clusters are the main determinant of the speed and not the small clusters. The wave appears, however, more smooth like the wave in the absence of the small clusters.

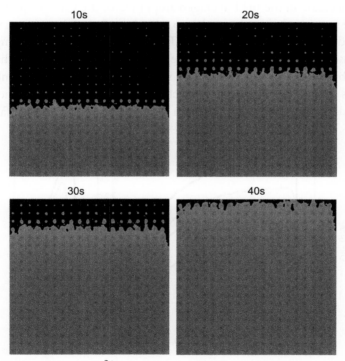

Figure 4.7 Snapshots of Ca^{2+} wave in egg (PMCA and SOCE absent, IP$_3$ affinity increased by a factor of 5, PLC$_\delta$ activated, large clusters with 980 IP$_3$Rs per cluster at a distance of 21 μm, small cluster with 20 IP$_3$Rs at a distance of 3 μm) at the indicated times after injecting 20 μM of IP$_3$ in 2 s at the bottom. The sheet has a size of 400 μm $\times$ 400 μm. Lighter gray scales indicate larger Ca^{2+} concentrations.

While we have shown above that spatial organization plays a major role for the speed of a calcium wave, it remains to be explored how spatial reorganization of channels affects the decay of intracellular calcium after a global stimulation. Specifically, how does spatial reorganization of channels influence the transition between a rapid decay in the oocyte to a plateau with slow decay in the egg (see Fig. 4.2). To this end, we simulate the decay of intracellular calcium for four additional cluster arrangements, that is, 20, 320, 500, and 980 channels per cluster, after globally injecting 10 μM of IP$_3$ for 1 s. In Fig. 4.8, we show the decay of the field-averaged cytosolic calcium with time at various cluster sizes.

Most importantly, the decay time of the intracellular calcium signal does not depend on the cluster organization. Hence, if these data where presented in a relative manner, normalized to, for example, the maximum, all curves would be close to identical. As the cluster size is increased, we observe, however, a decrease in the field-averaged amplitude of the Ca^{2+} response. This decrease provides an important clue to why the speed of the calcium wave is regulated by the cluster organization. The cause for the decreased amplitude is increased inhibition of IP$_3$Rs in the large clusters. We support this conclusion by the inset in Fig. 4.8 where we compare the fraction of inhibited channels of cluster of 20 channels and cluster of 980 channels. The fraction of inhibited channels in the cluster with 980 channels is larger by a factor of

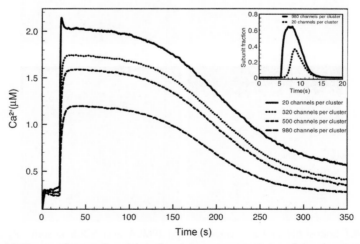

Figure 4.8 Decay of cytosolic calcium for various arrangements of the clusters of IP$_3$Rs (20, 320, 500, and 980 channels per cluster from top to bottom) in egg after globally injecting 10 μM of IP$_3$ for 1 s at time $t = 20$ s. The inset shows the fraction of inhibited subunits of 980 channels and 20 channels per cluster.

about 1.7 in comparison with that in cluster of 20 channels. Consistently, the maximum field-averaged calcium response of the cluster arrangement with 20 channels per cluster is larger by a factor of 1.7 than in the cluster arrangement with 980 channels per cluster. Channels in a large cluster are more likely inhibited because of their proximity to other channels and mutual inhibition, leading to a shorter open time and hence less calcium release.

4. DISCUSSION

The calcium–signaling machinery undergoes significant changes during oocyte maturation. These changes are manifest most obviously in the competence of (a) eliciting a sweeping Ca^{2+} wave upon local stimulation with IP$_3$ and (b) maintaining a high Ca^{2+} elevation for minutes in order to trigger embryonic development. On the scale of elemental Ca^{2+} release processes (puffs, blibs), the changes, while statistically significant, are more subtle.[3] The difference between the amount of Ca^{2+} released during a puff in an egg and an immature oocyte is statistically insignificant,[3] while the average puff duration is slightly shorter in the egg. During larger localized Ca^{2+}-release events through coalesced clusters (coined single release events in Ref. 3), a larger amount of Ca^{2+} is released in eggs than in oocytes. While numerous changes in the machinery (e.g., internalization of PMCAs, blocking of SOC, increased IP$_3$ affinity and remodeling of ER) are known, it is unknown what changes are responsible for the manifest changes of the cell's physiology. While it is very difficult to disentangle the effects of numerous simultaneous changes, mathematical modeling is perfectly suitable to address this question.

Differentiation of calcium signaling during oocyte maturation has been subject to previous modeling efforts. Ponce-Dawson *et al.* addressed this question using the fire–diffuse–fire model (FDF).[6] In FDF, discrete calcium-release sites, organized in a discrete array with a distance d, open for time τ if the local concentration exceeds a threshold C_T and release a fixed amount of calcium σ, leading to a local release concentration of σ/d^3. This concentration spreads out by diffusion and can cause the release of Ca^{2+} at another site if its local concentration exceeds the required threshold. Transition from saltatory Ca^{2+} waves, mimicking Ca^{2+} waves in immature oocytes, to smooth Ca^{2+} waves, mimicking Ca^{2+} waves in eggs, have been obtained with FDF by increasing the release time τ almost 200-fold, decreasing the amount of Ca^{2+} released about threefold and increasing basal Ca^{2+} levels eightfold. The velocities for the Ca^{2+} waves obtained by FDF[6] are consistent with typical experimental values of 20 μm/s in the immature oocyte and 5 μm/s in the egg. In contrast to FDF, our model is more detailed and microscopic, and the input parameters (such as τ and σ) emerge from receptor kinetics and calcium diffusion.

The assumed value for the open time τ of 9 s (in FDF) for the cluster release compares well with the typical duration of a release event of a cluster with about 1000 channels as obtained by our simulations. The typical duration of a release event in our oocyte model is about 100 ms, much shorter than for the egg, and of the same order of magnitude than the open time postulated in FDF.[6] The long event duration of the larger cluster in the egg model is caused mainly by the presence of PLC_δ and increased IP_3 affinity which reinforces channel opening but also due to the larger number of available channels. The predicted release concentration of a release site in the egg model (large clusters) is about 3 μM (with a decreasing temporal profile) and compares with the value of about 0.5 μM in Ref. 6. For the immature oocyte, the release concentration is about 1 μM, that is, one-third than in the egg, while it is about 0.6 μM in Ref. 6, that is, less in the egg. A larger release concentration is consistent with larger clusters of IP_3Rs in the mature egg.

There are other differences. While the overall change in the speed of the calcium wave during oocyte maturation is consistent between Ref. 6 and our model, the dependence of the speed on the release-site distance in our study is not as predicted as in Ref. 6. One reason could be that the FDF model does not include channel inhibition. As shown in the inset of Fig. 4.8, inhibition is substantially larger in channels within large clusters than small clusters. Hence, when we increase the cluster sizes during oocyte maturation, we increase inhibition.

Fertilization waves in eggs have been investigated through mathematical modeling and experimentation by Wagner and collaborators in Refs. 18,2. The numerical study in Ref. 18 focuses on the shape of an evolving Ca^{2+} wave in two-dimensional circular geometry and uses a continuous reaction-diffusion model with continuous fluxes from ER to cytosol through release and SERCA. The underlying kinetics of the model sports bistable Ca^{2+} dynamics, that is, the wave front separates low Ca^{2+} levels ahead of it and high Ca^{2+} on its tail, which remains high forever. The numerical study in Ref. 2, based on a three-dimensional (bistable) reaction–diffusion equation on a spherical domain with continuous fluxes, underscores the importance of local IP_3 generation through PLC_δ copropagating ahead of the Ca^{2+} wave front.

While these studies have contributed significantly to our understanding of the fertilization wave, none of these models are detailed enough to decipher the mechanism underlying calcium-signaling differentiation during oocyte maturation. Our model takes into account in addition to explicit expressions for Ca^{2+} release from stores, Ca^{2+} reuptake by stores, Ca^{2+} extrusion through PMCA, SOCE, the spatial organization of the calcium-release channels, which undergo significant redistribution through ER remodeling during oocyte maturation. In contrast to the study in Ref. 2, the calcium dynamics in our model is not bistable, that is, in the back of the wave front, the calcium concentration does not remain high indefinitely, but will decay. In the immature oocyte, in agreement with observations,[3] this decay happens on the order of seconds, while in the egg on the order of minutes, motivating the use of a bistable model for the characterization of the sweeping calcium wave front in the egg. In our study, however, we derive important clues for the changes in the calcium-signaling machinery during oocyte maturation exactly from the decay of the calcium in the wake of the wave front. Unlike the study in Ref. 2, where the role of PLC$_\delta$ was emphasized as an important ingredient for the sweeping calcium wave, we arrived at the conclusion that increased IP$_3$ affinity of the IP$_3$ plays a key role for the egg to generate a calcium wave sweeping self-sustained the entire cell.

Another key result of this study is that in order to explain the manifest physiological changes in calcium homeostasis during oocyte maturation, that is, increased duration of calcium transients and decreased wave speed, spatial arrangement of IP$_3$Rs plays an important role. Internalization of PMCAs, a known process during oocyte maturation, by itself gives rise to an increase of the wave speed, while most observations suggest a strong decrease during oocyte maturation. In a similar fashion, an increase in IP$_3$ affinity gives rise to a longer lasting calcium response upon stimulation as seen in the egg and also leads to a self-sustained Ca^{2+} wave. A decrease in IP$_3$ affinity, would result in a slower wave, as seen in the egg, but would also lead to a shorter duration of the overall transients, which is inconsistent with the observations.

The only possibility to accommodate both, the decrease of the wave speed and the increase of the duration of cytosolic calcium levels, is when the internalization of PMCA and the blockage of SOCE are accompanied by increased IP$_3$ affinity and a reorganization of the spatial distribution of IP$_3$Rs.

In summary, our model provides a microscopic understanding of the underlying changes in the open time, the release concentrations, and the wave speeds in terms of properties of the calcium-signaling machinery and its spatial organization.

ACKNOWLEDGMENT

We would like to thank the National Science Foundation for financial support under grant #IOS-0744798.

REFERENCES

1. Berridge M, Bootman MD, Lipp P. Calcium—a life and death signal. *Nature.* 1998;395:645–648.
2. Wagner J, Fall CP, Hong F, et al. A wave of IP3 production accompanies the fertilization Ca2+ wave in the egg of the frog, Xenopus laevis: theoretical and experimental support. *Cell Calcium.* 2004;35:433–447.
3. Machaca K. Increased sensitivity and clustering of elementary Ca^{2+} release events during oocyte maturation. *Dev Biol.* 2004;275:170–182.
4. El Jouni W, Jang B, Haun S, Machaca K. Calcium signaling differentiation during *Xenopus* oocyte maturation. *Dev Biol.* 2005;288:514–525.
5. Fontanilla RA, Nuccitelli R. Characterization of the sperm-induced calcium wave in Xenopus eggs using confocal microscopy. *Biophys J.* 1998;75:2079–2087.
6. Dawson SP, Keizer J, Pearson JE. Fire-diffuse-fire model of dynamics of intracellular calcium waves. *Proc Natl Acad Sci USA.* 1999;96:6060–6063.
7. Machaca K, Haun S. Store-operated calcium entry inactivates at the germinal vesicle breakdown stage of Xenopus Meiosis. *J Biol Chem.* 2000;275:38710–38715.
8. Hartzell HC. Activation of different C1 currents in Xenopus oocytes by Ca liberated from stores and by capacitative Ca influx. *J Gen Phys.* 1996;108:157–175.
9. Terasaki M, Rauft LL, Hand AR. Changes in organization of the endoplasmic reticulum during Xenopus oocyte maturation and activation. *Mol Biol Cell.* 2001;12:1103–1116.
10. Ullah G, Jung P, Machaca K. Modeling Ca(2+) signaling differentiation during oocyte maturation. *Cell Calcium.* 2007;42:556–564.
11. Sun XP, Callamaras N, Marchant JS, Parker I. A continuum of IP$_3$ mediated elementary Ca^{2+} signalling events in Xenopus oocytes. *J Physiol.* 1998;509:67–80.
12. Sneyd J, Dufour JF. A dynamic model of the type-2 inositol trisphosphate receptor. *Proc Natl Acad Sci USA.* 2002;99:2398–2403.
13. Sneyd J, Tsaneva-Atanasova K, Yule JI, Thompson JL, Shuttleworth TJ. Control of calcium oscillations by membrane fluxes. *Proc Natl Acad Sci USA.* 2004;101:1392–1396.
14. Stiber J, Hawkins A, Zhang ZS, et al. STIM1 signaling controls store-operated calcium entry required for development and contractile function in skeletal muscle. *Nat Cell Biol.* 2008;10:688–697.
15. Machaca K, Haun S. Induction of maturation-promoting factor during Xenopus oocyte maturation uncouples Ca^{2+} store depletion from store-operated Ca^{2+} entry. *J Cell Biol.* 2002;156:75–85.
16. Shuai JW, Jung P. Optimal ion channel clustering for intracellular calcium signaling. *Proc Natl Acad Sci USA.* 2003;100:506–510.
17. Snow P, Yim DL, Leibow JD, Saini S, Nuccitelli R. Fertilization stimulates an increase in inositol trisphosphate and inositol lipid levels in *Xenopus* eggs. *Dev Biol.* 1996;180:108–118.
18. Wagner J, Li XY, Pearson J, Keizer J. Simulation of the fertilization Ca^{2+} wave in Xenopus laevis eggs. *Biophys J.* 1998;75:2088–2097.
19. Kume S, Yamamoto A, Inoue T, Muto A, Okano H, Mikoshiba K. Developmental expression of the inositol 1,4,5-trisphosphate receptor and structural changes in the

endoplasmic reticulum during oogenesis and meiotic maturation of Xenopus laevis. *Dev Biol*. 1997;182:228–239.

20. Mehlmann LM, Terasaki M, Jaffe LA, Kline D. Reorganization of the endoplasmic reticulum during meiotic maturation of the mouse oocyte. *Dev Biol*. 1995;170:607–615.

21. Parys JB, McPherson SM, Mathews L, Campbell KP, Longo FJ. Presence of inositol 1,4,5-trisphosphate receptor, calreticulin, and calsequestrin in eggs of sea urchins and Xenopus laevis. *Dev Biol*. 1994;161:466–476.

22. Sun L, Yu F, Ullah A, et al. Endoplasmic reticulum remodeling tunes IP$_3$-dependent Ca^{2+} release sensitivity. *PLoS One*. 2011;6(11):e27928.

23. Shuai J, Rose H, Parker I. The number and spatial distribution of IP3 receptors underlying calcium puffs in Xenopus oocytes. *Biophys J*. 2006;91:4033–4044.

CHAPTER FIVE

Modeling Mitochondrial Function and Its Role in Disease

M. Saleet Jafri[*], Rashmi Kumar[†]

[*]School of Systems Biology and Department of Molecular Neuroscience, George Mason University, Fairfax, VA, USA
[†]School of Systems Biology, George Mason University, Fairfax, VA, USA

Contents

Abstract

Many neurodegenerative diseases involve defects in mitochondrial function. These defects often arise from mutations carried in the genes that code mitochondrial proteins. Many of these defective proteins are involved in mitochondrial energy metabolism which is one of the primary roles of the mitochondria. However, others proteins have other roles that are related to signaling or mitochondrial structure. The interaction of the mitochondrial proteins is complex. Understanding these complex dynamics requires the use of computational models. Studies have started to exploit such models, but much further work is necessary to understand mitochondrial function and its role in neurodegenerative disease.

Progress in Molecular Biology and Translational Science, Volume 123
ISSN 1877-1173
http://dx.doi.org/10.1016/B978-0-12-397897-4.00001-2

1. INTRODUCTION

Mitochondria are intracellular organelles that have multiple roles within cells. The mitochondria process metabolites to make ATP (adenosine triphosphate), the energy currency used by the cell, as is the case with cardiac muscle. In addition to this role, the mitochondria have a signaling role in cells. For example, they serve an essential role as a glucose sensor in pancreatic β-cells. Mitochondria also play a crucial role in apoptosis: programmed cell death. Finally, mitochondrial dysfunction has been implicated in a number of pathologies, including many neurodegenerative diseases such as Alzheimer's disease and Parkinson's disease. These diverse roles result from the complexity of the interaction of mitochondrial components which often cannot be directly measured by current experimental techniques. Hence, computational modeling of mitochondria has become an important tool to understand mitochondrial function. The complex physiology of the mitochondria and how modeling can be used to understand it will be discussed in this chapter.

Mitochondria are thought to have originated as a separate organism that was internalized early in evolution in a eukaryotic cell. Based on evolutionary and phylogenetic studies, many suggest that it was a protobacteria; however, there are some recent suggestions that it might have been a eukaryote.[1] As such, the mitochondria have an inner and outer membrane and mitochondrial DNA. The volume between the inner and outer membrane is the intermembrane space. The mitochondrial matrix is found within the inner membrane. Proteins in the mitochondrial matrix help to carry out mitochondrial function. The inner membrane forms tubular and lamellar structures called cristae. While the outer membrane is quite porous, the inner membrane is not. Instead, the large inner membrane surface area houses proteins that serve as transporters, exchangers, and ion channels that control the flow of ions, metabolites, and signaling molecules across it. The cristae structures are also not static, changing under different physiological conditions and during disease (including neurodegenerative diseases such as Parkinson's disease and Alzheimer's disease), suggesting that there are functional consequences of these structural changes.[2] In the following sections, some of the facets of mitochondrial function are described in more detail.

The approximately 1100 proteins that comprise the mitochondria arise from genes both in the nuclear and mitochondrial DNA. In most organisms, only a small number of mitochondrial proteins are encoded by the mitochondrial genome; whereas most are encoded by genes in the nucleus and transported to the mitochondria via the cytosol. The mitochondrial genome is small consisting of 37 genes. Of these, 13 code for proteins involved in the respiratory chain and oxidative phosphorylation and the remainder code for translation and transcription of the mitochondrial genome, RNA processing, protein import, and protein maturation.[1,3] These genes would follow a maternal inheritance pattern as the mitochondria in the oocyte all come from the mother. The remaining polypeptides that comprise the mitochondria come from the nuclear genome. These would be contributed by both parents. Tables 5.1 shows the mitochondrial genes involved in neurodegenerative disease and the effect they have on mitochondrial function. Table 5.2 shows the nuclear genes involved in neurodegenerative diseases and the effect they have on mitochondrial function.

Table 5.1 Summary of mitochondrial genes, their function, and their involvement in disease

Disease	Mitochondrial gene	Defect
Kearns–Sayre syndrome	Subunits of Complexes I, II, and IV	Respiratory chain dysfunction[4]
Leber heridary optic atrophy	Subunits of Complexes I, II, III, and IV	Impaired oxidative phosphorylation/electron transport chain
	MT-ATP6—ATP synthase subunit 6	Decreased ATP production
Leigh's syndrome	Subunits of Complex I	Decreased Complex I activity
Maternally inherited Leigh's syndrome (MILS)	MT-ATP6—ATP synthase subunit 6	Decreased ATP production
Mitochondrial Complex I deficiency	Many subunit of Complex I	Impaired Complex I function
Neuropathy, ataxia, and retinosa pigmentosa (NARP)	MT-ATP6—ATP synthase subunit 6	Decreased ATP production
Spinocerebellar ataxia	ATXN7—ataxin 7	Abnormally large mitochondria with irregular cristae

Table 5.2 Summary of nuclear genes, their function, and their involvement in disease

Disease	Gene	Defect
Alzheimer's disease	APP—amyloid β precursor protein	Impaired succinate dehydrogenase
Adrenoleukodystrophy	ABCD1—ATPase binding cassette I	Defect in β-oxidation of fatty acids
Amyotrophic lateral sclerosis	SOD1—Cu, Zn superoxide dismutase I	Increased cellular ROS
Encephalopathy (lethal)	DNM1L—dynamin 1 like	Mitochondrial and peroxisome fusion
Friedreich ataxia	FXN—frataxin	Increased ROS production through interaction at Complex II[5]
Hereditary spastic paraplegia	REEP1—receptor expression-enhancing protein-1	Abnormal mitochondrial network organization and impaired mitochondrial energy production[6]
	SPG7—paraplegin-7 and other SPG proteins	Abnormal mitochondria and dysfunction[7]
	HSPD1—heat-shock 60 KD protein-1 and other HSP proteins	Morphological changes in mitochondria, deficient ATP synthesis, defect in assembly of Complex III[7,8]
Huntington's disease	HTT—huntingtin	Mitochondria have lower membrane potential, have more ROS production, and are more sensitive to depolarization due to high-cellular calcium
Leigh's syndrome	Complex I subunits from nuclear DNA	Decreased oxygen consumption, decreased ATP production, and increased ROS and lactic acid production
	SDH1—succinate dehydrogenase complex subunit A in Complex II	Decreased succinate dehydrogenase and Complex II activity
	SURF1 SCO2 in Complex IV	Decreased cytochrome C oxidase activity
	Pyruvate dehydrogenase E1-α subunit	Decreased pyruvate dehydrogenase activity

Table 5.2 Summary of nuclear genes, their function, and their involvement in disease—cont'd

Disease	Gene	Defect
Mohr–Tranebjaerg syndrome	TIMM8A—translocase of the inner mitochondrial membrane	Decreased NADH levels by calcium binding aspartate–glutamate shuttles
Parkinson's disease	PARK2—parkin	Damaged mitochondria not degraded
	SNCA—α-synuclein	Mitochondrial fragmentation
	LRRK2—leucine-rich repeat kinase 2	Increases peroxiredoxin (PRDX3) phosphorylation which reduces its peroxidase activity in mitochondria
Spongiform encephalitis	PRNP—prion protein	Fewer mitochondria, with abnormal morphology, and increased ROS production
Supranuclear palsy	MAPT—microtubular associated protein Tau	Mitochondria do not enter neurites
Spinocerebellar ataxia	ATXN7—ataxin 7	Large mitochondria with no cristae

Data gathered from Refs. 3,9.

2. ENERGY METABOLISM

One of the primary functions of the mitochondria is catabolic energy metabolism; that is, substrates, such as carbohydrates, fatty acids, and proteins, are broken down to release energy that is stored in high-energy phosphate bonds in molecules such as ATP and CP (creatine phosphate). This occurs in multiple stages by multiple pathways. (1) The tricarboxylic acid (TCA) cycle breaks down small carbohydrates (acetyl-CoA and TCA cycle intermediates) to produce reducing equivalents, that store the released energy. (2) There are also pathways that bring energy substrates into the TCA cycle. (3) The reducing equivalents enter oxidative phosphorylation pathway to become ATP the energy substrate of the cell.

2.1. Tricarboxylic acid cycle

The TCA cycle is a set of eight catalyzed reactions and eight intermediates that break down hydrocarbon substrates into carbon dioxide (CO_2) and water (H_2O) using the energy released to protonate nicotinamide adenine dinucleotide converting from NAD^+ to NADH or flavin adenine dinucleotide from FADH to $FADH_2$ (Fig. 5.1). Acetyl-CoA enters the TCA cycle to combine with oxaloacetate in a reaction catalyzed by citrate synthase to form citrate. Citrate is converted to isocitrate by aconitase liberating water. ISO dehydrogenase catalyzes the next reaction converting isocitrate to α-ketoglutarate while reducing NAD^+ to NADH. This enzyme is activated by [Ca^{2+}] and inhibited by [NADH] and succinyl-CoA (SCoA). The α-ketoglutarate dehydrogenaseenzyme complex converst α-ketoglutarate to SCoA reducing NAD^+ in the process. This enzyme is activated by mitochondrial [Ca^{2+}] and [ADP] and inhibited by [NADH]. SCoAbecomes succinate through a reaction catalyzed by succinate-CoA ligase while forming guanosine triphosphate or ATP. Succinate dehydrogenase transforms succinate into fumarate and reduces FADH to $FADH_2$. This is followed by the conversion of fumarate to malate by fumarase. Completing the cycle, malate dehydrogenase creates oxaloacetate and NADH from malate and NAD^+ in a reaction inhibited by oxaloacetate.

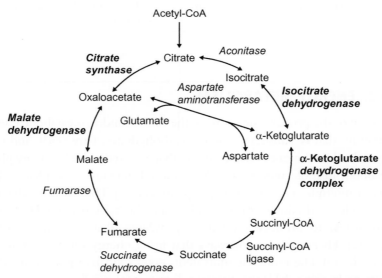

Figure 5.1 *Tricarboxylic acid cycle schematic.* The enzymes shown in bold text have been previously modeled in detail (see Dudycha and Jafri, 2001). The influx of acetyl-CoA (ACoA) to the system reflects the breakdown of substrate from the catabolism of glucose, fatty acids, and proteins.

As mentioned before, there are other biochemical pathways that allow the entry of substrate as intermediates of the TCA cycle. The cycle is split by a reaction catalyzed by aspartate amino transferase that converts oxaloacetate and glutamate to α-ketoglutarate and aspartate. In the brain under certain conditions, the top and bottom halves of the TCA cycle have actually been observed to occur at different rates.[10]

2.2. Substrate entry into the TCA cycle

There are several pathways by which different substrates enter the TCA cycle. For example, most carbohydrates are converted to glucose through digestion and postprocessing in the liver. The glucose absorbed by cells via the glucose transporter enters the glycolysis pathway in the cell cytoplasm to become pyruvate. Pyruvate enters the mitochondria via membrane transporters where it is metabolized by the pyruvate dehydrogenase complex to make acetyl-CoA. Pyruvate dehydrogenase complex is activated by mitochondrial calcium concentration ($[Ca^{2+}]$) and inhibited when the [ATP]/[ADP] ratio, $[NAD^+]$/[NADH] ratio, or [acetyl-CoA]/[CoA] ratio is increased. Acetyl-CoA enters the tricarbocylic acid cycle (also known as the TCA cycle, Krebs cycle, or the citric acid cycle).

Fatty acids in the cytoplasm are activated through a reaction catalyzed by fatty acid acyl-CoA synthetase and are then transported into the mitochondria across the inner membrane by a carnitine carrier system. Once in the mitochondrial matrix, β-oxidation of fatty acids catabolize the fatty acid into acetyl-CoA which then can enter the TCA cycle.

Proteins that are broken down into amino acids can enter the TCA cycle at different points depending upon the amino acid. Some glucogenic amino acids are deaminated and converted to glucose through gluconeogenic pathways. The glucose then enters glycolysis to form pyruvate and then acetyl-CoA. Other ketogenic amino acids are converted to acetyl-CoA or α-ketoglutarate which can enter the TCA cycle directly at different points of the cycle.

There are other anapleurotic pathways where substrates can enter the TCA cycle at additional loci. For example, pyruvate can be converted to oxaloacetate by pyruvate carboxylase found in mitochondria, aspartate can be converted to oxaloacetate through the transaminase reaction, glutamate can be converted to α-ketoglutarate by glutamate dehydrogenase, β-oxidation of fatty acids can result in SCoA, as well as others.

2.3. Oxidative phosphorylation

The reducing equivalents NADH and FADH$_2$ enter the electron transport chain to undergo a process called oxidative phosphorylation (Fig. 5.2). The steps of the electron transport chain are a series of reactions involving NADH-coenzyme Q oxidoreductase (Complex I), Succinate-Q oxidoreductase (Complex II), Electron transfer flavoprotein-Q oxidoreductase, Q-cytochrome c oxidoreductase (Complex III), Cytochrome c oxidase (Complex IV). This is followed by a final step that makes ATP by the F$_1$F$_0$-ATPase (Complex V). These protein complexes are found in the inner membrane distributed between the cristae membrane and the remaining inner membrane. In yeast, 60% of these proteins are in the cristae membrane while in cardiac mitochondria 94% are found in the cristae.[2] Note that the TCA cycle proteins are also clustered in a metabolon associated with the

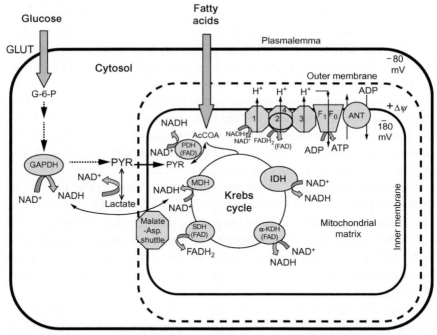

Figure 5.2 *Mitochondrial energy metabolism.* Energy substrates enter in the form of glucose or fatty acids that are converted to acetyl-CoA. The acetyl-CoA enters the tricarboxylic acid (TCA) or Krebs cycle to form NADH and FADH2 which enter the electron transport chain. The electron transport chain generates a membrane potential and proton gradient across the mitochondrial inner membrane. These gradients are then used to make ATP.

matrix side of the inner membrane so that isocitrate dehydrogenase can interact with Complex I and succinate dehydrogenase can interact with Complex II (Fig. 5.3).

Complex I is coupled to isocitrate dehydrogenase and is the point where NADH enters the electron transport change. The electron released when NADH is oxidized is passed through the complex resulting in the translocation or pumping of four protons (H^+) across the inner membrane out of the mitochondrial matrix. Ubiquinone (coenzyme Q10 or Q) is also reduced to ubiquinol (QH_2) as it gains two protons.

Complex II is coupled to succinate dehydrogenase and is the entry point for $FADH_2$ into the electron transport chain. $FADH_2$ is oxidized and the two protons are used to reduce ubiquinone.

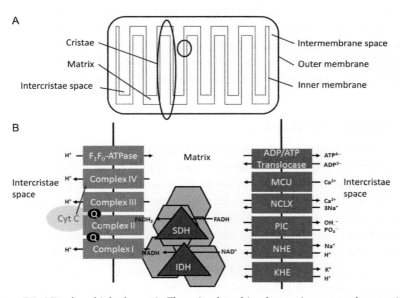

Figure 5.3 *Mitochondrial schematic.* The mitochondrion has an inner membrane with a large surface area that define the cristae (in oval) and mitochondrial matrix (top). The circled region (top) is enlarged in the bottom panel. The large surface area serves as the site for energy metabolism with the proteins for the respiratory chain and oxidative phosphorylation located in the membrane and the tricarboxylic acid (TCA) cycle proteins clustering on the matrix side of the membrane. There are also the proteins involved in transport of substances in and out of the mitochondria. (MCU, mitochondrial calcium uniporter; NCLX, mitochondrial sodium–calcium–lithium exchanger; PIC, phosphate carrier; NHE, sodium hydrogen exchanger; KHE, potassium hydrogen exchanger.) (For the color version of this figure, the reader is referred to the online version of this chapter.)

In Complex III, the ubiquinol is oxidized to reform ubiquinone. Four protons are extruded from the mitochondrial matrix across the inner membrane in this step. Energy stored as cytochrome C is reduced in this step.

In Complex IV, cytochrome C is oxidized to pump four protons across the inner membrane. Water is created in this step from protons and oxygen. This is the step that requires oxygen.

During electron transport, the respiration-drive proton pumps (the four steps) create a proton gradient across the mitochondrial inner membrane as well as a mitochondrial membrane potential ($\Delta\psi_m$). The membrane potential is extremely negative ranging from -160 to -180 mV. The proton gradient has been estimated to be between 0.1 and 0.5 pH units depending on the tissue type and conditions. This electrochemical gradient is then used by the F_1F_0-ATPase (ATP synthase) to make ATP. As protons enter through a pore in the ATP synthase, ATP is created from ADP and free inorganic phosphate. The typical ratio is three protons for one ATP molecule created, although this might vary under different conditions. The ATP synthase is thought to be activated by mitochondrial matrix calcium.

2.4. Substrate transport

In order for energy metabolism to work effectively, substrates, ADP, and phosphate need to enter the mitochondria and ATP needs to exit. To this end, there exists a series of transport proteins in the form of carrier and shuttles. The ATP/ADP translocase, located in the inner membrane, transports ATP out of the matrix and ADP into the matrix. As ATP^{4-} and ADP^{3-} have different charges there is a net charge transfer during the translocation process. This net movement of charge generates an ionic current that dissipates some of the mitochondrial membrane potential. Phosphate is transported into the mitochondrial matrix by the phosphate carrier that exchanges phosphate and hydroxide (OH^-) so that there is no net charge carried (electroneutral). As OH^- exits the matrix when phosphate enters, this dissipates some of the proton gradient.

Substrates enter the mitochondria through a number of pathways such as the malate–aspartate shuttle, the glutamine carrier, and the tricarboxylate carrier protein. The former exchanges malate for aspartate bringing malate into the matrix and extruding aspartate. The other two are unidirectional carriers of substrate.

2.5. Ionic and substrate homeostasis

As the mitochondria need to maintain homeostasis of its internal environment, the fluxes of ions and substrates across the inner membrane need to be balanced with the consumption or source fluxes. For example, the phosphate carrier flux must balance the consumption rate of phosphate by the F_1F_0-ATPase or the rate of ADP consumed or ATP produced needs to match the flux transported by the ATP/ADP translocase.

This is also true of ions such as protons (H^+), calcium (Ca^{2+}), and sodium (Na^+). The number of protons extruded by the respiration-drive proton pumps in the electron transport chain need to be balance by the entry through the F_1F_0-ATPase and the amount neutralizing the OH^- brought in by the phosphate carrier. This is complicated by proton buffering in the mitochondrial matrix.

This also extends to calcium which has thus far been presented as a regulator of energy metabolism. Calcium enters the mitochondrial matrix through the mitochondrial calcium uniporter (MCU). This ion channel open in response to extramitochondrial calcium (the outer membrane is leaky so the intermembrane space effectively has extramitochondrial calcium levels) and calcium enters driven by the electrochemical gradient created by the large mitochondrial membrane potential. The entry of positive charge generates a depolarizing current that partially dissipates the mitochondrial membrane potential. Calcium is buffered in the matrix by binding to proteins, ATP, and ADP. This calcium influx is balanced by the mitochondrial sodium–calcium (lithium) exchanger (NCLX) which extrudes one calcium ion for three sodium ions. This net charge difference creates a current that also helps to depolarize the mitochondrial membrane potential. The question of how much calcium is sequestered in the mitochondria and the resulting change in mitochondria is a topic of current debate. The interpretation of experiments from different groups suggest either little uptake of calcium by the mitochondria with a <100 nm change in mitochondrial $[Ca^{2+}]$ or a more significant contribution where mitochondrial $[Ca^{2+}]$ can change several micromolar.

3. MITOCHONDRIAL SIGNALING

The mitochondria also produce signaling molecules that have targets in the mitochondria as well as in the cell. One example of this is reactive oxygen species (ROS). ROS include superoxide ($O_2^{\cdot-}$), hydrogen peroxide (H_2O_2), and peroxynitrite ($ONOO^-$). Superoxide is created as byproduct of electron transport at Complex I and Complex III. Less than 1% of the electron transport flux results in the production of superoxide. It is highly reactive; hence, the mitochondria have a defense against it in the form of superoxide dismutase (SOD) which converts superoxide into hydrogen peroxide. In fact, the amount of mitochondrial superoxide dismutase is 4 times that found in the cytoplasm of heart. The hydrogen peroxide is converted to water and oxygen through the reaction with glutathions (GSH) to form glutathione disulfide (GSSG). The glutathione disulfide is reduced to glutathione by glutathione reductase replenishing the antioxidant system.

ROS play a signaling role in the cell acting upon many targets. For example, ROS does oxidative damage to DNA which is considered to be one of the mechanisms of aging. It can also damage lipids and proteins. On the other hand, it can serve as a reversible signaling molecule. For example, ROS target the ryanodine receptor, a calcium release channel in the ER and increases its open probability which might have implications for calcium signaling.

The mitochondria are also a target of signaling molecules from the cell. In the discussion above, calcium signaling in the cell cytoplasm can result in a rise in mitochondrial calcium that can stimulate energy metabolism. Increases in cytosolic ADP levels can translate into increases of mitochondrial ADP and stimulate energy metabolism. Hence, if the cell is active (lots of calcium signaling) and consuming ATP (increase in ADP) the mitochondria can respond by increasing energy metabolism.

Another example of the mitochondria being a target of intracellular signaling is apoptosis. Apoptosis is a complex signaling set of pathways that can be triggered through a variety of cell stress stimuli. This results in the induction of apoptosis and cell death. One of these pathways targets mitochondrial function. Typically, apoptosis seems to be regulated by the balance between proapoptotic factors such as Bax, Bid, Bad, and Bak and anti-apoptotic pathways such as the Bcl-2 protein family in the mitochondrial outer membrane. However, there are different pathways that alter this balance. When the balance is shifted toward apoptosis there is an increase in mitochondrial permeability that allows efflux of small mitochondria-derived activators of caspases (SMAC) and cytochrome c. SMAC binds to and deactivates mitochondrial inhibitor proteins allowing apoptosis to proceed. Cytochrome c binds with another protein apoptotic protease activating factor (APAF) to form the apoptosome which then bind to caspase-9 which starts the cascade to apoptosis.

4. MITOCHONDRIA IN DISEASE

The two main types of neurological diseases that involve mitochondria are ischemia and neurodegenerative diseases.

4.1. Ischemic disease

Ischemia is the deprivation of tissue of oxygen and metabolic substrate. This is in contrast to hypoxia which is simply depriving the tissue of oxygen. In the case of oxygen deprivation Complex IV activity ceases. The result is that all the cytochrome cremains in the reduced state. This impairs Complex III and prevents the replenishment of ubiquinone from ubiquinol. Without ubiquinone, Complex I and II cannot function and NADH and $FADH_2$ accumulate. The accumulation of these reducing equivalents will inhibit enzymes in the TCA cycle and energy metabolism ceases. As a result, the cell must rely on anaerobic metabolism for minimal ATP production and cell acidification from the lactic acid produced. As ATP falls, cellular processes such as ionic transport ceases resulting in dysfunctional or no action potentials and impaired neurotransmitter release. With ischemia, substrate is not available and the cell has to use substrates already available in the cell which quickly deplete. The result is that the crisis caused by low ATP occurs even sooner and is more severe. Prolonged ischemia/hypoxia can result in damage or death of neurons.

4.2. Neurodegenerative disease

Dysfunction of the mitochondria has been implicated in many neurodegenerative diseases, including Alzheimer's disease, Parkinson's disease, Huntington's disease, and amyotrophic lateral sclerosis (ALS). Accumulations of amyloid proteins have been found in the brain neurons of patients with Alzheimer's disease. However, symptoms are apparent before these plaques are observed. On the other hand, nonaggregated amyloid accumulation in mitochondria has been correlated with disease symptoms. Damage to the cell and mitochondria eventually lead to apoptosis.

In Parkinson's disease, Huntington's disease, and ALS excessive levels of ROS have been observed. In fact in some patients with these diseases, mutations in the ROS converting enzymes have been found. Excessive ROS levels can damage cell lipids, proteins, and DNA altering cell function and structure and eventually leading to cell death via apoptosis.

5. MODELS OF MITOCHONDRIAL ENERGY METABOLISM

Mitochondrial models have been an important tool in the understanding of mitochondrial function. The first computational model for mitochondrial energy metabolism was developed by Achs and Garfinkel[11,12] to study ischemic metabolism. This model included glycolysis, the TCA cycle and associated pathways. It demonstrated the changes in metabolite concentrations and pH during ischemia in dog heart as well as suggested the mechanism of metabolic oscillations. Metabolic oscillations are oscillations in energy metabolism observed during ischemia. It should be noted that other oscillations of energy metabolism based upon enzyme dynamics, for example, phosphofructokinase in glycolysis, have been described.[13] Metabolic oscillations caused by ischemia in heart and respiratory neurons do not rely on glycolytic oscillations. There have been other models based upon alternate mechanisms to explain this. These oscillations in heart have been shown to be blocked by antagonists of the mitochondrial inner membrane ion channel (IMAC). These two studies rely on two different mechanisms: The Jafri and Kotulska model relies on the regulation of IMAC by phosphate and magnesium while the model by Cortassa and coworkers predicts that redox sensitivity of the IMAC channel can allow for the oscillations.[14,15] Note that both of these models include descriptions of oxidative phosphorylation.

The first model for oxidative phosphorylation was developed by Korzeniewski[16] followed by improved versions. These models used empirical relations to describe the processed involved in oxidative phosphorylation and made predictions about the regulation of metabolism. One prediction was the idea of a "push–pull" mechanism in which energy metabolism was regulated both by the utilization of high-energy phosphates such as ATP and by factors that controlled the rate of energy production.

A model that tried to capture the mechanisms of energy metabolism and mitochondrial calcium dynamics in the pancreatic β-cell was developed by Magnus and Keizer.[17,18] This model predicted that calcium entry under certain conditions could be large enough to transiently depolarize the mitochondria and interfere with energy metabolism. A series of other models was developed based upon the Magnus–Keizer model. For example, Bertram[19] created a simplified version of this model that was more amenable to mathematical analysis. The Bertram model was later used by Deikman and coworkers[20] in a model that integrates astrocyte membrane potential, ionic homeostasis, mitochondrial ATP production and calcium handling by mitochondria, and the endoplasmic reticulum. This model predicted how changes in IP_3-mediated calcium release affects mitochondrial energy production under different conditions and how certain signaling molecules (P2Y1R) might protect astrocytes after ischemia.

Another series of models have integrated a model for the TCA cycle with the Magnus–Keizer model to describe energy metabolism in the cardiac myocyte.[21–23] The models of Cortassa and coworkers have also integrated cardiac myocyte action potential and calcium dynamics with the mitochondrial model. The Nguyen–Dudycha–Jafri model focuses more on the regulation of mitochondrial energy metabolism and mitochondrial ionic homeostasis. In the sections that follow, a detailed description of the Nguyen–Dudycha–Jafri model is given as an example of how mitochondria in neurons might be modeled.

Computational modeling of mitochondrial function involves similar principles used for modeling excitable cell membranes, ionic homeostasis, and biochemical pathways. The inner membrane supports a membrane potential ($\Delta\psi$) determined by the ionic currents across it. These currents include the respiration driven proton pumps of the electron transport chain (J_{resp}), the F_1F_0-ATPase (J_{F1F0}), the mitochondrial calcium uniporter (J_{MCU}), the proton leak (J_{Hleak}), the mitochondrial sodium–calcium exchanger (J_{NCLX}), and the ATP/ADP antiporter (J_{ANT}). Hence the membrane potential is described by the ordinary differential equation:

$$-C_m \frac{d\psi}{dt} = J_{resp} + J_{F1F0} + J_{Hleak} + J_{MCU} + J_{NCLX} + J_{ANT}$$

where C_m is the mitochondrial membrane capacitance.

Each of these fluxes should have a representation based on experimental data. For example, the respiration driven proton pumps have been described in different ways in previous modeling efforts. In some studies, the whole electron transport chain is lumped into one flux.[16,17] In the former, an empirical relation was derived for the flux through the electron transport chain. In the latter, a mechanistic state diagram of the electron transport chain is proposed and through mathematical analysis a flux equation is proposed. Others have developed descriptions of the individual complexes that form the electron transport chain.[24–26] Korzeniewski uses empirical descriptions of each complex. On the other hand, Beard uses thermodynamics

constraints to formulate descriptions of each complex based on first principles. Others have proposed detailed mechanistic models, such as Complex
III.[27] In this model, the individual entities that comprise Complex III are
represented with specific electron transfer between these entities described.
All these approaches have their applications depending upon the questions
being asked. It is important that any of these approaches be constrained by
available experimental data.

The chemical species must also be described by dynamic equations
as well. For example, since Complex I is the point of entry for NADH
into the respiratory chain, it must describe the consumption of NADH
($J_{ComplexI}$) produced by the TCA cycle via isocitrate dehydrogenase,
α-ketoglutarate dehydrogenase complex, and malate dehydrogenase (J_{IDH},
$J_{\alpha KDHC}$, and J_{MDH}, respectively).

$$\frac{d[\text{NADH}]}{dt} = J_{IDH} + J_{\alpha KDHC} + J_{MDH} - J_{Complex\ I}$$

The fluxes from the TCA cycle need to be described by principles from
chemical kinetics as they are simply a set of catalyzed chemical reactions.
There have been several efforts to model the cycle and its regulation by
metabolites and ions such as calcium, protons, ADP, NADH, NAD^+,
and ATP. The first such work by Dudycha and Jafri[22] included detailed
descriptions of the regulation of citrate synthase, isocitrate dehydrogenase,
α-ketoglutarate dehydrogenase complex, and malate dehydrogenase. These
enzymes were based upon Michaelis–Menten kinetics which assumes that
the reaction is at equilibrium. For example, the reaction velocity (v) of an
enzyme (E) and substrate (S) reaction that irreversibly yields a product

$$E + S \underset{k_b}{\overset{k_f}{\rightleftharpoons}} ES \overset{k_p}{\rightarrow} P$$

can be described by $v = \frac{V_{\max}[S]}{K_m + [S]}$, where $V_{\max} = k_p[\text{E}]_{total}$ and $K_m = \frac{k_b + k_p}{k_f}$.
$[\text{E}]_{total}$ is the total concentration of enzyme. Michaelis–Menten formalism
can also be used to derive expression for allosteric and competitive inhibitors
and activators for the reactions. The bisubstrate reactions catalyzed by
enzyme E with a random order of addition that can be used to describe
the TCA cycle dehydrogenases is shown in Fig. 5.4.[28] This yields the following net reaction velocity equation

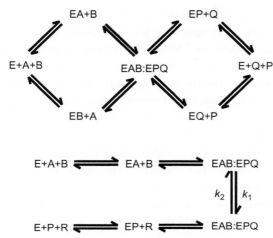

Figure 5.4 *Bisubstrate reaction schemes.* The top shows the binding of enzyme (E) to substrates A and B with random order of addition to form products P and Q. The bottom shows a scheme where A binds the enzyme first.

$$v = \frac{v_f \frac{[A]\,[B]}{K_A\,K_B} - v_r \frac{[P]\,[Q]}{K_P\,K_Q}}{1 + \frac{[A]}{K_A} + \frac{[B]}{K_B} + \frac{[A]\,[B]}{K_A\,K_B} + \frac{[P]}{K_P} + \frac{[Q]}{K_Q} + \frac{[P]\,[Q]}{K_P\,K_Q}}$$

where v_f and v_r are the maximal forward and reverse reaction rates, respectively, and the K_j where $j = A$, B, P, and Q are the Michaelis constants for the different reactants (A and B) and products P and Q.

The other reactions were described by the law of mass action which represents rate of change of the [A] (the concentration of A) governed by the chemical reaction equation

$$A + B \underset{k_b}{\overset{k_f}{\rightleftharpoons}} C + D$$

as

$$\frac{d[A]}{dt} = k_b[C][D] - k_f[A][B]$$

Similar equations can be derived for the concentration of the other chemical species involved.

This model set the stage for modeling by others who either adapted this model[21,23] or made their own new implementations.[29] The common feature is that these models used Michaelis–Menten formalism for bisubstrate reactions to describe the regulation of the key regulator steps in the TCA cycle. The parameters for the binding constants for these reactions were fit to experimental data on initial reaction rates for these reactions as that is what is typically measured experimentally. The caveat with this is that the reverse reaction is not well constrained. Experimental measurement of reaction kinetics measures initial reaction rates, that is, reaction rates in the absence of product. For example, the work by Dudycha and Jafri constrained the model with this data and hence did not include reverse reactions. This however, under physiological conditions in not an issue as the change in free energy for these reactions made the reverse reaction flux negligible.

The model of mitochondrial energy metabolism developed by Kumar and Jafri[26] will be used here to give an example of how some of the genetic defects in energy metabolism can be modeled and what insights might be gained about how these mutations affect energy metabolism. The Kumar and Jafri model improves upon the Nguyen–Dudycha–Jafri model[23] which combines the Dudycha and Jafri[22] model for the TCA cycle described above with the Magnus and Keizer model for mitochondrial energy metabolism and ionic homeostasis.[17] The improvements in the most recent version of the model include improved calcium, sodium, potassium, and proton homeostasis as well as a mechanistic description of the complexes in the respiratory chain and the F_1F_0-ATPase.[26] This detailed description of the respiratory chain and oxidative phosphorylation includes the kinetic reactions for Complex I, II, and III and the F_1F_0-ATPase instead of steady-state solutions used earlier. It therefore allows accurate modeling of laboratory experimental manipulations that target these mechanisms and provides a means of validating the model.

Genetic mutations lead to changes in protein function. These can simply be a reduction in the proteins function, complete loss of the protein function, or a change in the protein characteristic, for example, the shifting of a dose–response curve. In the model, a mutant of a protein/protein complex can be developed to fit the data from the mutant gene if such data exists. Unfortunately, in many of the cases of neurodegenerative diseases, detailed characterization is not yet available. Table 5.3 shows examples demonstrating the effects of mutations that simply cause a reduction of function for three cases: (1) Complex I, (2) Complex III, and (3) succinate

Table 5.3 Changes in mitochondrial species due to mutation

Mutation	$\Delta\psi$	ROS production	[NADH]	$[Ca^{2+}]$	ATP production
Complex I	−	−	+	−	−
Complex III	−	+	0	−	0
Succinate dehydrogenase	−	−	0	+	0

Key: +, indicates increase; −, indicates decrease; 0, indicates little change.

dehydrogenase. In all these cases, NADH-linked substrates are used, that is, substrate enters the cycle as acetyl-CoA.

When Complex I is blocked, NADH cannot enter the respiratory chain. As a result, the NADH accumulates, inhibiting the TCA cycle. Without the entry of reduction equivalents, electron transport slows and mitochondrial membrane potential ($\Delta\psi$) drops. With entry of NADH and reduced membrane potential, ROS production also drops. The reduced membrane potential is also accompanied by a reduction in ATP production and a reduction in calcium uptake by the mitochondria.

The defect in Complex III is simulated by blocking the oxidation of ubiquinol to ubiquinone. The extent of disruption of the Q-cycle depends upon the exact part of the enzyme blocked. The result is accumulation of ubiquinol and depletion of ubiquinone. There is some decrease in mitochondrial membrane potential and this has decreases in calcium uptake. In spite of the decrease in membrane potential, ROS production increases as electrons in Complex I cannot be transferred to ubiquinone.

When succinate dehydrogenase is blocked there are two results. The first is disruption of the bottom half of the TCA cycle. This results in reduced NADH and $FADH_2$ production. It also prevents entry of $FADH_2$ into the respiratory chain. The membrane potential drops resulting in a decrease in ROS production and calcium uptake.

The simulations above suggest how mitochondrial concentrations and fluxes change. These changes are likely to have impact on neuronal properties as well. For example, reduction in ATP production can affect ionic homeostasis and neurotransmitter release/reuptake. Further exploration of the impact of these mutations is possible when the mitochondrial model is coupled to a neuronal model. For example, if the aforementioned changes are coupled with a model for neuronal excitability that includes the ATP usage by pumps and other processes, the effect on neural excitability can be assessed. Coupling this to a model of apoptosis can show how this might make the neuron susceptible to programmed cell death.

6. MODELS OF MITOCHONDRIAL SIGNALING

Mitochondria are both the target of cellular signaling pathways, as in the case of apoptosis, and are the source of signaling molecules such as ROS. One computational study that explores the bistability caused by the interplay of Bax and Bcl-2, by Bagci and coworkers[30] suggested that the cooperative formation of the apoptosome is the key mechanism for achieving this bistability. The combination of this type of mechanistic model with other pathways will be an essential step for understanding apoptosis. One recent effort to do so by Hong and colleagues[31] developed a large-scale literature based model to describe the signaling pathways involved in cisplatin-induced apoptosis. The model suggests that the pathways involve activation of the ER-stress pathway, increased cytosolic calcium and activation of the TNF (tumor necrosis factor) pathway that reduces expression of Bcl-2. This tips the balance that allows Bax activation of the apoptosis. However, this result is controversial, because in experimental studies TNF increases Bcl-2 and Bax expression. This demonstrates the difficult and complexity of understanding the system and emphasizes the need for more computational modeling.

ROS are produced by the mitochondria as a byproduct of electron transport. ROS have many cellular targets such as the ryanodine receptor, DNA, RNA replication, and lipids. ROS also triggers apoptosis. One possible pathway is the increase in TNF expression resulting from high levels of ROS. TNF is known to be involved in the induction of apoptosis. ROS might also act directly on mitochondria to induce cytochrome c release. This area is mostly unexplored by computational modeling studies and presents an open area for such modeling work.

There are also other possible mechanisms for apoptosis. While calcium is an essential signaling molecule, prolonged levels of high calcium is toxic to the cell. Not only does calcium overload activate calcium dependent processes, it results in increased mitochondrial calcium. With mitochondrial calcium overload, permeability transition occurs resulting in membrane depolarization and release of cytochrome c into the cytosol. Cytochrome c release then leads to activation of apoptosis.

Computational modeling can help understand calcium overload and how it impacts mitochondrial calcium levels. Under normal conditions calcium transients of approximately 1.0 μM in amplitude occur. Modeling constrained by experiments indicates that the resulting mitochondrial calcium changes are relatively small.[23] However, for mitochondria near release sites that experience higher local calcium levels (~20 μM in amplitude) uptake can be much higher. These responses are transient, making them insufficient to cause mitochondrial dysfunction. Alzheimer's disease has been characterized by neurons with higher than normal calcium levels, presumable through calcium entry from the extracellular space. However, the mode of calcium entry is still an open question. Modeling can be used to explore the mode of calcium entry and how it must be localized with respect to the mitochondria to induce mitochondrial calcium overload.

These examples emphasize that mitochondria play an essential role in the induction of apoptosis. The resulting death of neurons, via apoptosis, results in permanent impaired function of neural networks, such as those involved in motor control, memory, and cognitive function. However, even prior to apoptosis, the changes to neural function by activation of signaling pathways that affect mitochondria impair energy metabolism resulting in neural dysfunction. It stresses the importance of developing an understanding of the pathways regulating and leading up to apoptosis which will require more advances in computational modeling in this area.

7. CONCLUDING REMARKS

Mitochondria play a critical role in the development and progression of neurodegenerative diseases through complex mechanisms. Computational modeling of mitochondria energy metabolism and signaling can be a powerful tool to understand these complex systems. Efforts have started to apply these models to neuroscience. However, there are many opportunities for the development and application of such models and their integration with experimental work. For example, connection of mitochondrial metabolism to signaling pathways will be an essential function of modeling. Modeling will also be crucial to explore functional implications of the high degree of spatial compartmentalization caused by complex cristae structure and their role in neurodegerative diseases.

ACKNOWLEDGMENTS

We would like to thank Aman Ullah and Hoang-Trong Minh Tuan for their helpful comments. This work was supported by the National Institutes of Health 5R01AR057348 and 5R01HL105239 and the National Science Foundation (DMS 0443843).

REFERENCES

1. Gray MW. Mitochondrial evolution. *Cold Spring Harb Perspect Biol.* 2012;4(9):a011403.
2. Mannella CA, Lederer WJ, Jafri MS. The connection between inner membrane topology and mitochondrial function. *J Mol Cell Cardiol.* 2013;62:51–57.
3. Schon EA, Manfredi G. Neuronal degeneration and mitochondrial dysfunction. *J Clin Invest.* 2003;111(3):303–312.
4. Tanji K, Vu TH, Schon EA, DiMauro S, Bonilla E. Kearns-Sayre syndrome: unusual pattern of expression of subunits of the respiratory chain in the cerebellar system. *Ann Neurol.* 1999;45(3):377–383.
5. Gonzalez-Cabo P, Palau F. Mitochondrial pathophysiology in Friedreich's ataxia. *J Neurochem.* 2013;126(Suppl. 1):53–64.
6. Goizet C, Depienne C, Benard G, et al. REEP1 mutations in SPG31: frequency, mutational spectrum, and potential association with mitochondrial morpho-functional dysfunction. *Hum Mutat.* 2011;32(10):1118–1127.
7. Fink JK. Hereditary spastic paraplegia: clinico-pathologic features and emerging molecular mechanisms. *Acta Neuropathol.* 2013;126:307–328.
8. Magnoni R, Palmfeldt J, Christensen JH, et al. Late onset motoneuron disorder caused by mitochondrial Hsp60 chaperone deficiency in mice. *Neurobiol Dis.* 2013;54:12–23.
9. OMIM. *Online MendelianInheritance in Man.* Baltimore, MD: The Johns Hopkins University; 2013.
10. Yudkoff M, Nelson D, Daikhin Y, Erecinska M. Tricarboxylic acid cycle in rat brain synaptosomes. Fluxes and interactions with aspartate aminotransferase and malate/aspartate shuttle. *J Biol Chem.* 1994;269(44):27414–27420.
11. Achs MJ, Garfinkel D. Metabolism of totally ischemic excised dog heart. I. Construction of a computer model. *Am J Physiol.* 1979;237(5):318–326.
12. Garfinkel D, Achs MJ. Metabolism of the acutely ischemic dog heart. II. Interpretation of a model. *Am J Physiol.* 1979;236(1):R31–R39.
13. Goldbeter A, Caplan SR. Oscillatory enzymes. *Annu Rev Biophys Bioeng.* 1976;5:449–476.
14. Jafri MS, Kotulska M. Modeling the mechanism of metabolic oscillations in ischemic cardiac myocytes. *J Theor Biol.* 2006;242(4):801–817.
15. Zhou L, Cortassa S, Wei AC, Aon MA, Winslow RL, O'Rourke B. Modeling cardiac action potential shortening driven by oxidative stress-induced mitochondrial oscillations in guinea pig cardiomyocytes. *Biophys J.* 2009;97(7):1843–1852.
16. Korzeniewski B, Froncisz W. An extended dynamic model of oxidative phosphorylation. *Biochim Biophys Acta.* 1991;1060(2):210–223.
17. Magnus G, Keizer J. Model of beta-cell mitochondrial calcium handling and electrical activity. I. Cytoplasmic variables. *Am J Physiol.* 1998;274(4 Pt 1):C1158–C1173.
18. Magnus G, Keizer J. Model of beta-cell mitochondrial calcium handling and electrical activity. II. Mitochondrial variables. *Am J Physiol.* 1998;274(4 Pt 1):C1174–C1184.
19. Bertram R, Gram Pedersen M, Luciani DS, Sherman A. A simplified model for mitochondrial ATP production. *J Theor Biol.* 2006;243(4):575–586.
20. Diekman CO, Fall CP, Lechleiter JD, Terman D. Modeling the neuroprotective role of enhanced astrocyte mitochondrial metabolism during stroke. *Biophys J.* 2013;104(8):1752–1763.
21. Cortassa S, Aon MA, O'Rourke B, et al. A computational model integrating electrophysiology, contraction, and mitochondrial bioenergetics in the ventricular myocyte. *Biophys J.* 2006;91(4):1564–1589.
22. Dudycha SJ. *A Detailed Model of the Tricarboxylic Acid Cycle in Heart Cells.* Baltimore, MD: Biomedical Engineering, The Johns Hopkins University; 2000.

23. Nguyen MH, Dudycha SJ, Jafri MS. Effect of Ca2 + on cardiac mitochondrial energy production is modulated by Na+ and H+ dynamics. *Am J Physiol Cell Physiol.* 2007;292(6):C2004–C2020.
24. Beard DA. A biophysical model of the mitochondrial respiratory system and oxidative phosphorylation. *PLoS Comput Biol.* 2005;1(4):e36.
25. Korzeniewski B. Regulation of oxidative phosphorylation through parallel activation. *Biophys Chem.* 2007;129(2–3):93–110.
26. Kumar R, Jafri MS. Mechanism of reactive oxygen species generation in cardiac mitochondria a computational approach. *Biophys J.* 2011;100(3 Suppl. 1):460.
27. Demin OV, Kholodenko BN, Skulachev VP. A model of O2.-generation in the complex III of the electron transport chain. *Mol Cell Biochem.* 1998;184(1–2):21–33.
28. Alberty RA. Enzyme kinetics. *Adv Enzymol Relat Areas Mol Biol.* 1956;17:1–64.
29. Wu F, Yang F, Vinnakota KC, Beard DA. Computer modeling of mitochondrial tricarboxylic acid cycle, oxidative phosphorylation, metabolite transport, and electrophysiology. *J Biol Chem.* 2007;282(34):24525–24537.
30. Bagci EZ, Vodovotz Y, Billiar TR, Ermentrout GB, Bahar I. Bistability in apoptosis: roles of bax, bcl-2, and mitochondrial permeability transition pores. *Biophys J.* 2006;90(5):1546–1559.
31. Hong JY, Kim GH, Kim JW, et al. Computational modeling of apoptotic signaling pathways induced by cisplatin. *BMC Syst Biol.* 2012;6:122.

CHAPTER SIX

Mathematical Modeling of Neuronal Polarization During Development

Honda Naoki*, Shin Ishii†

*Imaging Platform for Spatio-Temporal Information, Graduate School of Medicine, Kyoto University, Kyoto, Kyoto, Japan
†Graduate School of Informatics, Kyoto University, Uji, Kyoto, Japan

Contents

Abstract

During development of the brain, morphogenesis of neurons is dynamically organized from a simple rounded shape to a highly polarized morphology consisting of soma, one axon, and dendrites, which is a basis for establishing the unidirectional transfer of electric signals between neurons. The mechanism of such polarization is thought to be "local activation–global inhibition"; however, globally diffusing inhibitor molecules have not been identified. In this chapter, we present a theoretical modeling approach of such neuronal development. We first summarize biological research on neuronal polarization and then develop a biophysical model. Through mathematical analysis, principles of local activation–global inhibition are illustrated based on active transport, protein degradation, and neurite growth, but not on globally diffusing inhibitor.

Progress in Molecular Biology and Translational Science, Volume 123
ISSN 1877-1173
http://dx.doi.org/10.1016/B978-0-12-397897-4.00003-6

1. BIOLOGICAL BACKGROUND

Neurons extend several neurites that differentiate to axon and dendrites with distinct properties. Functionally, the axon carries spikes generated near the soma to its terminal, and then neurotransmitters are released onto the post-synaptic neuron, whereas the dendrites receive and integrate incoming pre-synaptic inputs and transfer the electrical signal to the soma. Structurally, the axon is a long, uniformly thin shaft, while the dendrites are relatively short and form a highly branched tree structure, whose shafts become thinner with distance from the soma. How is the polarity established? Here, we review prior research on the development of neuronal polarization.

Experimentalists frequently use cultured rat embryonic hippocampal neurons to observe the establishment of neuronal polarity. The morphological development of neurons is divided into five stages.[1,2] In stage 1, a spherical shaped cell first produces many thin extensions called filopodia. During stage 2, several filopodia are stabilized as short neurites of similar length (Fig. 6.1A). After several hours, the cell becomes polarized: one neurite abruptly begins to elongate and obtains the properties of an axon in stage 3 (Fig. 6.1B). Several days after polarization, other neurites acquire dendritic characteristics and begin to form premature synapses during stages 4 and 5.

Neurons spontaneously polarize in spite of uniform symmetrical distributions of extracellular chemical and adhesion factors,[1,3] indicating that

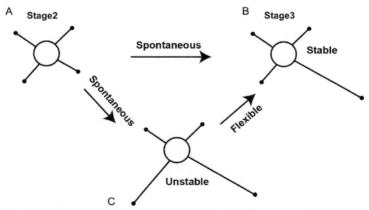

Figure 6.1 Flexible search for single-axon morphology.[11] (A) In stage 2, a neuron has several short neurites of similar length. (B) A developing neuron enters stage 3 if one neurite spontaneously elongates and develops to an axon. This established polarity is stably maintained. (C) Sometimes multiple neurites accidentally elongate, but this unfavorable pattern is flexibly reversed, leading to the single-axon pattern (B).

developing neurons have the ability to autonomously disrupt their own morphological symmetry. To achieve polarization, the elongation of a selected neurite must be spontaneously induced and accompanied by the suppression of the other neurites so that they do not extend further (Fig. 6.1B). How is such a *stable*, winner-takes-all mechanism achieved under symmetrical conditions? Sometimes multiple neurites accidentally elongate during the polarization process (Fig. 6.1C), but this unfavorable morphological pattern is eventually reversed so that a single axon remains, as though the cell is seeking a more favorable morphological pattern. How is such a *flexible* mechanism for finding a single-axon morphology achieved?

Neuronal polarization involves many molecular processes including intracellular signaling and active protein transport mediated by motor proteins, and it is accompanied by cellular morphological changes regulated by cytoskeletal rearrangements.[4-6] During the transition from stage 2 to 3, various molecules related to neurite extension are transported by motor proteins from the soma to the neurite tips, and then these molecules accumulate at neurite tips. The neurite extension could be regulated by switching molecules as observations reveal that the axonal neurite suddenly elongates in a switch-like manner.[1] In fact, fluorescence resonance energy transfer imaging[7] has also shown that a small GTPase, H-Ras, is significantly activated only in the axonal growth cone and not in the shorter neurites, which is evidence for the involvement of switching molecules. In addition, protein degradation is known to be involved in polarization as inhibition of the ubiquitin–proteasome system causes multiple axons.[8]

Theoretical studies have extensively examined polarized pattern formation with particular attention to reaction-diffusion systems with a local activation–global inhibition mechanism.[9,10] However, the molecules that mediate global inhibition have not been experimentally identified.[5] Thus, questions remain as to whether a local activation–global inhibition mechanism is plausible as a complete system without inhibitor molecules. In addition, polarization involves the active transport of various molecules and dynamic changes in morphology, further arguing against a local activation–global inhibition mechanism.

2. BIOPHYSICAL MODEL

We present a biophysical model[11] based on minimum, but essential assumptions: some factor, called X, which is actively transported from the soma to the growth cone, activates another factor, called Y, which directly

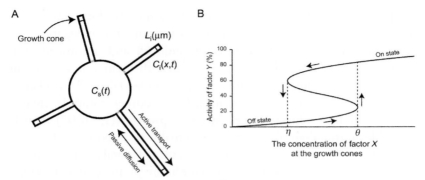

Figure 6.2 Model of neuronal polarization.[11] (A) The model neuron consists of one soma, a well-mixed compartment, and several neurites, each of which is represented as a continuous cable compartment. Each neurite has a growth cone at the tip. Factor X is produced by gene expression in the soma and is actively transported to each growth cone. Factor X is degraded throughout the cell and diffuses along the neurite shafts. $C_s(t)$ denotes the concentration of factor X in the soma at time t, and $C_i(x,t)$ denotes the concentration of factor X at x μm from the neck of neurite i at time t. L_i is the length of neurite i. (B) The axon is specified according to the activity of factor Y at the growth cone. Factor Y activity not only depends on the concentration of factor X but also has hysteresial behavior. Between the threshold concentrations of η and θ, the system exhibits bistability with one unstable and two stable states (on and off states). When factor Y is in the on state, the neurite elongates, and when factor Y is in the off state, the neurite shrinks.

controls axon growth. In this section, we explain the characteristics of these factors X and Y. In the discussion, we present a few molecules that may play the role of factors X and Y.

The model neuron consists of $N+1$ compartments: one soma and N neurites ($3 \leq N \leq 6$) (Fig. 6.2A). The somatic cytoplasm was modeled as a well-mixed compartment with a volume of V, and each neurite shaft was considered to be a continuous compartment with a cross-sectional area A and length L_i, where the subscript i indexes a neurite.

2.1. Model for active transport, diffusion, and degradation of factor X

Factor X is involved as a key regulator of axon specification and is produced by gene expression in the soma at rate G. Factor X is degraded (or inactivated) at rate k; it diffuses at rate D throughout the entire neuron and is transported stochastically by motor proteins and/or waves that depend on actin filaments[12,13] from the soma to each growth cone. This transport is assumed to be Poissonian, and the rate is proportional to the somatic concentration of factor X. No time delay in transport is considered in the model because the time required for the motor proteins to travel to the growth cone is negligible in comparison to the time scale of polarization ($\sim$36 h); the kinesin motor protein moves with a velocity of 0.2–1.5 μm/s along neurites, which are approximately 10 μm long.[1,14] The dynamics of the concentration of factor X in the soma, $C_s(t)$, and along each neurite, $C_i(x,t)$, are described as:

$$\frac{dC_s}{dt} = G - kC_s + \frac{1}{V}\sum_{i}^{N}\left[DA\frac{\partial C_i(0,t)}{\partial x} - \alpha\frac{d\mathbf{n}_i(t)}{dt}\right], \qquad (6.1)$$

$$\frac{\partial C_i}{\partial t} = D\frac{\partial^2 C_i}{\partial x^2} - kC_i + \frac{\alpha}{A\Delta L}\frac{d\mathbf{n}_i(t)}{dt}H(x - L_i + \Delta L)H(L_i - x), \qquad (6.2)$$

with boundary conditions $C_i(0,t) = C_s(t)$, where x represents the distance from the neck of each neurite and α is a positive constant representing the amount of factor X per single transport. The variable $\mathbf{n}_i(t)$ designates the number of active transport events to each neurite tip at time t and is an independent random variable sampled from a Poisson probability distribution with the frequency $\lambda C_s(t)$. With each transport event, factor X is released at the tip of a neurite of length ΔL. $H(x)$ is a step function in which $H(x) = 1(x \geq 0)$ or $0(x < 0)$. Initially, the concentration of factor X is 0 in the entire cell and the length of every neurite is L_o. The diffusion and degradation of proteins are usually fast and slow, respectively; the proteins in the neurons disperse with diffusion coefficients of 0.23–44 μm^2/s,[15,16] and they have a half-life of approximately 1 day.[17,18]

2.2. Model for regulation of neurite growth by factor Y

Axon specification in stage 2, during which one of the neurites extends suddenly, is considered to be a threshold phenomenon, and its stable nature suggests the existence of bistability. In the model, we assume that a certain molecule, factor Y, plays an essential role in axon specification at the growth cones. Factor Y is activated by factor X, but its activation is characterized by hysteresis (Fig. 6.2B), implying that factor Y works as a bistable switch; factor Y takes the binary activity, for example, an "on" and an "off" state, and factor Y in the off (on) state suddenly transitions to an on (off) state when the concentration of factor X exceeds the threshold θ (decreases below the threshold η). Such hysteresis is observed in many dynamic systems with nonlinear positive feedback loops. In fact, positive feedback loops have been found in growth cone signaling,[5] and one of small GTPases, H-Ras, is

suddenly activated in only one growth cone, in accordance with axon specification.[7] Our model assumes that if factor Y is in the on (or off) state, the
neurite elongates (or shrinks to L_o) as follows:

$$\frac{dL_i}{dt} = v_{growth}\delta(Y_i - 1) - v_{shrink}\delta(Y_i),\qquad(6.3)$$

where v_{growth} and v_{shrink} are the rates of growth and shrinkage of the growth
cone. Y_i represents the binary activity of factor Y in the growth cone of neurite i (1 for the on state and 0 for the off state of factor Y). $\delta(x)$ is a delta
function in which $\delta(x) = 1 (x=0)$ or $0 (x \neq 0)$. Once the on state is realized,
growth continues robustly due to hysteresis (Fig. 6.2B).

3. MATHEMATICAL ANALYSIS

3.1. Factor X along neurite

In the model, the factor X is actively transported from the soma and released
at the axonal tips, and is then distributed due to its diffusion along the neurites. Here, we mathematically derive the spatial distribution of factor X
along neurites. Equation (6.4) is approximated by assuming point transport
to neurite tips at $x = L_i$, and taking a temporal average with respect to
Poissonian transport variables $\mathbf{n}_i(t)$,

$$\frac{\partial\langle C_i\rangle}{\partial t} = D\frac{\partial^2\langle C_i\rangle}{\partial x^2} - k\langle C_i\rangle + \frac{\alpha\lambda C_s}{A}\delta(x - L_i),\qquad(6.4)$$

The concentration distribution along the neurite can be assumed to be in
steady state, because the dynamics are dominated by rapid diffusion, compared with the slow rate of degradation within the soma. The general
steady-state solution of Eq. (6.2) can be expressed by

$$\langle C_i(x)\rangle = A\cosh\left(\sqrt{k/D}x\right) + B\sinh\left(\sqrt{k/D}x\right).\qquad(6.5)$$

A solution with boundary conditions: $\langle C_i(0)\rangle = 0$ and $\langle C_i(L_i)\rangle = C_o$
becomes

$$C_i^T(x) = \frac{C_o}{\sinh\left(\sqrt{k/DL_i}\right)}\sinh\left(\sqrt{k/D}x\right),\qquad(6.6)$$

which reflects the distribution due to transport to the neurite tips. C_o can be
determined by satisfying the detailed balance, $\alpha\lambda C_s = DAdC_i^T(L_i)/dx$.
A solution with boundary conditions: $\langle C_i(0)\rangle = C_s$ and $d\langle C_i(L_i)\rangle/dx = 0$ is

$$C_i^L(x) = \frac{C_s}{\cosh\left(\sqrt{k/DL_i}\right)} \cosh\left(\sqrt{k/D}(L_i - x)\right), \qquad (6.7)$$

which reflects the distribution due to leakage from the soma into the neurites. The sum of these solutions is the distribution realized along the neurites:

$$C_i(x) = C_i^L(x) + C_i^T(x). \qquad (6.8)$$

In addition to the distribution, we can calculate the fraction of the transported factor X returned back to the soma without degradation as follows:

$$R(L_i) \equiv \frac{DA\frac{\partial C_i^T(0)}{\partial x}}{\alpha \lambda C_s} = \frac{1}{\cosh\left(\sqrt{k/DL_i}\right)}, \qquad (6.9)$$

where $C_i^T(x)$ is the distribution due to transport at the neurite tip. This fraction was calculated as the diffusion influx rate into the soma at the neck of the neurite divided by the transportation rate.

3.2. Factor X in the soma

Each neurite initially is short ($L_i = L_o$), such that almost all of the factor X that is transported to the growth cone can rapidly diffuse back to the soma. Thus, the third term of Eq. (6.1) can be dropped. The concentration of factor X in the soma increases according to

$$C_s(t) = G/k\left(1 - e^{-kt}\right), \qquad (6.10)$$

with a relatively large time constant of degradation.

To evaluate the somatic concentration dependent on neurite length, Eq. (6.1) was approximately expressed as

$$\frac{d\langle C_s \rangle}{dt} \simeq G - k\langle C_s \rangle + \frac{1}{V}\sum_i^N \left[DA\frac{\partial C_i^L(0)}{\partial x} + \alpha\lambda\langle C_s \rangle\{R(L_i) - 1\} \right], \qquad (6.11)$$

here, we took a temporal average over the processes by which the transported factor X diffuses back to the soma using Eq. (6.9). This equation is nothing but a simple linear differential equation:

$$\frac{d\langle C_s \rangle}{dt} = -\frac{\langle C_s \rangle - C_s^\infty}{\tau} \qquad (6.12)$$

with an equilibrium concentration $C_s^\infty = GQ^{-1}$ and a time constant $\tau = Q^{-1}$, where

$$Q(\{L_i\}) = k + V^{-1}\sum_i^N \left[A\sqrt{kD}\tanh\left(\sqrt{k/D}L_i\right) + \alpha\lambda\{1 - R(L_i)\}\right] \quad (6.13)$$

The first and second terms in the summation represent the rate of leakage from the soma to each neurite shaft and the rate of degradation of factor X along each neurite per single unit of transport, respectively. When the neurites are kept short before polarization ($V \gg NAL_o$), these two terms can be ignored so that $Q \cong k$. Once a neurite elongates, Q increases, causing the concentration of factor X in the soma to decrease.

4. MECHANISM OF NEURONAL POLARIZATION

4.1. Local activation

The gene expression and active transport results in abundant transport to and accumulation of factor X in the growth cones. The mean concentration in the growth cone can be expressed as $C_i(L_i)$:

$$\langle C_i(L_i)\rangle = C_s\left[\frac{\alpha\lambda}{A\sqrt{kD}}\tanh\left(\sqrt{k/D}L_i\right) + 1/\cosh\left(\sqrt{k/D}L_i\right)\right]. \quad (6.14)$$

Eventually, the concentration of factor X, $C_i(L_i)$, in one of the growth cones reaches a threshold θ (Fig. 6.2B) because the time necessary to reach the threshold θ varies among the different growth cones. The variations in time are due to the inherent stochasticity of the transport process.[6,18,19] Once the threshold θ is reached, factor Y changes from the off state to the on state and the corresponding neurite begins to elongate. According to Eq. (6.14), as the neurite elongates, factor X accumulates at the neurite tip (Fig. 6.3A) under the condition that the transport is sufficiently strong. This property suggests a local activation (local positive feedback) mechanism because elongation of a neurite leads to a greater accumulation of factor X, which stably maintains the activity of factor Y to facilitate further elongation of the neurite.

4.2. Global inhibition

In long neurites, most of the transported factor X cannot return to the soma because the retrograde diffusion back takes so long that the factor is degraded along the way. The proportion of factor X that returns to the soma is described by Eq. (6.9). As L_i increases, the proportion substantially decreases (Fig. 6.3B), even if there is little degradation. This result indicates

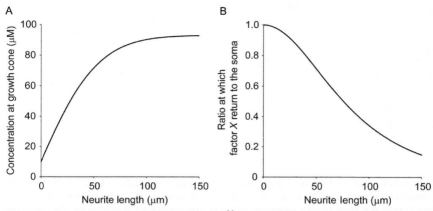

Figure 6.3 Local activation–global inhibition.[11] (A) Relationship between neurite length and factor X concentration at the growth cone at steady state. This curve is mathematically obtained by Eq. (6.14). (B) Relationship between neurite length and the proportion of factor X diffusing back to the soma without being degraded. This curve is mathematically obtained by Eq. (6.9).

a global inhibition (global negative feedback) mechanism. With such a mechanism, a decrease in the proportion caused by neurite elongation induces a decrease in the somatic concentration, resulting in a decrease in the amount of factor X transported to all the neurites. The equilibrium concentration in the soma, which depends on the length of the neurites, was approximately calculated as

$$C_s^\infty(\{L_i\}) = \frac{G}{k + V^{-1}\sum_i^N \left[A\sqrt{kD}\tanh\left(\sqrt{k/D}L_i\right) + \alpha\lambda\{1 - R(L_i)\}\right]}.$$

$$(6.15)$$

4.3. Logic of neuronal polarization

From the mathematical analyses, we obtained a scenario of spontaneous neuronal polarization based on the phase diagram in Fig. 6.4. First, the somatic concentration of factor X increases due to gene expression (black line in Fig. 6.4A). The concentration of factor X in the growth cones increases according to Eq. (6.14). The concentration in only one of the neurites happens to reach the threshold θ due to stochastic variations in transport. In that specific growth cone, factor Y switches from the off state to the on state and the neurite elongates (a in Fig. 6.4B). The concentration of factor X in the remaining neurites remains below the threshold θ (b in Fig. 6.4B). The

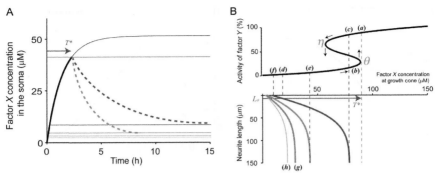

Figure 6.4 Logic of neuronal polarization.[11] (A) Expected time course of the somatic concentration of factor X during the establishment of polarization. The black line shows the time course when no neurites are elongated, mathematically obtained as solutions to Eq. (6.10). The expected time courses with one and two elongated neurites are depicted by the blue and red dashed curves, respectively. The solid blue, red, cyan, and purple lines depict the equilibrium concentrations when one, two, three, and four neurites, respectively, elongate (100 μm), as obtained from Eq. (6.15). The green line indicates the threshold at which factor Y is activated to the on state in the growth cone of a neurite of length L_0. (B) Activity of factor Y (upper panel) and concentration of factor X (lower panel) in the growth cone. The blue, red, cyan, and purple lines in the lower panel indicate the concentration of factor X given as a function of neurite length when the somatic concentration is in equilibrium, as depicted by the solid blue, red, cyan, and purple lines in (A), respectively. Points (a–f) indicate states when the concentration of factor X in growth cones is above (a) or below (b) the threshold θ, when the somatic concentration is in equilibrium as depicted by the solid blue (c and d) and red (e and f) lines in (A) and when the neurite length is infinite (c and e) and L_0 (d and f). The green arrows in (A) and (B) indicate the time that elapses until the concentration of factor X in one growth cone reaches the threshold θ. The parameters used in this analysis were $G = 10^{-2}\,\mu M/s$, $k = 7.5 \times 10^{-5}\,s^{-1}$, $D = 0.25\,\mu m^2/s$, $\alpha = 4 \times 10^{-2}\,\mu m^3/s$, $\lambda = 10\,s^{-1}\mu M^{-1}$, $L_0 = 7.5\,\mu m$, $V = 300\,\mu m^3$, $A = 10\,\mu m^2$, $N = 4$, and $\theta = 90\,\mu M$. A neurite was assumed to elongate to 100 μm if factor Y switches to the on state in the corresponding growth cone. (See the color plate.)

somatic concentration then starts to decrease to the level given by Eq. (6.15) (solid blue line in Fig. 6.4A), which decreases the concentration of factor X in the growth cone of the elongating neurite, but factor Y remains in the on state due to hysteresis (c in Fig. 6.4B). The concentration of factor X in the remaining growth cones also decreases (d in Fig. 6.4B), suggesting that the concentration of factor X in these growth cones will never reach the threshold θ; thereby stabilizing the disruption of symmetry (Fig. 6.1B).

To validate such a scenario, we performed Monte Carlo simulations of the stochastic transport process and successfully reproduced the expected

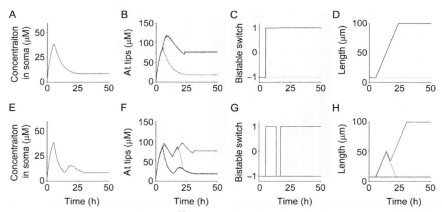

Figure 6.5 Computer simulations.[11] The biophysical model is simulated under single-axon conditions, that is, the threshold η is between the points indicated by (c) and (e) in Fig. 6.4 ($\lambda = 60\,\mu$M). The rates of neurite growth and shrinkage are set to $v_{growth} = 5\,\mu$m/h and $v_{shrink} = 5\,\mu$m/h, respectively. (A–D) Typical simulation results showing a direct transition from an initial state to a single-axon state, depicted as a transition from state (A) to state (B) in Fig. 6.1. (E–H) Typical simulation result showing an indirect transition from an initial state to a single-axon state via a two-axon state, depicted as a transition from state (A) to state (B) via state (C) in Fig. 6.1. Time courses of the concentration of factor X in the soma (A and E), at growth cones (B and F), state of factor Y (C and G) and neurite lengths (D and H) are shown. Note that four curves with different colors are plotted in (B–D, F–H) to correspond to four neurites, although distinguishing them in some cases is difficult. The other parameters used in this simulation were the same as those in Fig. 6.4. (See the color plate.)

behaviors in most trials (Fig. 6.5A–D). In some trials, two neurites simultaneously elongate, as suggested by Fig. 6.1B, because the concentrations in the two growth cones almost coincidentally reach the threshold θ (Fig. 6.5E–H). After a certain period of time, one of these two neurites shrinks so that the other neurite remains to develop into an axon, indicating that favorable morphology with a single axon is flexibly sought (Fig. 6.6A).

To understand the logic of this flexibility, we show that the multiple-axon state is unstable according to our phase diagram (Fig. 6.4). When two neurites elongate, the somatic concentration decreases to a lower level than when a single neurite elongates (red solid line in Fig. 6.4A). The factor X concentration in the growth cones of the long neurites likely falls below the threshold η (e in Fig. 6.4B). Consequently, factor Y easily returns from the on state to the off state, causing the neurites to shrink to length L_o. Thus, the automatic selection of a single future axon among multiple elongating neurites occurs flexibly by means of the regularization of the

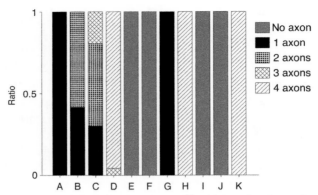

Figure 6.6 Distribution of the numbers of axons.[11] Each bar plot shows the distribution of the number of axons among 500 simulation runs for each condition. (A) Single-axon condition in Fig. 6.5. (B) Two-axons possible conditions, that is, the threshold η rests between the points indicated by (e) and (g) in Fig. 6.4 ($\eta = 35\,\mu$M). (C) Three-axons-possible condition, that is, the threshold η rests between the points indicated by (e) and (g) in Fig. 6.4 ($\eta = 25\,\mu$M). Conditions with a low (D) and high (E) degradation rates k for factor X, that is, the k value here was one-fifth of and five times larger than that in Fig. 6.5, respectively. Conditions with low (F) and high (G) rates of transport λ, that is, the λ value here was one-fifth of and five times larger than that in Fig. 6.5, respectively. Conditions with lower ($\theta = 40\,\mu$M, $\eta = 10\,\mu$M) (H) and higher ($\theta = 130\,\mu$M, $\eta = 100\,\mu$M) (I) thresholds than those in Fig. 6.5. Conditions with a low (J) and high (K) rates of production G of factor X, that is, the G value here was one-fifth of and five times larger than that in Fig. 6.5, respectively.

somatic pool of factor X. Based on the phase diagram in Fig. 6.4B, we see that robust neuronal polarization is realized when the threshold η is between the points indicated by (c) and (e). When the threshold η is lower than the point indicated by (e), multiple axons can develop (Fig. 6.6B–C).

5. DISCUSSION

5.1. Candidates for two factors

Several experimental studies have shown that inositol phospholipid signaling and small GTPases are involved in a positive feedback loop during axon specification.[20–23] Therefore, one possible candidate for factor Y is PIP3. Factor X may be the PIP3 inducer, PI3K or a PI3K–activating factor such as HRas,[7,21] Rap1B (Ras–related protein 1B),[23] Crmp2 (collapsing response mediator protein 2),[24,25] Shootin1,[13,18] Grb2,[26] or NudE–like (NUDEL).[27] One possible candidate for factor X, HRas, is significantly activated only in the axonal growth cone and not in shorter neurites.[7] That study also showed that there is a mutual activation of HRas and PI3K, which is required to configure the bifurcation diagram with hysteresis as shown in Fig. 6.2B. Axonal elongation is induced by the local application of actin–depolymerizing reagents to neurite tips, suggesting that factor Y regulates axon specification via actin depolymerization.[4]

5.2. Comparisons with experiments

Our model can explain a number of experimental observations that are reproduced by simulations:

1. Inhibition of the ubiquitin–proteasome system induces the localization of Akt, a protein downstream of PIP3 (a factor Y candidate) in all of the neurite tips and causes multiple axons.[8] If the model degradation rate k is low, the simulation reproduces this phenomenon with no selected neurite (Fig. 6.6D). When k is small, most of the transported factor X returns to the soma, preventing significant decreases in the somatic concentration. Therefore, the winning neurite cannot easily suppress the elongation of other neurites.

2. The motor protein kinesin-1, comprised of kinesin heavy chains (KIF5) and kinesin light chains (KLC), regulates active transport. Introducing a dominant-negative KIF5 or depleting KLC by RNAi prevents axon formation.[28] In the simulation, if the transport rate λ is low, no neurite is selected as an axon (Fig. 6.6F). As Eq. (6.14) indicates, a low transport rate makes the concentration of factor X lower than the threshold θ in all of the growth cones, thus preventing axon formation. Our simulation predicts that polarization with a single axon is robustly achieved while increasing the transport rate λ (Fig. 6.6G). The concentration of factor X at the tips (Eq. 6.14) is almost independent of λ and this contributes to the robustness. The bracketed term in Eq. (6.14) is almost proportional to λ, whereas the somatic concentration C_s is almost inversely proportional to λ, as can be seen in Eq. (6.15) when the transport rate is large.

3. Overexpression of the phosphate and tension homolog protein (PTEN) disrupts the establishment of polarity, whereas the inhibition of PTEN expression by siRNA increases the number of axons.[29] According to our previous model of inositol phospholipid signaling,[11] increasing PTEN elevates both the θ and η thresholds, whereas decreasing PTEN lowers both of the thresholds. Simulations using the present model show that high and low PTEN levels induce the loss of axons and the development of multiple axons, respectively (Fig. 6.6H–I).

4. The up- or downregulation of factor X candidates, including HRas, Rap1B, Crmp2, and Shootin1, induces multiple axons or suppresses axonal formation, respectively.[13,20,21,23] Accordingly in our model, if the expression rate G is high (low), multiple (no) axons are selected in simulations (Fig. 6.6J–K). Equations (6.14) and (6.15) indicate that the concentration of factor X at the growth cones varies proportionally to the expression rate G. Therefore, if factor X is upregulated, the concentrations at all of the growth cones surpass the threshold θ even after axon formation, whereas if factor X is downregulated, the concentrations at the growth cones never reach the threshold θ.

5. If the axon is transected to a shorter length than other neurites, the longest neurite becomes the axon. If the axon is cut but remains longer than the other neurites, the cut axon remains the axon.[18,30] Equation (6.14) indicates that, as a neurite grows in length, the concentration of factor X at the corresponding growth cone increases, indicating that the factor X concentration in the longest neurite is closest to the threshold θ. Therefore, the longest neurite will most likely become the axon even after it is cut. Note that factor Y is assumed to be constantly present at the tip of the transected neurites.

ACKNOWLEDGMENTS

Supported by the Next Generation Supercomputing Project and the Strategic Research Program for Brain Sciences, both from the MEXT (Ministry of Education, Culture, Sports, Science, and Technology) of Japan. We thank Drs. Kozo Kaibuchi and Shinichi Nakamuta for valuable discussions.

REFERENCES

1. Dotti CG, Sullivan CA, Banker GA. The establishment of polarity by hippocampal neurons in culture. *J Neurosci*. 1988;8:1454–1468.
2. Caceres A, Ye B, Dotti CG. Neuronal polarity: demarcation, growth and commitment. *Curr Opin Cell Biol*. 2012;24:547–553. http://dx.doi.org/10.1016/j.ceb.2012.05.011.
3. Craig AM, Banker G. Neuronal polarity. *Annu Rev Neurosci*. 1994;17:267–310.
4. Bradke F, Dotti CG. The role of local actin instability in axon formation. *Science*. 1999;283:1931–1934.
5. Arimura N, Kaibuchi K. Neuronal polarity: from extracellular signals to intracellular mechanisms. *Nat Rev Neurosci*. 2007;8:194–205.
6. Bradke F, Dotti CG. Neuronal polarity: vectorial cytoplasmic flow precedes axon formation. *Neuron*. 1997;19:1175–1186.
7. Fivaz M, Bandara S, Inoue T, Meyer T. Robust neuronal symmetry breaking by Ras-triggered local positive feedback. *Curr Biol*. 2008;18:44–50.
8. Yan D, Guo L, Wang Y. Requirement of dendritic Akt degradation by the ubiquitin-proteasome system for neuronal polarity. *J Cell Biol*. 2006;174:415–424.

9. Turing AM. The chemical basis of morphogenesis. *Bull Math Biol*. 1953;52: 153–197, discussion 119-152 (1990).

10. Meinhardt H, Gierer A. Pattern formation by local self-activation and lateral inhibition. *Bioessays*. 2000;22:753–760.

11. Naoki H, Nakamuta S, Kaibuchi K, Ishii S. Flexible search for single-axon morphology during neuronal spontaneous polarization. *PLoS One*. 2011;6:e19034. http://dx.doi. org/10.1371/journal.pone.0019034.

12. Ruthel G, Banker G. Actin-dependent anterograde movement of growth-cone-like structures along growing hippocampal axons: a novel form of axonal transport? *Cell Motil Cytoskeleton*. 1998;40:160–173.

13. Toriyama M, et al. Shootin1: a protein involved in the organization of an asymmetric signal for neuronal polarization. *J Cell Biol*. 2006;175:147–157.

14. Hirokawa N. The mechanisms of fast and slow transport in neurons: identification and characterization of the new kinesin superfamily motors. *Curr Opin Neurobiol*. 1997;7:605–614.

15. Popov S, Poo MM. Diffusional transport of macromolecules in developing nerve processes. *J Neurosci*. 1992;12:77–85.

16. Naoki H, Sakumura Y, Ishii S. Local signaling with molecular diffusion as a decoder of Ca2+ signals in synaptic plasticity. *Mol Syst Biol*. 2005;1:2005.0027.

17. Droz B, Leblond CP. Axonal migration of proteins in the central nervous system and peripheral nerves as shown by radioautography. *J Comp Neurol*. 1963;121:325–346.

18. Toriyama M, Sakumura Y, Shimada T, Ishii S, Inagaki N. A diffusion-based neurite length-sensing mechanism involved in neuronal symmetry breaking. *Mol Syst Biol*. 2010;6:394.

19. Jacobson C, Schnapp B, Banker GA. A change in the selective translocation of the Kinesin-1 motor domain marks the initial specification of the axon. *Neuron*. 2006;49:797–804.

20. Menager C, Arimura N, Fukata Y, Kaibuchi K. PIP3 is involved in neuronal polarization and axon formation. *J Neurochem*. 2004;89:109–118.

21. Yoshimura T, et al. Ras regulates neuronal polarity via the PI3-kinase/Akt/GSK-3beta/ CRMP-2 pathway. *Biochem Biophys Res Commun*. 2006;340:62–68.

22. Shi SH, Jan LY, Jan YN. Hippocampal neuronal polarity specified by spatially localized mPar3/mPar6 and PI 3-kinase activity. *Cell*. 2003;112:63–75.

23. Schwamborn JC, Puschel AW. The sequential activity of the GTPases Rap1B and Cdc42 determines neuronal polarity. *Nat Neurosci*. 2004;7:923–929.

24. Inagaki N, et al. CRMP-2 induces axons in cultured hippocampal neurons. *Nat Neurosci*. 2001;4:781–782.

25. Arimura N, Menager C, Fukata Y, Kaibuchi K. Role of CRMP-2 in neuronal polarity. *J Neurobiol*. 2004;58:34–47.

26. Shinoda T, et al. DISC1 regulates neurotrophin-induced axon elongation via interaction with Grb2. *J Neurosci*. 2007;27:4–14.

27. Taya S, et al. DISC1 regulates the transport of the NUDEL/LIS1/14-3-3epsilon complex through kinesin-1. *J Neurosci*. 2007;27:15–26.

28. Kimura T, Watanabe H, Iwamatsu A, Kaibuchi K. Tubulin and CRMP-2 complex is transported via Kinesin-1. *J Neurochem*. 2005;93:1371–1382.

29. Jiang H, Guo W, Liang X, Rao Y. Both the establishment and the maintenance of neuronal polarity require active mechanisms: critical roles of GSK-3beta and its upstream regulators. *Cell*. 2005;120:123–135.

30. Goslin K, Banker G. Experimental observations on the development of polarity by hippocampal neurons in culture. *J Cell Biol*. 1989;108:1507–1516.

CHAPTER SEVEN

Multiscale Modeling of Cell Shape from the Actin Cytoskeleton

Padmini Rangamani[*,†], **Granville Yuguang Xiong**[†], **Ravi Iyengar**[†,‡]

[*]Department of Molecular and Cell Biology, University of California, Berkeley, California, USA
[†]Department of Pharmacology and Systems Therapeutics, Mount Sinai School of Medicine, New York, USA
[‡]Systems Biology Center New York, Mount Sinai School of Medicine, New York, USA

Contents

Abstract

The actin cytoskeleton is a dynamic structure that constantly undergoes complex reorganization events during many cellular processes. Mathematical models and simulations are powerful tools that can provide insight into the physical mechanisms underlying these processes and make predictions that can be experimentally tested. Representation of the interactions of the actin filaments with the plasma membrane and the movement of the plasma membrane for computation remains a challenge. Here, we provide an overview of the different modeling approaches used to study

Progress in Molecular Biology and Translational Science, Volume 123
ISSN 1877-1173
http://dx.doi.org/10.1016/B978-0-12-397897-4.00002-4

143

cytoskeletal dynamics and highlight the differential geometry approach that we have used to implement the interactions between the plasma membrane and the cytoskeleton. Using cell spreading as an example, we demonstrate how this approach is able to successfully capture in simulations, experimentally observed behavior. We provide a perspective on how the differential geometry approach can be used for other biological processes.

1. INTRODUCTION

Cell shape and structure are controlled by the actin cytoskeleton, a self-assembled polymeric system that is dynamic. In addition to maintaining or changing cell shape, the actin cytoskeleton is also required for sensing environmental cues, aiding processes such as exo- and endocytosis, cell motility, and cell division. The actin cytoskeleton is a structural polymeric system that is dynamic: monomers of actin assemble into filaments and disassemble on a continuing basis. The energy for this process is provided by ATP hydrolysis. The actin cytoskeleton can exist in different structural configurations: lamellipodium, filopodium, and stress fibers. The actin cytoskeleton structure is used in different cell types for different purposes. In neurons, the varied uses include driving axon growth at the growth cone[1] and changing size of the spine upon synaptic transmission.[2] The subcellular mechanisms that are operational in neurons are largely the same as in other cell types. The modeling approaches described here can be used to model cellular behaviors as varied as differentiation and shape changes in neurons, movement of fibroblasts, and regulation of foot process interactions in kidney podocytes. The different structural configurations of the actin cytoskeleton result from different biochemical and mechanical cues that control a range of actin-associated regulatory proteins. Mathematical modeling of the actin cytoskeleton has provided unique insights into the regulation, growth, and dynamics of these structures. In this chapter, we consider the reorganization of the actin cytoskeleton as a multiscale process in both time and space and outline the different computational modeling approaches that can be used in understanding each step of the process. We focus specifically on a stochastic approach combining both discrete biochemical kinetics and evolving differential geometry that we used to computationally model the interaction between the actin cytoskeleton and the plasma membrane.

2. CELL SPREADING

The ability to move and migrate is very important for the various cell types to perform different physiological functions. White blood cells move freely in the blood stream, neutrophils migrate to sites of injury in response to cytokines, and fibroblasts migrate within connective tissue to wound sites. Furthermore, cell migration is fundamental to embryonic development.[3] The motile behavior of cells is made possible by the dynamic reorganization of the underlying cytoskeleton. The coupling of actin filament reorganization with the movement of the membrane has been studied extensively.[4–9] It was shown that the filament reorganization events alone can generate sufficient force to push the leading edge of the membrane forward.[5,10,11]

Cell motility is a complex process, dependent on the reorganization of the underlying actin cytoskeleton. There are three main steps in motility: protrusion, attachment, and traction.[12–14] Each of these steps recruits different sets of coordinated signaling molecules that in concert with the actin cytoskeleton reorganization events allow the cell to change shape and move forward. This multistep-integrated process of motility is observed in embryonic development, tissue repair and wound healing, immune response, and growth cones in neurons.

The steps involved in cell motility have been studied in depth experimentally.[15–17] Cell adhesion to a substrate or an extracellular matrix plays a critical role in the regulation of downstream signaling pathways via connections to the cytoskeleton.[14–23] Once the cell forms adhesive contacts with the substrate, it starts spreading on the surface. These contacts, termed *focal contacts*, are the signaling centers that connect the signaling pathways to the actin cytoskeleton. Cell spreading is dependent on actin filament dynamics and force generation by the focal contacts on the surface.[14]

Plasma membrane protrusion is driven by actin filament reorganization; the protrusion can take on different forms such as filopodia, lamellipodia, and pseudopodia.[24] While all three structures are made of the core actin cytoskeleton, different arrangements of the actin polymers result in different shapes. In a filopodium, the filaments are long and bundled to form finger-like protrusions. Lamellipodia on the other hand are characterized by sheet–like structures and contain a cross–linked meshwork of actin filaments. Pseudopodia have been observed in ameobae and neutrophils. Among these three structures, lamellipodia are best understood.

The lamellipodium contains all the machinery necessary for cell motility. In order to understand how the protrusion of the lamellipodium occurs and aids cell motility, it is important to understand the dynamics of the underlying actin cytoskeleton.

3. ACTIN CYTOSKELETON

Actin cytoskeletal reorganization occurs during cell membrane protrusion and retraction.[16,25,26] Actin is a globular protein that exists in the monomeric form (G–actin) and polymeric filamentous form (F–actin). G–actin is an ATPase, containing a deep cleft where the adenosine nucleotide binds in the presence of divalent magnesium. F–actin is polarized with preferred monomer addition occurring at the barbed end and monomer depolymerization occurring at the pointed end. ATP-bound G–actin rather than ADP-bound G–actin is favorable for filament elongation and branching reactions.[27] F–actin is spatially organized into a variety of structures, such as stress fibers, cortical actin networks, surface protrusions, and the contractile ring formed during cell division.[28,29] Precise temporal and spatial control of the actin cytoskeleton is required for these activities, but it is not clear how the changes are mediated. The biochemistry of actin reorganization in response to external cues includes filament elongation, branching, capping, and depolymerization.

4. BIOCHEMICAL SIGNALING TO THE ACTIN CYTOSKELETON

Actin cytoskeleton reorganization is mediated by four major proteins: (i) profilin, (ii) cofilin, (iii) Arp2/3, and (iv) gelsolin. The dendritic nucleation model outlines the sequence of events that take place during actin reorganization at the leading edge.[30] Profilin mediates the fast exchange of ATP for ADP in G–actin molecules, favoring filament elongation.[29,31,32] The exchange product, profilin–ATP–actin complex, leads to filament elongation at the barbed or growing end of the filament. Cofilin is a depolymerization factor which mediates actin depolymerization by sequestering ADP-bound G–actin tightly, thereby regulating the amount of free monomer available for filament elongation.[29,33–35] Arp2/3 is important for filament branching at the leading edge of the cell and acts by nucleating new filaments from existing filaments by binding to seven actin molecules on an existing filament and initiating a new branch formation at a 70°

angle.[28,36–38] The concentration of Arp2/3 is regulated by WASP in response to extracellular signals.[38] The capping protein gelsolin negatively regulates filament growth by binding to the barbed end of a growing filament, thereby preventing further monomer addition.[39,40] Actin filaments are thus regulated by multiple proteins, each of which separately and collectively modulates cytoskeletal reorganization. Recently, Urban *et al.* showed using electron tomography that the actin filaments at the leading edge may be predominantly unbranched.[41] This suggests that there may be multiple mechanisms of actin polymerization at the leading edge, resulting in different configurations of actin filaments.

4.1. Regulatory proteins for actin filament reorganization

4.1.1 Arp2/3

Arp2/3 is responsible for actin filament branching. Arp2/3 is a seven–protein complex containing Arp2, Arp3, and five unique polypeptides.[42,43] Arp2/3 is activated by WASP (Wisckott–Aldrich syndrome protein), which exists in an autoinhibited conformation. Binding of Cdc42 (GTP bound) and PI(4,5)P_2 relieves the inhibitory conformation of WASP and allows Arp2/3 complex formation and subsequently actin filament branching. WASP is phosphorylated by Src, and this phosphorylation event leads to WASP ubiquitination and degradation.[44–46] WAVE is another protein that activates Arp2/3. Unlike WASP, WAVE is not autoinhibited and is activated by Rac and PI(3,4,5)P_3 binding.[42] Thus, WASP and WAVE are important signal transducers that convert signals from protein–protein and protein–membrane interactions to actin polymerization.

4.1.2 Profilin

Profilin binds to ATP-bound G–actin and promotes filament elongation. Binding of VASP to profilin facilitates actin polymerization.[47] Ligand-bound activated integrin receptors form a complex with α–actinin, zyxin, and VASP that is responsible for profilin activation. PI(4,5)P_2 on the other hand, binds profilin and sequesters it. PI(4,5)P_2-bound profilin is incapable of promoting actin filament elongation.[48] Phosphorylation of VASP reduces its actin-binding capability. VASP and profilin interaction promotes actin elongation by interacting with barbed ends, shielding them from capping protein.[49] This effect is also modulated by PI(4,5)P_2. PI(4,5)P_2 binding to gelsolin inactivates it, and profilin and gelsolin compete to bind with PI(4,5)P_2, antagonizing each other.

4.1.3 Capping protein/gelsolin

Capping protein binds to the barbed ends of actin filaments and caps them, thus preventing filament elongation. Capping can be considered to be a terminal event in filament dynamics because the off-rate constant for capping protein to fall off a barbed end is slow, predicting the half-life of a capped filament to be at least 30 min.[50] PI(4,5)P$_2$ and PI(4)P cause rapid and efficient dissociation of capping protein from capped filaments.[48] Binding of phosphoinositides to the capping protein gelsolin prevents it from capping actin filaments.[50,51]

4.1.4 Cofilin

Cofilin depolymerizes the pointed, slow-growing ends of actin filaments. Cofilin phosphorylation by LIM kinase renders it incapable of depolymerization activity. LIM kinase is activated by p65PAK and p160ROCK, downstream of cdc42, Rac, and Rho, respectively.[52,53] PI(4,5)P$_2$ can sequester cofilin and prevent phosphorylation by LIMK.[48] Cofilin is the only known substrate of LIM kinase.[52,53]

Cell motility is a process governed by both spatial and temporal variations in the various regulators of the actin cytoskeleton.[21] Processes involved in cell migration and morphology depend on the spatial regulation of both cytoskeletal activity and the upstream signaling network. This spatial regulation controls the appropriate polar response of the cell to extracellular signals, and the control of membrane extension and retraction. Spatial regulation is controlled mainly by the polarized nature of F-actin and the binding properties of the intermediary proteins to F-actin.[54] Temporal regulation controls when the cell responds to the extracellular signals and is determined by the kinetic parameters of the biochemical reactions.

5. DIFFERENT APPROACHES FOR COMPUTATIONALLY MODELING THE ACTIN CYTOSKELETON

Many informative computational models of actin polymerization–depolymerization cycles have been developed.[5,6,8,9,55–57] Often these models are in one spatial dimension and analyze the cytoskeletal reorganization process in an abstracted cytoskeletal structure at steady state. These models have yielded substantial insight into the dynamics of the underlying actin cytoskeleton and enable the development of models that explore the relationship between actin cytoskeleton dynamics and whole cell behavior such as cell spreading. Brownian ratchet models provide mechanisms by which actin polymerization can drive the motility of a load, in this case, the plasma membrane.[4,6] The Brownian ratchet model[4] posits that if the membrane undergoes Brownian motion, then occasionally the distance between the barbed end of the actin filament and the membrane is large enough to allow the addition of a new monomer. The elastic Brownian ratchet model is a modification of the original model where the random bending of the filament provides space for the addition of new monomer.[6] The membrane is then pushed forward because of the elastic energy stored in the filament. Recently, Schaus *et al.* developed a computational model of actin filament orientation in the dendritic nucleation model.[58] This model provides insight into how the steady-state actin filament patterns emerge using stochastic simulations. Using these observations, other groups have developed models of populations of actin filaments and analyzed the work required to push a flexible membrane forward.[10,59]

5.1. Kinetic modeling of the cytoskeleton

A large number of experimental studies of the cytoskeleton focus on purification of the cytoskeletal components and studying their interactions and effects in cell-free systems. These experiments provide us with kinetic rate constants and have identified the important components required for the cytoskeletal reorganization events to occur. These experiments can be used to build kinetic models of the actin cytoskeleton. The premise here is that in a well-mixed system of actin, reactions can be treated as occurring spatially uniformly, and the concentrations of the different biochemical species are changing only in time and not in space. The mathematical modeling effort gives us the concentrations of the different actin species in time and the equilibrium concentrations. We demonstrate this with an example. Consider the actin reactions required for filament growing, branching, and capping. The rate of filament elongation depends on the concentration of monomeric actin (G-ATP–actin) and the concentration of filamentous actin (F-actin). The addition of a monomer to an "n-mer" of filamentous actin results in a filament that is of length "$n+1$-mer." The rate of filament branching depends on the concentration of Arp2/3 and the amount of filamentous actin, and the rate of capping of filaments depends on the amount of capping protein and filamentous actin available. As in Refs. 27,60, we assume that this set of reactions is sufficient to capture the fundamental dynamics of growing actin filaments. Further complexity can be added by adding filament depolymerization and severing, but the fundamental events of the reorganization can be captured by these three reactions of growing, branching, and capping.

In a well-mixed system, the rate of polymerization is a direct measure of F-actin concentration in the system and an indirect measure of spread cell area and therefore size. In this model, the actin reactions are treated as irreversible and the reaction rates are written as mass–action laws.

The elongation reaction and the corresponding reaction rate are given by

$$\text{G-ATP−actin} + \text{F-actin}_n \xrightarrow{k_{\text{elongation}}} \text{F-actin}_{n+1}$$
$$R_{\text{elongation}} = k_{\text{elongation}} [\text{F-actin}_n][\text{G-ATP−actin}] \tag{7.1}$$

The branching reaction results in the addition of a new filament along the side of an existing filament. Two G-ATP–actin monomers and one Arp2/3 molecule participate in this reaction:

$$\text{Arp2/3} + 2\text{G-ATP−actin} + \text{F-actin}_n \xrightarrow{k_{\text{branching}}} \text{F-actin}_n + \text{F-actin}_2$$
$$R_{\text{branching}} = k_{\text{branching}} [\text{F-actin}_n][\text{G-ATP−actin}]^2[\text{Arp2/3}] \tag{7.2}$$

The capping reaction results in the capping of a filament; only existing filaments and the capping protein participate in this reaction:

$$\text{Gelsolin} + \text{F-actin}_n \xrightarrow{k_{\text{capping}}} \text{F-actin}_n\text{ -capped}$$
$$R_{\text{capping}} = k_{\text{capping}} [\text{F-actin}_n][\text{gelsolin}] \tag{7.3}$$

The net rate of polymerization is now given by

$$R_{\text{polymerization}} = \frac{\text{d}[\text{F-actin}]}{\text{d}t} = R_{\text{elongation}} + R_{\text{branching}} - R_{\text{capping}}$$

$$\frac{\text{d}[\text{F-actin}]}{\text{d}t} = k_{\text{elongation}}[\text{G-ATP−actin}][\text{F-actin}_n]$$
$$+ k_{\text{branching}}[\text{F-actin}_n][\text{G-ATP−actin}]^2[\text{Arp2/3}]$$
$$- k_{\text{capping}}[\text{F-actin}_n][\text{gelsolin}] \tag{7.4}$$

$$\frac{\text{d}(\ln[\text{F-actin}])}{\text{d}t} = k_{\text{elongation}}[\text{G-ATP−actin}]$$
$$+ k_{\text{branching}}[\text{G-ATP−actin}]^2[\text{Arp2/3}]$$
$$- k_{\text{capping}}[\text{gelsolin}] = R_{\text{free}} \tag{7.5}$$

Therefore, the net rate of filament growth can be written as a function of the actin-regulating proteins alone. This type of calculation can allow us to generate time plots of F– and G–actin concentrations in the system as observed in cell-free experiments.[33,57,61] This approach has been used to explain the role of actin in endocytosis.[62] Including diffusion of reacting species in addition to reaction rates alone has allowed researchers to model the spatial distribution of actin components. An elegant example of this is seen in the modeling studies of keratocyte movement.[63,64]

In cellular systems, such as endocytosis and spreading, the regulation of the actin proteins is not controlled by experimentally adding Arp2/3 or WASP to the system but by an upstream signaling system in response to an extracellular signal. In that case, a model of the biochemical signaling network is required to capture the dynamics of the system. We developed a model of the key signaling events for actin in Ref. 65. These reactions then provide additional equations to the set of differential equations shown above and can be integrated simultaneously to obtain the time courses of the concentrations of the signaling proteins.

5.2. Interaction of actin filaments with the plasma membrane
5.2.1 Local membrane regulation of cell spreading
While the deterministic models described above provide insight into the dynamics of the actin reorganization events, within cells, the actin cytoskeleton reactions take place in the vicinity of the plasma membrane. The actin filament reorganization reactions are modulated by the interaction of the cytoskeleton with the plasma membrane. The membrane has been studied as a smooth surface, allowing continuum mechanics approaches to develop partial differential equations for the shape of the membrane.[66] The actin filament network can be treated as a viscous gel, also a continuum formulation.[67] However, the interface between the membrane and the actin filaments is based on the elastic Brownian ratchet, a principle based on thermal fluctuations.[4–6]

The elastic Brownian ratchet model proposed by Mogilner and Oster[4–6] is adapted in three dimensions to model the filament–membrane interactions. The actual reaction velocity in the presence of a load is less than the reaction velocity for a freely growing filament without any resistance and is dependent on the probability that a gap of width (δ) is created between the filament tip and the load (in this case, the membrane). The modified rate constant is given by

$$k'_{on} = k_{on}e^{-\Delta E/k_B T} \tag{7.6}$$

where ΔE is the energy change required to push the membrane forward by a distance δ. ΔE is a local parameter and depends on the location of the growing filament and the area of the membrane it is pushing. ΔE is computed as follows.

There are three main contributions to the energy change—membrane surface energy, filament flexibility, and membrane bending.[5,68] Here, we treat the actin filaments as rigid filaments based on the assumption that filament bending undulations are much faster than the polymerization kinetics.[6-8] We include two major contributions to the membrane energy change—the membrane surface energy and the membrane-bending energy.

The membrane surface energy characterizes the work required by the filament to push an area ΔA (nm^2) of the membrane forward to accommodate an actin monomer of length ($\delta = 0.275$ nm). We characterize the membrane surface resistance by a pressure p (pN/nm^2). This is the load offered by the plasma membrane, similar to the definitions in Refs. 6,69. The energy contribution from the membrane surface term is given by

$$\Delta E_{\text{surface}} = p\mathrm{d}A\delta \tag{7.7}$$

The other important contribution comes from membrane bending. The membrane-bending coefficient is a physical property that characterizes the flexibility of the membrane K_b (pN nm). By incorporating this term, we are accounting for a bendable rather than rigid membrane. The bending energy contribution is then given by[66]

$$\Delta E_{\text{bending}} = K_b \int H^2 \mathrm{d}A \tag{7.8}$$

where H is the local membrane curvature (μm^{-1}). Therefore, the net change in energy that affects the biochemical rates is

$$\Delta E = \Delta E_{\text{bending}} + \Delta E_{\text{surface}} \tag{7.9}$$

In the presence of the membrane, each reaction experiences resistance offered by the membrane by a combination of surface load and bending rigidity (Eq. 7.13). Then, the observed average rate of polymerization is now given by

$$\frac{d(\ln[F\text{-actin}])}{dt} = \left(k_{\text{elongation}}[\text{G-ATP-actin}] + k_{\text{branching}}\right.$$
$$[\text{G-ATP-actin}]^2[\text{Arp2/3}] - k_{\text{capping}}[\text{gelsolin}]\right)e^{(-\Delta E/k_B T)} = R_{\text{observed}} \tag{7.10}$$

The first observation is that the concentration of F-actin is influenced mainly by the amount of G-ATP–actin present. Sensitivity of the time evolution of F-actin to Arp2/3 concentration (F-actin/Arp2/3=d

$(\ln[\text{F-actin}])/\text{d}[\text{Arp2/3}])$ is independent of Arp2/3 concentration and depends on G-actin concentration alone.

$$\frac{\text{d}\Phi_{\text{F-actin/Arp2/3}}}{\text{d}t} = k_{\text{branching}}[\text{G-ATP}-\text{actin}]^2 e^{-\Delta E/k_\text{B}T} \tag{7.11}$$

Similarly, the sensitivity of temporal evolution of F-actin to gelsolin concentration $(\Phi_{\text{F-actin/gelsolin}} = \text{d}(\ln[\text{F-actin}])/\text{d}[\text{gelsolin}])$ is a constant.

$$\frac{\text{d}\Phi_{\text{F-actin/gelsolin}}}{\text{d}t} = -k_{\text{capping}} e^{-\Delta E/k_\text{B}T} \tag{7.12}$$

These relationships highlight the fact that while Arp2/3 and gelsolin are required for the maintenance of filament branching and polymerization, within a reasonable concentration range, the actual value of Arp2/3 and gelsolin is less important than the amount of monomeric actin present for polymerization to proceed.

In the foregoing discussion, we have assumed that each point along the membrane exerts the same energy penalty on the growing actin filaments. However, actin filament reactions are stochastic in time and space. How do we implement the growth of the actin filament network in the presence of the membrane along with the spatial and temporal stochasticity? As we noted above, the membrane alone can be characterized by the bending energy and surface tension, and continuum models including the Helfrich model[66] can capture the change of membrane shape. On the other hand, the actin filament network can be treated as a viscous gel and treated as a continuum. However, putting the two pieces together and capturing the dynamics of cytoskeletal reorganization is something that requires more than a continuum approach. In Section 6, we outline a computational geometry method for implementing the actin cytoskeleton reorganization events and its interaction with the cytoskeleton.

6. COMPUTATIONAL GEOMETRY APPROACH FOR MODELING CELL SPREADING

How can we model the interaction of the membrane with the growing actin filaments and track the change in cell shape during active motile events? Here, we elaborate on a computational geometry approach to develop an interface between the ends of the actin filament and the membrane. The concept behind this model is simple: rather than represent the membrane as a smooth manifold, we represent the membrane as a triangulated mesh (Fig. 7.1). Each vertex on this mesh represents the end of an uncapped actin filament. When a filament grows by the addition of an actin monomer, then the vertex moves forward by the length of the monomer. When a filament is capped, the vertex is fixed in space and may be removed from further calculations as described below, and when a new filament branches from an existing filament, a new vertex is added to the mesh surface. As a result, we have a realistic, three-dimensional model of the actin cytoskeleton and the plasma membrane.

During cell spreading, the absolute area of the membrane surface at the leading edge increases as the leading edge protrudes outward. Such area increase is realized by the removal of invaginations in the plasma membrane or fusion of inner membrane reservoir to cell surface to meet the need of cell

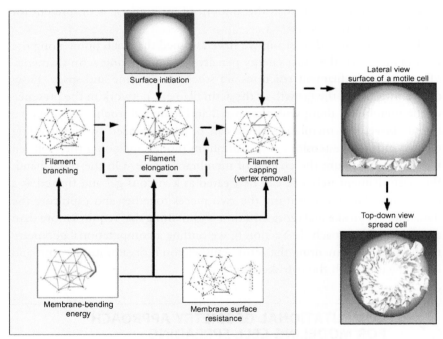

Figure 7.1 Differential geometry approach for interfacing the membrane with the cytoskeleton. A differential geometry approach is used to represent the membrane and its interaction with the actin filament. Each vertex on the triangulated surface represents an end of one actin filament. The vertices move according to the underlying growth of actin filaments in response to biochemical reactions modulated by the energy penalty imposed by membrane bending and surface resistance. As these events proceed in space and time, we are able to simulate cell spreading behavior. (See the color plate.)

spreading. When uncapped filaments elongate and push the cell membrane, the mechanical energy associated with the change of surface geometry is represented as a change in surface curvature. This energy change regulates filament growth negatively and is incorporated as a feedback feature in our model, where change in energy becomes the negative regulator of reactions underlying filament growth. This energy change of the cell membrane is estimated by the work that actin filaments have to do in order to break and reform the surface. These concepts are implemented in the model using computational geometry methods linking the actin filament biochemistry to membrane biophysics.

7. COMPUTATIONAL IMPLEMENTATION

This computational core of the model is composed of four parts: dynamic filament network, dynamic cell surface, membrane energy feedback, and stochastic reaction machinery. Each part describes one aspect of the model: dynamic filament network is modified in response to individual actin filament reactions, dynamic cell surface represents the changing cell geometry, membrane energy-based filament growth reflects the interaction between biochemical actin filament reactions and biophysical cell membrane mechanics, and stochastic reaction machinery determines the temporal dynamics of the system. These parts are integrated into autonomous and functional machinery that drives cell motility. The computational program, written in C++, contains about 10,000 lines of code.

7.1. Dynamic filament network

This module forms the basis of the actin filament network structure at the leading edge. The actin filament network is initialized as a set of seeding filaments (composed of one Arp2/3 and two prepolymerized actin monomers) evenly distributed in the leading edge beneath the spherical cell surface (molecule reservoir). The filaments in this module are initiated and connected with other filaments by branches originating from Arp2/3 binding sites on the actin filament. The dynamics growth of this network is modulated by the iterative occurrences of the three actin biochemical reactions as specified by a modified Gillespie's algorithm.[70]

The barbed ends point radially outward toward the cell membrane. The occurrence of the elongation, branching, and capping reactions leads to filament growth and branching. One challenge in validating this model is that the actual number of filaments at the leading edge of a cell is not known. We conducted parameter variation for initial values of actin filament density.[71] Setting values too low causes the rate of filament polymerization to be slow and prevents branching reactions from occurring within the timescale of the observed experimental effect. High initial densities resulted in reaction rates that caused clashes of filaments and terminated cell spreading in the simulations. Based on these observations, we selected an initial condition of 4000 actin filaments distributed evenly around the periphery of the leading edge.

From the simulation program, we collect the barbed and pointed end location of every filament in the cytoskeleton in the x, y, and z dimensions. We then plot the filament network as shown in Fig. 7.2 as x–y projection for 0.2-μm thick slices. Over the duration of the simulation, the filament network shows an increase in elongation and branch density. We also show an electron micrograph provided by Dr. Tatyana Svitkina (University of Pennsylvania) and trace the outermost edge of the filament network in a representation of the membrane. The comparison between the micrograph and the simulation filament network shows that the model assumptions are reasonable and the stochastic spatiotemporal reactions are able to qualitatively capture the filament reorganization events. A sequence of steps on the growth of the actin cytoskeleton in our spreading model is shown in Fig. 7.2A.

7.2. Dynamic cell surface

The dynamic cell surface keeps track of the exact location of the leading edge. Based on the nature of actin-based motility machinery used by the model, the surface of a motile cell is constructed from the underlying actin filament network. Changes in the biochemical reactions lead to the change of cell surface and consequently lead to cell spreading. The cell surface is constructed by a series of adjacent triangular polyhedrons enclosed by a triangulated surface embedded in a three-dimensional space, resulting in a triangularized sphere (Fig. 7.1). This is similar to the experimentally observed round fibroblast cell as its starts to spread on a fibronectin-coated glass surface.[14] As the underlying filament reactions progress, the cell surface is actively updated, such that the dynamics of the filament network directly changes the location of the cell surface, therefore, representing the experimentally observed movement of the leading edge.

Surface construction using triangulation algorithms and computational geometry is not new.[72] However, using these methods to simulate changes in cell shape has not been done before. In this representation, each vertex of

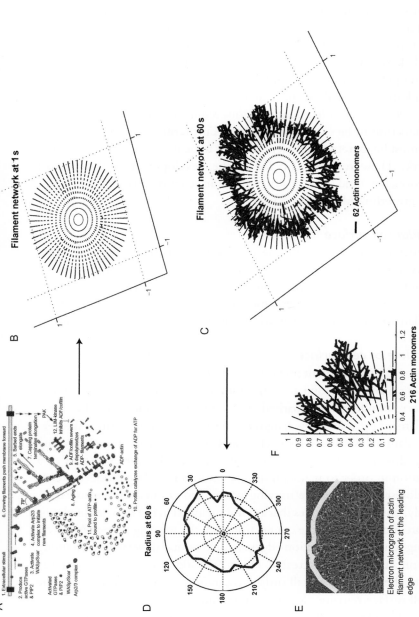

Figure 7.2 Growth of the actin cytoskeleton from the spreading model. The filament network in the model is initiated as a number of seed filaments, which can elongate, branch or cap, based on the actin reactions (A). The resulting actin network (B and F) looks similar to the cross-linked actin network at the leading edge (E). Tracing the periphery of the filaments (C) allows us to construct the shape of cell (D) as a function of time. *Figure S2 from Xiong et al.,* [71] . (See the color plate.)

the cell surface corresponds to the barbed end of an actin filament beneath the cell surface, and the edges connecting all vertices define the triangulation of the surface. Thus, the membrane surface results from the underlying actin cytoskeleton. Because we are not simulating the membrane independent of the cytoskeleton, we do not consider situations where there are no filaments (similar to lipid vesicles). As a result, the cell surface is a direct consequence of the underlying actin filament structure.

The large number of filament reactions requires each step of the Monte Carlo simulation to be extremely efficient. Furthermore, the surface triangulation must be extremely efficient in order to be included in the stochastic framework. These requirements led to the development of the original computational method to reconstruct dynamic cell surface based on filament network growth used in the model.[71]

7.2.1 Initial cell surface

The geometry of cell surface at the leading edge is initialized to a triangulated surface. As mentioned before, the vertex of surface triangles represents the barbed end of underlying actin filaments. Each initial actin filament is composed of one Arp2/3 and two polymerized actin monomers, and the direction of filament growth is initialized to the radial orientation of the sphere.

7.2.2 Update of cell surface

The occurrence of a filament polymerization reaction increases the length of the existing filament by one actin monomer. Thus, the vertex on the cell surface corresponding to this barbed end moves its location along the direction of filament growth by the length of one actin monomer. Since the vertex of a filament is always connected with the vertex of its neighboring filaments, this connectivity must be updated once the vertex moves to a new location in order to maintain surface smoothness. The model searches for the closest vertices to the moving vertex and connects them together. Previously, connected vertices that are not included in the new set of closest vertices are removed from the connection with the moving vertex. All facets containing removed connection edges are deleted and new facets containing added connection edges are created. During the addition and removal of surface facets, the topology of cell surface is carefully maintained such that the entire cell surface remains closed in three dimensions.

The update method when a filament capping reaction occurs is similar to the case of filament polymerization reaction except that the molecule added to the barbed end of a filament is a capping protein. The only difference is that the current filament is now capped and is not able to grow anymore. Thus, the vertex corresponding to the capped filament is now incapable of moving.

The filament branching reaction creates a new actin filament from an existing filament by the binding of Arp2/3 to the side of the existing filament. Therefore, a new vertex representing the barbed end of the new filament must be created and added into the cell surface (Fig. 7.1). The spatial location of the new filament is determined by the binding site of Arp2/3 on the existing filament, the branching orientation, and the initial length of the new filament. Because Arp2/3 is activated by the upstream signaling molecules that are attached to the cell membrane, the model assumes that Arp2/3 binds to the membrane facing side of the existing filament, usually three to four actin monomers away from the barbed end of the existing filament. The angle between the new filament and the existing filament is 70°, and the number of initially polymerized actin monomers is 2. Based on these constraints, the new filament can be created from the Arp2/3 binding site on the existing filament and its possible orientations form a conic surface around the existing filament. The resulting cytoskeletal structure from these simulations is shown in Fig. 7.2 and compared against electron micrographs of the actin network at the leading edge of cells.

7.3. Membrane energy-based feedback to the biochemical reaction rate

The filament network has to overcome the change in the mechanical energy associated with the forward movement of the membrane. This results in a reduction of the rate of the reaction as shown in Eq. (7.10). The concept of the elastic Brownian ratchet is explained in Ref. 5. In our model, we consider two contributions to the change in membrane energy—the surface-resistance pressure p and the membrane-bending stiffness K_b.

The energy computation in the simulation characterizes the resistance imposed on the growing filaments. In our model, we use the energy–velocity dependence defined in Eq. (7.6) to compute the effective rate constants for all three biochemical reactions. The main challenge here lies in the computation of dA and H for the triangular facets while keeping a closed smooth surface. We turn to discrete differential geometry framework for computing dA and H. The geometrical framework for calculating the area and the curvature integral for a given filament is shown in Fig. 7.3.

The calculation of local area of a facet and local curvature is well established for triangular meshes in discrete differential geometry. We use

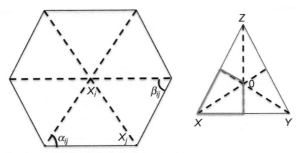

Figure 7.3 Computation of area element to obtain membrane resistance. In order to calculate the Voronoi area of the vertex X in the triangle XYZ, we find the circumcenter O. The Voronoi area associated with the vertex X lies within the triangle (marked by the blue lines) for nonobtuse angles and is $(1/8)(|XY|^2 \cot \angle Z + |XZ|^2 \cot \angle Y)$. Using this method, we then sum these areas for a vertex x_i as a function of the neighbors x_j and obtain the area associated with each filament. *Xiong et al.,* [71] *supplemental figure 7.*

the method proposed in Ref. 72 to calculate the area of a facet and to obtain the mean curvature integral of the surface. The complete mathematical framework for this is presented in Ref. 72.

Using the triangulations shown in Fig. 7.3, for a vertex x_i on the surface, the area as a function of its neighbors x_j is given by

$$dA_{ij} = \frac{1}{8} \sum (\cot(\alpha_{ij}) + \cot(\beta_{ij})) \| x_i - x_j \|^2 \qquad (7.13)$$

The pressure term is used to compute the vectorial force acting on each filament during the motion of the surface. In order to compute the curvature integral, we use the following equation:

$$U = \int H^2 dA = \frac{1}{2A_{\text{mixed}}} \sum (\cot(\alpha_{ij}) + \cot(\beta_{ij}))(x_i - x_j) \qquad (7.14)$$

The vertices and angles in Eqs. (7.13) and (7.14) also correspond to Fig. 7.3. A_{mixed} is the cumulative area of the individual triangles on the surface. The complete derivation of this relationship between the curvature integral U and the individual angles can be found in Ref. 72.

7.4. Membrane surface resistance

The membrane surface resistance is characterized by a pressure term p (pN/nm^2). The energy change associated with moving a membrane facet of area dA by a length (one actin monomer) is given by $\Delta E_{\text{surface}} = p dA \delta$.

The force associated with this energy change is pdA. The resistance force f imposed on growing filaments gets updated as soon as the local membrane geometry changes throughout the simulation process. The resistance force f imposed on a growing filament is calculated by the following formula:

$$f = \hat{n} \cdot \sum_{i=1}^{5} \overrightarrow{f_i} \tag{7.15}$$

where $\hat{n}$ is a unit vector pointing to the inverse direction of filament growth, $\overrightarrow{f_i}$ is the resistance force generated by the neighboring triangular facet $O-V_{1,i-1}-V_{1,i}$, and its magnitude is calculated by

$$f_i = p \cdot dA_{O-V_{1,i-1}-V_{1,i}} \tag{7.16}$$

where p is the resistance pressure of membrane surface and $dA_{O-V_{1,i-1}-V_{1,i}}$ is the area of the triangular facet $O-V_{1,i-1}-V_{1,i}$. The direction of $\overrightarrow{f_i}$ points from the vertex O to the center of the triangular facet $O-V_{1,i-1}-V_{1,i}$. Therefore, using the cotangents calculated for the Voronoi area, the integral can be computed directly for the surface. For further details, please see Ref. 71. Based on these two membrane energy contributions, we compute the energy change associated with moving the membrane forward and suitably modify the reaction rate of the actin filament reactions.

7.5. Surface clashes

The model only selects the orientation that makes the new filament intersect with the cell surface in order to avoid the formation of a concave surface around the newly added vertex. For filaments that have been capped, no filament reactions can occur on them and they cannot grow any more. Then as the neighboring filaments keep growing, the local surface around the barbed ends of these filaments becomes concave or even invaginated. Such concave surface may cause the clash of local surfaces. In order to avoid this, once the spatial location of the new filament is determined, the vertex corresponding to the barbed end of new vertex is added into the cell surface and the vertex connectivity in this local surface area is updated by following the same principle as that used in filament polymerization reaction.

In order to determine when to remove the attachment of a capped filament from the cell surface, we have used the resistance force imposed on this filament by the cell membrane as the criterion: if the resistance force on the capped filament disappeared based on the geometry of the local surface around this filament, then the barbed end of this capped filament should be removed from cell surface by removing the corresponding vertex from cell surface. As the removal of a vertex leaves a three-dimensional polygon hole on the cell surface, the surface area within this region must be retriangulated to maintain the closed topology of cell surface. A straightforward method is used to triangulate this polygon hole by connecting neighboring edges of the polygon to form new triangular facets.

The mechanism of alternating branching and capping reactions keeps filaments short during filament network growth, and the filament network remains rigid enough to be able to push the cell membrane forward. On the other hand, this mechanism causes the cross growth of uncapped filaments which leads to the clash of cell surface. Surface clash not only causes a jagged cell surface but also makes the filament network have equal chance to grow both forward and backward such that there is no net forward growth for the leading edge for a spreading cell. This problem is avoided by the nature of the membrane: rigidity (such as membrane elasticity and membrane-bending resistance) and merging intracellular membranes into the cell surface membrane during the spreading process. The computational model uses the concept of protrusion guidance to direct the branching direction of filament branching reaction based on the experimental observations of fibroblast cell spreading. Protrusion guidance posits that the branching direction that deviates the least from the protrusion direction of the leading edge is preferred. Together with the intersection requirement of creating new filaments, the branching reaction that creates a new filament that intersects with the cell surface and has the least deviation angle from the protrusion direction of the leading edge is preferred.

7.6. Dynamic dependency graph

A dynamic dependency graph is built on the graph data structure. Each node of the dependency graph contains three dependency lists: the list of other nodes that this node can modify, the list of other nodes that this node can destroy, and the list of other nodes that can modify or destroy this node. At the beginning of the spreading simulation, this dependency graph is initialized to reflect the interaction relationships among the initial filament reactions. At each step of simulation: (1) when a filament polymerization reaction occurs, the filament reactions affected by the current reaction are updated based on their interaction relationships in the dependency graph; (2) when a filament branching reaction occurs, three associated filament reactions are created and their interaction relationships are added into the dependency graph; and (3) when a filament capping reaction occurs, the filament reactions associated with the capped filament are destroyed and their dependency relationships are removed from the graph.

The dynamic dependency graph is then integrated with Gillespie's stochastic framework, specifically the first reaction method, by introducing the sorted list of the waiting time of all filament reactions into the model. This waiting time list updates the waiting time of a filament reaction if this reaction is changed indicated by the dynamic dependency graph and uses a dichotomy sorting algorithm to move the updated waiting time to a new position to keep the list sorted. Since this waiting list is always sorted from the minimum to the maximum, the filament reaction to occur at the next simulation step is always at the front of this list.

8. COMPARISON OF EXPERIMENTS AND SIMULATED CELL SPREADING

One application of our model has been to understand how cell spreading during the isotropic phase is regulated. Using this spreading model, we were able to capture key aspects of cell spreading during the isotropic phase. Our model also identified that the shape of the spreading cell is maintained primarily by the membrane properties and the interaction of the growing cytoskeleton with the plasma membrane.[65,71] While signaling and the concentrations of actin-regulating proteins are required for initiation of cell spreading, they do not control the shape evolution of the spreading cell. These observations were validated experimentally using fibroblasts spreading on fibronectin-coated surfaces. Thus, a simple model of three actin reactions, coupled with a unique representation of membrane geometry was sufficient to model the spreading behavior of fibroblasts during the isotropic phase. Our approach of including the membrane interactions with the actin not just allowed us to capture experimentally observed behavior in a complex cellular system but also provided some fundamental insights into how membrane properties regulate actin reactions.

9. CONCLUSIONS AND PERSPECTIVES

In this chapter, we describe a distinctive approach to modeling the cytoskeletal changes during cell spreading. Borrowing from techniques for chemical reactions, mechanics of actin filaments and the plasma membrane, and implementation techniques from computational geometry allowed us to build a three-dimensional model of cell spreading. This approach can be extended to study contractile processes in cells, neurite outgrowth, and reorganization during synaptic transmission, with suitable modifications.

Mathematical and computational modeling compliment experimental studies in biology. This is particularly true in complex biological systems where multiple variables may be affected by a single change in an experimental setting. Modeling tools allow us to distill complex processes into a set of equations that explicitly describe multiple relationships. Solutions to these equations can provide deep insight into the mechanisms that regulate the process of interest. Since most biological models are solved numerically, the simulations often provide explicit predictions that can be tested experimentally. There is the ever-present danger of the urge to oversimplify biological processes in modeling; however, the ability of such models to predict the role of regulatory molecules in controlling the dynamics of the process and comparison of the simulations with experimental results as a gold standard can help us build models with the correct level of detail that provide deep mechanistic insight not obtained solely from experiments.

ACKNOWLEDGMENTS

This work was supported by NIH Grant GM 072853 and a Systems Biology Center Grant GM-071558.

REFERENCES

1. Vitriol EA, Zheng JQ. Growth cone travel in space and time: the cellular ensemble of cytoskeleton, adhesion, and membrane. *Neuron.* 2012;73(6):1068–1081.
2. Shirao T, González-Billault C. Actin filaments and microtubules in dendritic spines. *J Neurochem.* 2013;126(2):155–164.
3. Bray D. *Cell Movements: From Molecules to Motility.* 2nd ed. New York: Garland Science; 2000.
4. Peskin CS, Odell GM, Oster GF. Cellular motions and thermal fluctuations: the Brownian ratchet. *Biophys J.* 1993;65(1):316–324.
5. Mogilner A, Oster G. Force generation by actin polymerization II: the elastic ratchet and tethered filaments. *Biophys J.* 2003;84(3):1591–1605.
6. Mogilner A, Oster G. Cell motility driven by actin polymerization. *Biophys J.* 1996;71(6):3030–3045.
7. Carlsson AE. Growth of branched actin networks against obstacles. *Biophys J.* 2001;81:1907–1923.
8. Mogilner A, Edelstein-Keshet L. Regulation of actin dynamics in rapidly moving cells: a quantitative analysis. *Biophys J.* 2002;83(3):1237–1258.
9. Zhu J, Carlsson AE. Growth of attached actin filaments. *Eur Phys J E Soft Matter.* 2006;21(3):209–222.

10. Atilgan E, Wirtz D, Sun SX. Mechanics and dynamics of actin-driven thin membrane protrusions. *Biophys J.* 2006;90(1):65–76.

11. Lee KC, Liu AJ. Force-velocity relation for actin-polymerization-driven motility from Brownian dynamics simulations. *Biophys J.* 2009;97(5):1295–1304.

12. Döbereiner HG, Dubin-Thaler B, Giannone G, Xenias HS, Sheetz MP. Dynamic phase transitions in cell spreading. *Phys Rev Lett.* 2004;93(10):108105.

13. Dubin-Thaler BJ, Hofman JM, Cai Y, et al. Quantification of cell edge velocities and traction forces reveals distinct motility modules during cell spreading. *PLoS One.* 2008;3(11):e3735.

14. Dubin-Thaler BJ, Giannone G, Döbereiner HG, Sheetz MP. Nanometer analysis of cell spreading on matrix-coated surfaces reveals two distinct cell states and STEPs. *Biophys J.* 2004;86(3):1794–1806.

15. Aplin AE, Howe A, Alahari SK, Juliano RL. Signal transduction and signal modulation by cell adhesion receptors: the role of integrins, cadherins, immunoglobulin-cell adhesion molecules, and selectins. *Pharmacol Rev.* 1998;50:197–263.

16. Sheetz MP, Felsenfeld D, Galbraith CG, Choquet D. Cell migration as a five-step cycle. *Biochem Soc Symp.* 1999;65:233–243.

17. Ridley AJ, Schwartz MA, Burridge K, et al. Cell migration: integrating signals from front to back. *Science.* 2003;302(5651):1704–1709.

18. Miyamoto S, Teramoto H, Coso OA, et al. Integrin function: molecular hierarchies of cytoskeletal and signaling molecules. *J Cell Biol.* 1995;131(3):791–805.

19. Geiger B, Bershadsky A, Pankov R, Yamada KM. Transmembrane crosstalk between the extracellular matrix–cytoskeleton crosstalk. *Nat Rev Mol Cell Biol.* 2001;2(11):793–805.

20. Zamir E, Geiger B. Molecular complexity and dynamics of cell-matrix adhesions. *J Cell Sci.* 2001;114(20):3583–3590.

21. Cohen M, Joester D, Geiger B, Addadi L. Spatial and temporal sequence of events in cell adhesion: from molecular recognition to focal adhesion assembly. *Chembiochem.* 2004;5(10):1393–1399.

22. Bershadsky AD, Balaban NQ, Geiger B. Adhesion-dependent cell mechanosensitivity. *Annu Rev Cell Dev Biol.* 2003;19:677–695.

23. Zaidel-Bar R, Cohen M, Addadi L, Geiger B. Hierarchical assembly of cell-matrix adhesion complexes. *Biochem Soc Trans.* 2004;32(3):416–420.

24. Alberts B, Johnson A, Lewis J, Raff M, Roberts K, Walter P. *Molecular Biology of the Cell.* 4th ed. New York: Garland Science; 2002.

25. Carlier MF, Pantaloni D. Control of actin dynamics in cell motility. *J Mol Biol.* 1997;269(4):459–467.

26. Le Clainche C, Carlier MF. Regulation of actin assembly associated with protrusion and adhesion in cell migration. *Physiol Rev.* 2008;88(2):489–513.

27. Pantaloni D, Le Clainche C, Carlier MF. Mechanism of actin-based motility. *Science.* 2001;292(5521):1502–1506.

28. Pollard TD, Blanchoin L, Mullins RD. Actin dynamics. *J Cell Sci.* 2001;114(1):3–4.

29. Pollard TD, Blanchoin L, Mullins RD. Molecular mechanisms controlling actin filament dynamics in nonmuscle cells. *Annu Rev Biophys Biomol Struct.* 2000;29:545–576.

30. Blanchoin L, Amann KJ, Higgs HN, Marchand JB, Kaiser DA, Pollard TD. Direct observation of dendritic actin filament networks nucleated by Arp2/3 complex and WASP/Scar proteins. *Nature.* 2000;404(6781):1007–1011.

31. Gutsche-Perelroizen I, Lepault J, Ott A, Carlier MF. Filament assembly from profilin-actin. *J Biol Chem.* 1999;274(10):6234–6243.

32. Kovar DR, Wu JQ, Pollard TD. Profilin-mediated competition between capping protein and formin Cdc12p during cytokinesis in fission yeast. *Mol Biol Cell.* 2005;16(5):2313–2324.

33. Carlier MF, Ducruix A, Pantaloni D. Signalling to actin: the Cdc42-N-WASP-Arp2/3 connection. *Chem Biol.* 1999;6(9):R235–R240.
34. Ressad F, Didry D, Egile C, Pantaloni D, Carlier MF. Control of actin filament length and turnover by actin depolymerizing factor (ADF/cofilin) in the presence of capping proteins and ARP2/3 complex. *J Biol Chem.* 1999;274(30):20970–20976.
35. Lin T, Zeng L, Liu Y, et al. Rho-ROCK-LIMK-cofilin pathway regulates shear stress activation of sterol regulatory element binding proteins. *Circ Res.* 2003;92(12):1296–1304.
36. Higgs HN, Pollard TD. Regulation of actin polymerization by Arp2/3 complex and WASp/Scar proteins. *J Biol Chem.* 1999;274(46):32531–32534.
37. Higgs HN, Pollard TD. Regulation of actin filament network formation through ARP2/3 complex: activation by a diverse array of proteins. *Annu Rev Biochem.* 2001;70:649–676.
38. Miki H, Takenawa T. Regulation of actin dynamics by WASP family proteins. *J Biochem (Tokyo).* 2003;134(3):309–313.
39. Cooper JA, Schafer DA. Control of actin assembly and disassembly at filament ends. *Curr Opin Cell Biol.* 2000;12(1):97–103.
40. Krause M, Dent EW, Bear JE, Loureiro JJ, Gertler FB. Ena/VASP proteins: regulators of the actin cytoskeleton and cell migration. *Annu Rev Cell Dev Biol.* 2003;19:541–564.
41. Urban E, Jacob S, Nemethova M, Resch GP, Small JV. Electron tomography reveals unbranched networks of actin filaments in lamellipodia. *Nat Publ Group.* 2010;12(5):429–435.
42. Weaver AM, Young ME, Lee WL, Cooper JA. Integration of signals to the Arp2/3 complex. *Curr Opin Cell Biol.* 2003;15(1):23–30.
43. Millard TH, Sharp SJ, Machesky LM. Signalling to actin assembly via the WASP (Wiskott-Aldrich syndrome protein)-family proteins and the Arp2/3 complex. *Biochem J.* 2004;380(1):1–17.
44. Takenawa T, Miki H. WASP and WAVE family proteins: key molecules for rapid rearrangement of cortical actin filaments and cell movement. *J Cell Sci.* 2001;114(10):1801–1809.
45. Torres E, Rosen MK. Protein-tyrosine kinase and GTPase signals cooperate to phosphorylate and activate Wiskott-Aldrich syndrome protein (WASP)/neuronal WASP. *J Biol Chem.* 2006;281(6):3513–3520.
46. Takenawa T, Suetsugu S. The WASP-WAVE protein network: connecting the membrane to the cytoskeleton. *Nat Rev Mol Cell Biol.* 2007;8(1):37–48.
47. Reinhard M, Jarchau T, Walter U. Actin-based motility: stop and go with Ena/VASP proteins. *Trends Biochem Sci.* 2001;26(4):243–249.
48. Niggli V. Regulation of protein activities by phosphoinositide phosphates. *Annu Rev Cell Dev Biol.* 2005;21:57–79.
49. Bear JE, Svitkina TM, Krause M, et al. Antagonism between Ena/VASP proteins and actin filament capping regulates fibroblast motility. *Cell.* 2002;109(4):509–521.
50. Schafer DA, Jennings PB, Cooper JA. Dynamics of capping protein and actin assembly in vitro: uncapping barbed ends by polyphosphoinositides. *J Cell Biol.* 1996;135(1):169–179.
51. Janmey PA, Stossel TP. Gelsolin-polyphosphoinositide interaction. Full expression of gelsolin-inhibiting function by polyphosphoinositides in vesicular form and inactivation by dilution, aggregation, or masking of the inositol head group. *J Biol Chem.* 1989;264(9):4825–4831.
52. Arber S, Barbayannis FA, Hanser H, et al. Regulation of actin dynamics through phosphorylation of cofilin by LIM-kinase. *Nature.* 1998;393(6687):805–809.
53. Raftopoulou M, Hall A. Cell migration: Rho GTPases lead the way. *Dev Biol.* 2004;265(1):23–32.

54. Sheetz MP, Sable JE, Döbereiner HG. Continuous membrane-cytoskeleton adhesion requires continuous accommodation to lipid and cytoskeleton dynamics. *Annu Rev Biophys Biomol Struct.* 2006;35:417–434.

55. Edelstein-Keshet L, Ermentrout GB. A model for actin-filament length distribution in a lamellipod. *J Math Biol.* 2001;43(4):325–355.

56. Carlsson AE. Growth velocities of branched actin networks. *Biophys J.* 2003;84(5):2907–2918.

57. Brooks FJ, Carlsson AE. Actin polymerization overshoots and ATP hydrolysis as assayed by pyrene fluorescence. *Biophys J.* 2008;95(3):1050–1062.

58. Schaus TE, Taylor EW, Borisy GG. Self-organization of actin filament orientation in the dendritic-nucleation/array-treadmilling model. *Proc Natl Acad Sci USA.* 2007;104(17):7086–7091.

59. Atilgan E, Wirtz D, Sun SX. Morphology of the lamellipodium and organization of actin filaments at the leading edge of crawling cells. *Biophys J.* 2005;89(5):3589–3602.

60. Loisel TP, Boujemaa R, Pantaloni D, Carlier MF. Reconstitution of actin-based motility of Listeria and Shigella using pure proteins. *Nature.* 1999;401(6753):613–616.

61. Huber F, Käs J, Stuhrmann B. Growing actin networks form lamellipodium and lamellum by self-assembly. *Biophys J.* 2008;95(12):5508–5523.

62. Berro J, Sirotkin V, Pollard TD. Mathematical modeling of endocytic actin patch kinetics in fission yeast: disassembly requires release of actin filament fragments. *Mol Biol Cell.* 2010;21(16):2905–2915.

63. Novak IL, Slepchenko BM, Mogilner A. Quantitative analysis of G-actin transport in motile cells. *Biophys J.* 2008;95(4):1627–1638.

64. Mogilner A, Keren K. The shape of motile cells. *Curr Biol.* 2009;19(17):R762–R771.

65. Rangamani P, Fardin M-A, Xiong Y, et al. Signaling network triggers and membrane physical properties control the actin cytoskeleton-driven isotropic phase of cell spreading. *Biophys J.* 2011;100(4):845–857.

66. Helfrich W. Elastic properties of lipid bilayers: theory and possible experiments. *Z Naturforsch C.* 1973;28(11):693–703.

67. Jean-François Joanny JP. Active gels as a description of the actin-myosin cytoskeleton. *HFSP J.* 2009;3(2):94.

68. Grimm HP, Verkhovsky AB, Mogilner A, Meister JJ. Analysis of actin dynamics at the leading edge of crawling cells: implications for the shape of keratocyte lamellipodia. *Eur Biophys J.* 2003;32(6):563–577.

69. Mogilner A, Rubinstein B. The physics of filopodial protrusion. *Biophys J.* 2005;89(2):782–795.

70. Gillespie DT. Exact stochastic simulation of coupled chemical reactions. *J Phys Chem.* 1977;81(25):2340–2361.

71. Xiong Y, Rangamani P, Fardin M-A, et al. Mechanisms controlling cell size and shape during isotropic cell spreading. *Biophys J.* 2010;98(10):2136–2146.

72. Meyer M, Desbrun M, Schroeder P, Barr AH. Discrete differential-geometry operators for triangulated 2-manifolds. *Visualization and Mathematics.* 2002;3(2):52–58.

CHAPTER EIGHT

Computational Modeling of Diffusion in the Cerebellum

Toma M. Marinov, Fidel Santamaria
UTSA Neurosciences Institute, University of Texas at San Antonio, San Antonio, Texas, USA

Contents

Abstract

Diffusion is a major transport mechanism in living organisms. In the cerebellum, diffusion is responsible for the propagation of molecular signaling involved in synaptic plasticity and metabolism, both intracellularly and extracellularly. In this chapter, we present an overview of the cerebellar structure and function. We then discuss the types of diffusion processes present in the cerebellum and their biological importance. We particularly emphasize the differences between extracellular and intracellular diffusion and the presence of tortuosity and anomalous diffusion in different parts of the cerebellar cortex. We provide a mathematical introduction

Progress in Molecular Biology and Translational Science, Volume 123
ISSN 1877-1173
http://dx.doi.org/10.1016/B978-0-12-397897-4.00007-3

169

to diffusion and a conceptual overview of various computational modeling techniques. We discuss their scope and their limit of application. Although our focus is the cerebellum, we have aimed at presenting the biological and mathematical foundations as general as possible to be applicable to any other area in biology in which diffusion is of importance.

1. INTRODUCTION

Diffusion takes place at all spatial and temporal scales in the brain: from the movement of lipids on a membrane to the diffusion of water and biomolecules in the extracellular space, over large volumes. Diffusion is a measurement that describes how difficult it is for molecules to move in space. Thus, depending on the viscosity of the medium and the structural properties of the neuropil, the same molecule could experience very different diffusion processes. These properties determine how far and how fast biochemical signals propagate inside neurons and in the extracellular space. Diffusion can also be used to measure the complexity of the nervous system in healthy and pathological conditions. Overall, diffusion determines the propagation of biochemical signals and can be used to determine the nervous system structure at multiple spatial scales.

We have divided this chapter in three sections that treat diffusion from different perspectives: biological, mathematical, and practical modeling. The first part provides an intuitive explanation of diffusion, how it relates to the cerebellum, types of diffusion processes encountered extracellularly and intracellularly, and a discussion on the biological importance of diffusion. This section contains very little mathematics and is intended to provide an overview of the problem from a biological perspective. The second section of the chapter contains a formal mathematical definition of the diffusion equation and its solutions. The purpose of this section is to provide a formal introduction to those interested in learning the mathematical foundations of diffusion. The last section provides an overview of the techniques to model diffusion in neurons. Our objective is not only to provide the mathematical foundations on how to model a diffusion process but also to identify the conceptual and technical problems that any modeler will encounter. Although our focus is the cerebellum, we have aimed at presenting the biological and mathematical foundations as general as possible to be applicable to any other area in biology in which diffusion is of importance.

2. DIFFUSION IN THE CEREBELLUM

We will start by providing a brief description of the anatomy of the cerebellum; then, we will describe the types of diffusion processes that take place in the extracellular and intracellular volumes. It is important to note that the structure of the brain or individual neurons affects the diffusion of molecules. How it affects the diffusion depends on multiple factors such as the diffusion coefficient of the molecule, size of the molecule, and the presence and arrangement of obstacles. At such small volumes, as those found in the nervous system, volume–exclusion effects could have a strong influence on the diffusion process. For example, if a molecule is 5 nm in diameter and the pore size of a highly crowded environment is 4 nm, then the molecule will not be able to move, while other smaller molecules would be hampered but would still diffuse. Such a process can give rise to a process known as anomalous diffusion. Also, the intricate arrangement of dendrites and glia generate a mesh that increases the actual distance a particle has to travel to diffuse along the neuropil, this net effect, reflected in a reduced diffusion coefficient, is known as tortuosity. Both anomalous diffusion and tortuosity have been found in the cerebellum.

2.1. The anatomy of the cerebellum

The cerebellum is traditionally divided into the cerebellar cortex and the deep cerebellar nuclei. Inputs to the cerebellum arrive via two axonal bundles. The first one is from mossy fibers that carry sensory and motor information from the periphery and the cerebral cortex. The second one is from the climbing fibers that originate in the inferior olive. There are three cerebellar nuclei (fastigal, interpositus, and dentate). Both axonal input bundles send collaterals to the deep cerebellar nuclei and the cerebellar cortex (Fig. 8.1A).

The widely studied cerebellar cortex is composed, mainly, of four types of neurons: granule cells, Purkinje cells, inhibitory molecular layer interneurons (stellate and basket), and Golgi cells (also inhibitory). The arrangement of the cerebellar cortical neurons is regular across the cortex. Briefly, mossy fibers contact granule cells and Golgi cells; granule cells send axons into the molecular layer. This section of the axon is called the ascending segment. The ascending segments then bifurcate and the bundle of axons traverse the cerebellar cortex in what is known as the parallel fibers. Parallel fibers

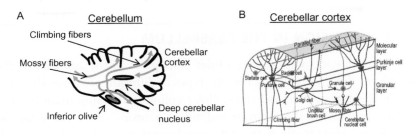

Figure 8.1 The cerebellum and cerebellar cortex. (A) The overall structure of the cerebellum. (B) Basic architecture and principal neurons of the cerebellar cortex. (See the color plate.)

contact Golgi cells, inhibitory molecular layer interneurons, and Purkinje cells. Purkinje cells are contacted by synapses from a single climbing fiber, although a single climbing fiber can contact several Purkinje cells (Fig. 8.1B).

The output of the cerebellar cortex is provided by Purkinje cells that integrate the synaptic activity of climbing fibers, granule cells, and inhibitory interneurons. Golgi cells provide inhibitory feedback to the granule cell layer. The deep cerebellar nuclei are the final integrator of cerebellar information. Cells in the deep cerebellar nuclei receive input from Purkinje cells and mossy fibers.[1]

Functionally, the cerebellum processes sensory and motor information.[2-4] However, there is evidence suggesting cerebellar processing of higher brain activity.[5,6] In any case, the general consensus is that the basic architecture of the cerebellar cortex is regular across the entire cerebellum. Thus, all the different functions are processed by the same network structure.

2.2. Extracellular and intracellular diffusion

Diffusion is a process in which a molecule randomly moves at every time step. The direction of each step is independent of the direction of any previous steps. This is called a Markov process or Brownian motion.[7] As shown elsewhere,[8] the mean square displacement (MSD) over time (t) of a particle moving in one dimension is described as

$$MSD = 2Dt \qquad (8.1)$$

where D is the diffusion coefficient (Fig. 8.2). For two and three dimensions, the relationship is $MSD = 4Dt$ and $6Dt$, respectively. The diffusion coefficient depends on the viscosity of the medium, the diameter of the particle,

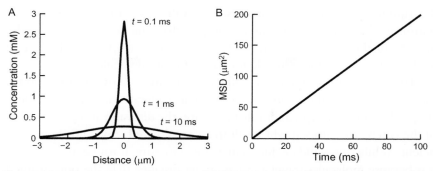

Figure 8.2 Diffusion in one dimension. (A) Three spatial profiles of the solution of the diffusion equation at $t=0.1$, 1, and 10 ms. A 1 mM initial condition is released at $x=0$ at $t=0$. The solution of the diffusion equation $\partial C/\partial t = D\partial^2 C/\partial x^2$ is $C=1/\sqrt{\pi 4Dt}e^{-x^2/4Dt}$, with the diffusion coefficient $D=0.1\ \mu m^2/ms$. (B) Mean square displacement (MSD) calculated from the spatial profiles in (A). $MSD(t)=\sum(\bar{x}-x)^2 C(x,t)$, with $\bar{x}$ the mean position of the distribution (in this case, $\bar{x}=0$).

and temperature. The electrical charge of a molecule does not influence the diffusion process.

Clearly, Eq. (8.1) has a constant D only in the case where the entire space where the particle is moving is homogeneous. However, the different concentration of dendrites, glia, vascularities, and changes in neuron densities make the neuropil a highly anisotropic medium.[9] For molecules that cannot go through the lipid bilayer, each structure acts as an obstacle or barrier that hampers diffusion. Furthermore, the viscosity of the medium can change either intracellularly or extracellularly.[10] Thus, a molecule diffusing in one direction might experience a different diffusion coefficient than moving in the perpendicular direction.

2.3. Extracellular diffusion in the cerebellum

Diffusion in the extracellular space is responsible for delivering oxygen and glucose from the vascular system to the cells, as well as for molecular signaling substances between cells in the brain tissue. This is known as volume transmission.

Extracellular space diffusion takes place in the spaces between cells and can be modeled as diffusion in a porous medium with two phases—an impermeable and a permeable phase. Given a sufficiently large volume of brain tissue, one can consider the ratio between the extracellular space volume V_o and the total volume V. Such a parameter $\rho=V_o/V (0\leq\rho\leq 1)$ is called volume fraction (or void fraction), and it gives a quantitative description of the porosity of the system. In the case of $\rho < 1$, the concentration in the vicinity of a source is higher than the concentration in the vicinity of the same source in a nonporous medium ($\rho = 1$). This is due to the reduced volume available for the diffusing molecules.

Tortuosity arises when the effective distance between two points increases because there are obstacles that generate an intricate mesh (Fig. 8.3). Thus, an experiment that measures the diffusion along the surface of the cerebellar cortex might result in a very slow effective diffusion coefficient. While this could be due to increased viscosity, it has been found that it is because molecules travel a much longer distance than the one assumed from the simplified geometry. Using experimental and computational techniques, it is possible to determine the correction factor necessary to account for the tortuosity of the environment. Usually, this has been shown in the extracellular diffusion of ions. Tortuosity is defined as

$$\lambda = \sqrt{\frac{D}{D^*}} \tag{8.2}$$

where D is the diffusion coefficient of a given substance in free medium (or cerebro spinal fluid) and D^* is the diffusion coefficient of the same substance in the brain. It is an averaging parameter reflecting the medium

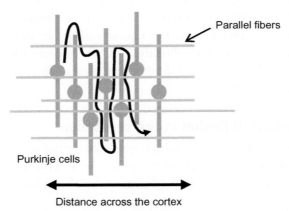

Figure 8.3 Tortuosity is a measurement of the effective free volume in the brain. The schematic shows a top view of the cerebellar cortex with Purkinje cells and parallel fibers. Although the measured distance along the cortex is short, the effective path that a diffusing molecule, with diffusion coefficient in free medium D, has to travel could be much larger, thus resulting in a much smaller diffusion coefficient (D^*). In order to measure tortuosity, it is necessary that the intricate neuronal and network structure can be averaged in a constant tortuosity parameter $\lambda = \sqrt{D/D^*}$.

properties such as pore geometry and size, as well as its connectivity. In the case when the medium is anisotropic (different values along different axes, usually due to anatomy), λ is a tensor. Physically, the tortuosity λ can be interpreted as the factor of increase in distance a given substance travels in the brain due to obstacles compared to free medium. Of importance is that, although the diffusion process is slower than the one expected from diffusion in free medium, the movement of the molecules follows the same classical equation (Eq. 8.1) (Fig. 8.4).

Additionally, the diffusion parameters can be inhomogeneous, that is, they can vary with the location in the tissue along the same axis. The cerebellum exhibits such an inhomogeneity between the molecular and the granule cell layers manifesting itself with distinct values of both the volume fraction and the tortuosity in both layers (Fig. 8.5).

In the cerebellum, there are two main types of phenomena where extracellular diffusion is present. The first one is spreading depression. This is a neurophysiological phenomenon in which neural activity is reduced. The spreading depression moves as a wave across the cortex accompanied by a wave of vasoconstriction and subsequent vasodilation.[11] Spreading depression is observed both in the cerebral and the cerebellar cortices; however, the cerebellum seems to be more resistant to it.

Cerebellar spreading depression can occur by raising the extracellular K^+ and reducing Cl^- concentrations. It is characterized by an extracellular Ca^{2+} concentration decrease due to Ca^{2+} influx in the cell. This increase in intracellular Ca^{2+} could even lead to cell death. Cerebellar spreading depression

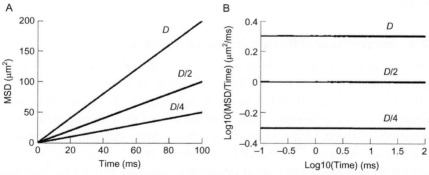

Figure 8.4 Tortuosity follows the classical diffusion equation. (A) Plots of the mean square displacement (MSD) against time for a diffusion in which the diffusion coefficient decreases by a factor of 2 and 4. (B) A logarithmic (base 10) transformation of the data in (A) shows that the diffusion coefficient is constant (compare to anomalous diffusion).

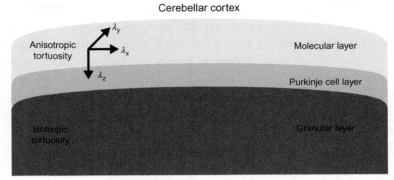

Figure 8.5 Anisotropy in the cerebellar cortex. The value of the diffusion coefficient depends on the specific layer of the cerebellum.

can occur without blood flow, implying that unlike the cerebral case, it does not have a vascular component.[12,13]

Another type of wave-spreading phenomenon is spreading acidification and depression. This type of depression of neuronal activity has been observed in the cerebellar cortex after electrical stimulation. Spreading acidification and depression and spreading depression are different in their rate and direction of propagation. While spreading acidification propagates faster and perpendicularly to the parallel fibers, spreading depression propagates radially. Also, unlike spreading depression, spreading acidification does not induce vascular constriction or dilation. Additionally, spreading acidification depends on extracellular Ca^{2+}, while it has been demonstrated that spreading depression can occur in calcium-free-incubated hippocampal tissue.[13,14]

2.4. Intracellular diffusion in the cerebellum

Although much is known about how the structure of dendrites affects their electrophysiological properties, little is known about how the shape and content of neurons affect biochemical signaling. In the cerebellar cortex dendritic trees can be small, such as in the case of granule cells with a few dendrites of about 15 μm in length, to the Purkinje cell with hundreds of microns of dendrites.[15] Purkinje cell dendrites branch, taper, and are covered with dendritic spines, small cytosolic protrusions that are the location of parallel fiber synapses. The dendritic structure of Purkinje cells spans three orders of magnitude, from the dendritic spines (less than a cubic micron) to thick dendrites that are almost 10 μm in diameter. Dendrites contain organelles (endoplasmic reticulum, mitochondria, Golgi apparatus, lysosome, and vesicles), the cytoskeleton (actin fibers and microtubules), and the cytosol. The cytosol of a dendrite is not a homogeneous fluid; instead, it contains a large concentration of macromolecules that could hamper the diffusion of small and large molecules.[16]

Assume that molecules are moving along the intracellular space of a dendrite. These molecules would follow Eq. (8.1) if the dendrites were hollow tubes. This relationship would still hold even in the case when the dendrites branch and taper.[17] Now assume the same basic process of diffusion in which the diffusing particles are trapped in the same place for a random period of time. The source of the trapping could be due to a mesh or macroproteins in the cytosol or due to entering a dendritic spine (Fig. 8.6). If the traps are heterogeneous and randomly distributed over the dendritic volume, then the molecules experience random trapping periods over a wide range of values. This intuitive description results in an ever slowing diffusion process that follows a power law

$$MSD = 2Dt^{\propto} \tag{8.3}$$

where α is between 0 and 1 and is called the anomalous exponent.[18]

We have found anomalous diffusion not only in Purkinje cells but also in hippocampal pyramidal cells caused by spine trapping.[17,19] Molecules moving along dendrites can enter a spine and remain in that small volume for a random period of time that depends on the diffusion of the molecule and the bottle neck generated by the head-to-neck spine radius ratio. Research in

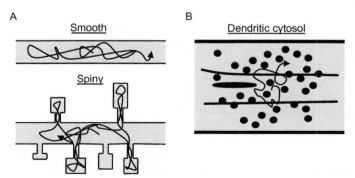

Figure 8.6 Trapping of molecules diffusing inside neurons. (A) Diffusion of a particle in a dendrite without spines (smooth) or with spines (spiny). (B). Trapping of molecules due to collisions with macromolecules or intracellular structures.

other cell types has found that the intricate mesh created by the high density of macro proteins in the cytosol also causes anomalous diffusion.[16,20–22]

Our work has shown that the anomalous exponent is a function of the spine density, with increasing spine density resulting in lower values of α[19] (Figs. 8.7 and 8.8). Since dendritic shape and density is a function of neuronal activity, then the diffusional environment of the dendrite can be modulated by the same processes believed to underlie learning and memory. However, not all molecules undergo anomalous diffusion. We have shown that calcium ions locally released in any part of the dendritic tree are removed from the cytosol in about 120 ms. This short window of time does not allow calcium ions to be trapped by spines and thus calcium does not undergo anomalous diffusion. At the same time, the slow degradation of inositol-1,4,5-triphoshpate (IP$_3$) allows it to linger in the cytosol for several seconds. Consequently, this molecule undergoes anomalous diffusion in spiny dendrites. Calcium ions and IP$_3$ are two molecules essential for the expression of long-term synaptic plasticity.[23]

2.5. Differences between tortuosity and anomalous diffusion

Tortuosity and anomalous diffusion describe two different processes. In general, tortuosity applies when the obstacles can be averaged out. For example, if obstacles are equally spaced, then there is a specific factor that decreases the spread of the diffusing molecule. Even when the obstacles shapes and positions are random, if they can be described by a mean (e.g., average diameter)

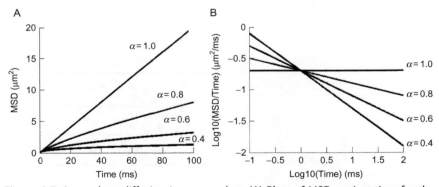

Figure 8.7 Anomalous diffusion is a power law. (A) Plots of MSD against time for the anomalous diffusion equation $MSD = 2Dt^{\alpha}$, where α is the anomalous exponent and is less than 1. (B) Anomalous diffusion is different than tortuosity. Performing a logarithmic transformation shows that the diffusion process is slowing down, thus the diffusion coefficient depends on time (compare to tortuosity).

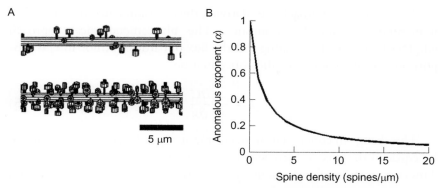

Figure 8.8 The anomalous exponent depends on spine density. (A) Rendering of structural models of Purkinje cell dendrites with different spine densities (1 spine/μm, top; 5 spines/μm, bottom). (B) Plot of the experimentally determined relationship between the anomalous exponent and spine density, $\propto = 2/(2 + 1.6\,s_d)$, with S_d being the spine density. For details, see Ref.9.

and a finite standard deviation, then it is expected that the diffusion will be described with the classical diffusion equation with tortuosity.

Anomalous diffusion emerges when the obstacles are heterogeneous and broadly distributed. For example, if the density or shapes of dendritic spines shows a long-tail distribution (meaning that there are cases of very long spines or very high head-to-neck diameter ratios).[17,19,24] It is in these cases that using an average pore size or spine shape will not model the diffusion process appropriately. A single molecule can undergo anomalous diffusion at some period of time and tortuosity in another.[25] This depends on the lifetime of the diffusing molecule, the structure in which it is moving, its diffusion coefficient, and the time constant of the biochemical mechanism in which it is involved.

3. DIFFUSION: THEORY

3.1. The classical diffusion equation

The physical process of diffusion in the absence of any chemical reactions is described by the diffusion equation:

$$\frac{\partial C(x, t)}{\partial t} = \nabla \cdot (D(x, t)\nabla C(x, t)) \qquad (8.4)$$

where D is the diffusion coefficient and C is the concentration.

There are two approaches to deriving the diffusion equation—the phenomenological and the atomistic.[8] The phenomenological approach employs Fick's first law, stating that the flux of particles J (mol/m^{2}s) is proportional to the gradient of the concentration

$$J = -D\frac{\partial C}{\partial x} \qquad (8.5)$$

The change in concentration over time at a specific point in space has to be equal to the flux of particles exchanged with the neighboring regions. This conservation of mass statement is described in the continuity equation

$$\frac{\partial C}{\partial t} + \nabla \cdot J = 0 \qquad (8.6)$$

Combining Fick's first law with the continuity equation yields

$$\frac{\partial C}{\partial t} = \nabla \cdot (D\nabla C)$$

Another way of deriving the diffusion equation is by introducing the concept of the random walk. Consider a particle moving along the real line starting at $t = 0$ and $x = 0$. At each time step Δt, the particle can make a jump left or right at a distance Δx with equal probability. Then, the probability for the particle to be at position x at time t is

$$p(x, t) = \frac{1}{2}(p(x - \Delta x, t - \Delta t) + p(x + \Delta x, t - \Delta t)) \qquad (8.7)$$

A simple manipulation results in

$$\frac{p(x, t) - p(x, t - \Delta t)}{\Delta t} = \frac{(\Delta x)^2}{2\Delta t} \frac{p(x - \Delta x, t - \Delta t) - 2p(x, t - \Delta t) + p(x + \Delta x, t - \Delta t)}{(\Delta x)^2}$$

$$(8.8)$$

Defining $D \equiv \frac{(\Delta x)^2}{2\Delta t}$, then in the limit as Δx and Δt go to zero then

$$\frac{\partial p(x, t)}{\partial t} = D\frac{\partial^2 p(x, t)}{\partial x^2}$$

Using the same basic principle of random walk, it is possible to derive the MSD relationship shown in Eq. (8.1).

The solution for the diffusion equation assuming a continuous homogeneous media, infinite boundary conditions, and a point source at $t=0$ is given by (in 1D)

$$C(x, t) = \frac{M}{\sqrt{4\pi Dt}} e^{-\frac{x^2}{4Dt}} \tag{8.9}$$

where $M = \int_{-\infty}^{+\infty} C dx$.

Analogously, the diffusion equation in 2D and 3D are described by

$$\frac{\partial C}{\partial t} = D \left(\frac{\partial^2 C}{\partial x^2} + \frac{\partial^2 C}{\partial y^2} \right)$$

$$\frac{\partial C}{\partial t} = D \left(\frac{\partial^2 C}{\partial x^2} + \frac{\partial^2 C}{\partial y^2} + \frac{\partial^2 C}{\partial z^2} \right)$$

and their solutions are

$$C(x, y, t) = \frac{M}{4\pi Dt} e^{-\frac{x^2 + y^2}{4Dt}}$$

$$C(x, y, z, t) = \frac{M}{(4\pi Dt)^{3/2}} e^{-\frac{x^2 + y^2 + z^2}{4Dt}}$$

where $M = \iint_{-\infty}^{+\infty} C dx dy$ and $M = \iiint_{-\infty}^{+\infty} C dx dy dz$, respectively.

The point-source solutions can be employed for modeling both intracellular as well as interstitial diffusion in the case of point-source release. This simplified approach often requires the substitution of the free-diffusion coefficient D with an effective diffusion coefficient D^*. In the case of interstitial diffusion, this is due to the porosity of the medium (see Section 3.2) as well as obstacles in the extracellular matrix. In the case of intracellular diffusion, geometry, molecular crowding, and trapping in microdomains (e.g., dendritic spines) could also require employment of an effective diffusion coefficient. In both cases, whenever the dwell time due to trapping is significant, one can use an alternative formulation of the problem (see Section 3.3).

3.2. Tortuosity

Modeling diffusion in the extracellular space requires modifications of the diffusion equations due to the fact that the brain acts as a porous medium. Such a generalized modification of the diffusion equation is given by

$$\frac{\partial C}{\partial t} = D^*\nabla^2 C + \frac{Q}{\rho} - v \cdot \nabla C - \frac{f(c)}{\rho} \tag{8.10}$$

where $D_x^* = D/\lambda_x^2$ is the effective diffusion coefficient in the brain, D is the free-diffusion coefficient. The second term is a source Q reduced by the volume fraction ρ in order to reflect the reduced available volume for diffusion.[35] The third term is the bulk flow, where v is the velocity vector. The last term represents loss of material from the extracellular space due to permeation in cells or degradation. This term can be nonlinear, which makes the solution of the equation nonstandard.

In the case of anisotropy, the tortuosity becomes a second-order tensor and the diffusion equation can be written as

$$\frac{\partial C}{\partial t} = D_x^*\frac{\partial^2 C}{\partial x^2} + D_y^*\frac{\partial^2 C}{\partial y^2} + D_z^*\frac{\partial^2 C}{\partial z^2} + \frac{Q}{\rho} \tag{8.11}$$

where $D_x^* = D/\lambda_x^2$, $D_y^* = D/\lambda_y^2$, and $D_z^* = D/\lambda_z^2$.

3.3. Anomalous diffusion

Random walks with biased jump probabilities give rise to the so-called anomalous diffusion. In this case, the MSD of the diffusing particles is a power law (Eq. 8.3). The regime with $0 < \alpha < 1$ is called anomalous subdiffusion.[26]

Anomalous diffusion can be described in terms of fractional calculus

$$\frac{\partial^\alpha C}{\partial t^\alpha} = D\frac{\partial^2 C}{\partial x^2} \tag{8.12}$$

where the fractional derivative of the Riemann–Liouville type for $0 < \alpha < 1$ is given by

$$\frac{\partial^\alpha C(t)}{\partial t^\alpha} = \frac{1}{\Gamma(1-\alpha)}\frac{\mathrm{d}}{\mathrm{d}t}\int_0^t \frac{C(\tau)}{(t-\tau)^\alpha}\mathrm{d}\tau \tag{8.13}$$

where Γ is the Gamma function. Using the fractional diffusion equation provides a framework to understand implications of highly correlated phenomena that arise from complex systems. The *memory kernel* $(t-\tau)^{-\alpha}$ indicates how much the present value of the concentration depends on the initial condition.[27] As α decreases, then the previous history of the concentration influences more and more the present result. Thus, α is a measurement of the correlations imposed by the system on the diffusing particles. Analysis of the fractional diffusion equation is done with Fourier and Laplace transformations and requires the use of Fox-H functions.[26]

3.4. Numerical techniques

The analytical solutions given above usually are restricted to simple geometries and constant diffusion coefficients. Thus, for modeling diffusion in more complicated systems, it is often necessary to employ numerical techniques.[28] The two most widely used methods are the finite difference method and its variation—the Crank–Nickolson method. Usually (but not always), the diffusion equation is nondimensionalized first. The convenience of the nondimensionalization is that various solutions for different parameters can be obtained from the nondimensional solution by scaling. For example, in the 1D case one can introduce nondimensional variables:

$$X = \frac{x}{L}; T = \frac{D}{L^2}t; \text{and } c = \frac{C}{C_0}$$

where L is a length typical for the system and C_o is some standard concentration for the problem. Note that if we choose L to be the length of the system, then $0 \leq X \leq 1$. Thus, the nondimensional 1D diffusion equation becomes

$$\frac{\partial c}{\partial T} = \frac{\partial^2 c}{\partial X^2}. \tag{8.14}$$

An alternative approach to solving the classical diffusion equation is the stochastic approach. The motion of the particles of substance A is tracked through a Monte Carlo simulation of particles, each of which is involved in a random walk. Depending on the choice of the jump probabilities of the particles, one can simulate both regular and anomalous diffusion. In the case of unbiased probability, the population of particles released results in position distributions described by Gaussian curves centered at $x = 0$ and $\text{MSD} \sim t$. This is consistent with the solution to the classical point-source macroscopic diffusion equation given above.

Numerical solutions of the fractional diffusion equation are computationally intensive. This is due to the fact that for each spatial point at each time step, we have to integrate the entire history. This is a direct consequence from the existence of the kernel $(t-\tau)^{-\alpha}$ in the definition of the fractional derivative and reflects the non-Markovian nature of the system. As time goes by, the length of the sequences to be integrated increases one term per time step. We have recently released a toolbox that can be used to efficiently integrate fractional derivatives and efficiently solve the fractional diffusion equation.[29]

Anomalous diffusion can also be simulated using Monte Carlo processes by using a random number that determines how long a particle remains in its current position. The probability distribution can have a long-tailed waiting time with an asymptotic behavior ($t^{-(1+\alpha)}$, where $0 < \alpha < 1$; the value of α is the same that then appears in the fractional diffusion equation).

4. MODELING DIFFUSION

Diffusion can be modeled with multiple strategies using stochastic processes and differential equations and with different degrees of freedom (from one to three dimensions). It is important to define the spatial and temporal scales of the problem in order to identify the best computational strategy.

4.1. Choosing between differential and stochastic simulations

Diffusion is one of the biophysical problems in neuroscience that is modeled using two strategies: stochastic and deterministic. Choosing between these two modeling frameworks in which one uses differential equations while the other uses stochastic processes depends on the nature of the problem. The differential equation model is amenable to be applied to compartments as those used in modeling the electrical activity of neurons. The fundamental properties of the classical diffusion equation are identical to those of the cable equation. However, a reality check has to be performed when using differential equations, as the volumes being investigated could be very small and the concentrations could be in the nanomolar range. If the concentrations being modeled are equivalent to fractions of molecules, then that would call into question the results of the simulations, especially if there are biochemical reactions involved.

The stochastic simulation strategy is used when the problem consists of tracking the position of individual molecules or small groups of molecules. In this case, boundary conditions are solved by assuming that particles collide and are reflected. Although highly accurate, this process could become computationally intractable for large sections of dendritic tree being modeled and is not easily integrated into compartmental models of electrical activity.

4.2. Compartmental approximation

In compartmental modeling, the electrical variables (conductances, voltages, and currents) are homogeneous through the compartment. Following this modeling strategy, a single compartment, representing a section of dendrite, is considered to have a homogeneous concentration. Radial diffusion can be incorporated into this framework by assuming a 2D diffusion process. Thus, the volume is divided into concentric shells allowing diffusion between contiguous shells with each shell potentially having mobile or immobile reactants. Diffusion from one compartment to its neighbors has been implemented on shells of the same level across compartments. Thus, each compartment has the same number of shells but not all shells have the same thickness due to variations in the diameter of each compartment.[30]

4.3. Grid and continuous random walk modeling

When modeling diffusion of single molecules, there are two strategies: modeling the displacement of a molecule along a mesh (grid) and modeling the movement as a continuous process in space.

The grid modeling strategy requires defining a space in which the molecule can move. For example, diffusion on two dimensions of a molecule with a diffusion coefficient D_1 and a time step Δt then results in a grid spacing of $\Delta x = \sqrt{4 D_1 \Delta t}$. At every time step of the simulation then the molecule takes a random walk along the X and Y axis. An obstacle is modeled as a position in the grid that is forbidden for the molecules, thus it is reflected to its original position in the previous time step. If the molecule moves to the position of a reactant, then another random number determines their probability of binding and unbinding. Based on the assumptions of the model, the molecule could then diffuse in any direction after unbinding. This is a good strategy for simulations of molecules with identical diffusion coefficients.[31]

A more general method to simulate the diffusion of molecules in space is to do it in a mesh–less framework. In this case, a random number selects the direction of the movement and another random number determines the length of the jump. On average, the jump length will be determined by the diffusion coefficient. With this strategy, it is possible to model multiple molecules diffusing with different diffusion coefficients. Collisions are treated as reflections. This type of simulation can model any structure and could make use of reconstructed electron microscopy images of neuropil.[32,33]

4.4. Tortuosity and anomalous diffusion

For a given volume fraction ρ, it is possible to calculate the tortuosity λ and equivalently the effective diffusion coefficient D^*; however, they will depend on how the extracellular space is modeled. There is a great variety of geometries that could be chosen: stacking elementary unit cells with space between them to account for the extracellular space (Fig. 8.9A), tubular channels inserted in a volume (Fig. 8.9B), octahedral cells, cells with asymmetric two- or three-dimensional smooth surfaces (Fig. 8.9C), impermeable wrapping (representing glial cells) around elementary cubic cells creating trapping microdomains (Fig. 8.9D), etc.[35] Depending on the complexity of the model and the initial and boundary conditions, one can choose between continuous (solving the DE) or the stochastic (Monte Carlo simulation) approaches.[25,34–38]

As explained above, anomalous diffusion can be modeled stochastically (using $MSD \sim t^{\alpha}$) or deterministically (using the fractional diffusion equation). Anomalous diffusion could also make use of the same strategies used to model tortuosity. In this case, the diffusing molecule follows a random walk. The obstacles, either modeling intracellular or extracellular space, then affect the diffusion of the molecules causing anomalous diffusion.[39]

4.5. Modeling tools

Selection of a modeling tool is dependent on the type of modeling strategy required for the scientific question being asked. For Monte Carlo simulations to track single molecules, the most widely used tool is MCell.[33] A more recent tool is STEPS.[39] For diffusion in cylinder-like structures compatible with compartmental models is Neuron.[41]

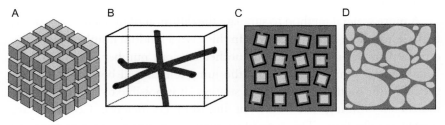

Figure 8.9 Simplified models of tortuosity. (A) Regular cubic stacks. (B) Diffusion only allowed through tubes. (C) Random obstruction. (D) Impermeable wrapping around cells with a small opening. (See the color plate.)

5. SUMMARY

Diffusion is a fundamental process in biology that underlies the propagation of nutrients and biochemical signals intracellularly or through the interstitial space. As a technique, diffusion measurements can be used to determine how difficult it is for a substance to move in the brain. Diffusion can be studied at multiple scales, from single ions to large concentrations of proteins. The modeling of diffusion poses a computational challenge to study neuronal structure at multiple scales. Furthermore, the study of diffusion introduces a theoretical area of study in which tortuosity and anomalous diffusion could coexist and participate in the processing of biochemical signals. Since the obstacles for diffusion are due to the structure of the brain, from molecular crowding to the anisotropy of the neuropil, then there is tremendous potential to study how brain structure in healthy and pathological cases affect the spread of biochemical signals in the nervous system.

ACKNOWLEDGMENTS

NSF IOS-1208029, NSF-EF 1137897, and NSF-HDR 0932339.

REFERENCES

1. D'Angelo E, Mazzarello P, Prestori F, et al. The cerebellar network: from structure to function and dynamics. *Brain Res Rev.* 2011;66(1–2):5–15.
2. Manto M, Bower J, Conforto A, et al. Consensus paper: roles of the cerebellum in motor control—the diversity of ideas on cerebellar involvement in movement. *Cerebellum.* 2012;11(2):457–487.
3. Barlow JS. *The Cerebellum and Adaptive Control.* Cambridge, UK; New York: Cambridge University Press; 2002, xi, 340 p.
4. Gao Z, van Beugen BJ, De Zeeuw CI. Distributed synergistic plasticity and cerebellar learning. *Nat Rev Neurosci.* 2012;13(9):619–635.
5. Schmahmann JD. The cerebellum and cognition. In: *International Review of Neurobiology.* San Diego: Academic Press; 1997, xxxi, 665 p.
6. Strata P, Scelfo B, Sacchetti B. Involvement of cerebellum in emotional behavior. *Physiol Res.* 2011;60(Suppl. 1):S39–S48.
7. Codling EA, Plank MJ, Benhamou S. Random walk models in biology. *J R Soc Interface.* 2008;5(25):813–834.
8. Koch C. Biophysics of computation: information processing in single neurons. In: *Computational Neuroscience.* New York: Oxford University Press; 1999, xxiii, 562 p.
9. Sykova E, Nicholson C. Diffusion in brain extracellular space. *Physiol Rev.* 2008;88(4):1277–1340.
10. Kuimova MK. Mapping viscosity in cells using molecular rotors. *Phys Chem Chem Phys.* 2012;14(37):12671–12686.
11. Ayata C. Spreading depression and neurovascular coupling. *Stroke.* 2013;44(6 Suppl. 1): S87–S89.

12. Case GR, Lavond DG, Thompson RF. Cortical spreading depression and involvement of the motor cortex, auditory cortex, and cerebellum in eyeblink classical conditioning of the rabbit. *Neurobiol Learn Mem.* 2002;78(2):234–245.

13. Vincent M, Hadjikhani N. The cerebellum and migraine. *Headache.* 2007;47(6):820–833.

14. Ebner TJ, Chen G. Spreading acidification and depression in the cerebellar cortex. *Neuroscientist.* 2003;9(1):37–45.

15. Shepherd GM. *The Synaptic Organization of the Brain.* 5th ed. Oxford; New York: Oxford University Press; 2004, xiv, 719 p.

16. Dix JA, Verkman AS. Crowding effects on diffusion in solutions and cells. *Annu Rev Biophys.* 2008;37:247–263.

17. Santamaria F, Wils S, De Schutter E, Augustine GJ. Anomalous diffusion in Purkinje cell dendrites caused by spines. *Neuron.* 2006;52(4):635–648.

18. Ben-Avraham D, Havlin S. *Diffusion and Reactions in Fractals and Disordered Systems.* Cambridge; New York: Cambridge University Press; 2000, xiv, 316 p.

19. Santamaria F, Wils S, De Schutter E, Augustine GJ. The diffusional properties of dendrites depend on the density of dendritic spines. *Eur J Neurosci.* 2011;34(4):561–568.

20. Mika JT, Poolman B. Macromolecule diffusion and confinement in prokaryotic cells. *Curr Opin Biotechnol.* 2010;22(1):117–126.

21. Gal N, Weihs D. Experimental evidence of strong anomalous diffusion in living cells. *Phys Rev E Stat Nonlin Soft Matter Phys.* 2010;81(2 Pt. 1):020903.

22. Guigas G, Weiss M. Sampling the cell with anomalous diffusion—the discovery of slowness. *Biophys J.* 2008;94(1):90–94.

23. Tanaka K, Khiroug L, Santamaria F, et al. Ca2+ requirements for cerebellar longterm synaptic depression: role for a postsynaptic leaky integrator. *Neuron.* 2007;54(5):787–800.

24. Harris KM, Stevens JK. Dendritic spines of rat cerebellar Purkinje cells: serial electron microscopy with reference to their biophysical characteristics. *J Neurosci.* 1988;8(12):4455–4469.

25. Lacks DJ. Tortuosity and anomalous diffusion in the neuromuscular junction. *Phys Rev E Stat Nonlin Soft Matter Phys.* 2008;77(4 Pt. 1):041912.

26. Metzler R, Klafter J. The random walk's guide to anomalous diffusion: a fractional dynamics approach. *Phys Rep.* 2000;339(1):1–77.

27. Heymans N, Podlubny I. Physical interpretation of initial conditions for fractional differential equations with Riemann-Liouville fractional derivatives. *Rheol Acta.* 2006;45(5):765–771.

28. Crank J. *The Mathematics of Diffusion.* Oxford: Clarendon Press; 1975.

29. Marinov T, Ramirez N, Santamaria F. Fractional integration toolbox. *Fract Calc Appl Anal.* 2013;16(3):670–681.

30. De Schutter E, Smolen P. Calcium dynamics in large neuronal models. In: Koch C, Segev I, eds. *Methods in Neuronal Modeling.* Cambridge: MIT press; 1998:211–250.

31. Santamaria F, Gonzalez J, Augustine GJ, Raghavachari S. Quantifying the effects of elastic collisions and non-covalent binding on glutamate receptor trafficking in the postsynaptic density. *PLoS Comput Biol.* 2010;6(5):e1000780.

32. Lopreore CL, Bartol TM, Coggan JS, et al. Computational modeling of threedimensional electrodiffusion in biological systems: application to the node of Ranvier. *Biophys J.* 2008;95(6):2624–2635.

33. Stiles J, Bartol TM. Monte Carlo methods for simulating realistic synaptic microphysiology using MCell. In: De Schutter E, ed. *Computational Neuroscience.* Boca Raton: CRC; 2000:87–127.

34. Hrabetova S, Hrabe J, Nicholson C. Dead-space microdomains hinder extracellular diffusion in rat neocortex during ischemia. *J Neurosci.* 2003;23(23):8351–8359.

35. Sykova E, Nickolson C. Diffusion in brain extracellular space. *Phisiol Rev.* 2008;88(4):1277–1340.
36. Sykova E, Vargova L. Extrasynaptic transmission and the diffusion parameters of the extracellular space. *Neurochem Int.* 2008;52(1–2):5–13.
37. Saftenku EE. Modeling of slow glutamate diffusion and AMPA receptor activation in the cerebellar glomerulus. *J Theor Biol.* 2005;234(3):363–382.
38. Saftenku EE. Determinants of slowed diffusion in the complex space of the cerebellar glomerulus. *Neurophysiology.* 2004;36(5–6):371–384.
39. Nicolau Jr DV, Hancock JF, Burrage K. Sources of anomalous diffusion on cell membranes: a Monte Carlo study. *Biophys J.* 2007;92(6):1975–1987.
40. Wils S, De Schutter E. STEPS: modeling and simulating complex reaction-diffusion systems with python. *Front Neuroinformatics.* 2009;3:15.
41. Hines ML, Carnevale NT. NEURON: a tool for neuroscientists. *Neuroscientist.* 2001;7(2):123–135.

CHAPTER NINE

Astrocyte–Neuron Interactions: From Experimental Research-Based Models to Translational Medicine

Marja-Leena Linne*, Tuula O. Jalonen†
*Computational Neuroscience Group, Department of Signal Processing, Tampere University of Technology, Tampere, Finland
†Department of Physiology and Neuroscience, St. George's University, School of Medicine, Grenada, West Indies

Contents

Abstract

In this chapter, we review the principal astrocyte functions and the interactions between neurons and astrocytes. We then address how the experimentally observed functions have been verified in computational models and review recent experimental literature on astrocyte–neuron interactions. Benefits of computational neuroscience work are highlighted through selected studies with neurons and astrocytes by analyzing the existing models qualitatively and assessing the relevance of these models to experimental data. Common strategies to mathematical modeling and computer simulation in neuroscience are summarized for the nontechnical reader. The astrocyte–neuron interactions are then further illustrated by examples of some neurological and neurodegenerative diseases, where the miscommunication between glia and neurons is found to be increasingly important.

Progress in Molecular Biology and Translational Science, Volume 123
ISSN 1877-1173
http://dx.doi.org/10.1016/B978-0-12-397897-4.00005-X

1. INTRODUCTION

The human brain is a complex structure containing multiple cell types, each of which to some extent can control and regulate the other cells' functions. Glial cells (microglia, oligodendrocytes, and especially astrocytes) play a critical role in the central nervous system (CNS) by affecting in various, yet specific, ways the neuronal single cell level interactions as well as connectivity and communication at the network level, both in the developing and mature brain.[1–4] Numerous studies (see, e.g., Refs. 5,6) show an important modulatory role of astrocytes, in general, in brain homeostasis but most specifically in neuronal metabolism, plasticity, and survival. Information processing locally in synapses and in extrasynaptic areas of the brain is strongly regulated by astrocytes (see, e.g., Ref. 7). One future challenge will be to combine the knowledge of the roles of astrocytic and neuronal networks with the information processing in human cognition and behavior in both health and disease (see Ref. 8). This will involve integration of the most relevant computational models of neuronal networks with models of astrocyte networks.

The astrocyte is the most numerous type of glia in the mammalian brain, and its number in the brain is greater than that of neurons.[9,10] Human astrocytes have been shown to form a structurally and functionally heterogeneous group.[11–13] For a long time, it was believed that the only role of all glial cells in brain functions is to provide structural support for the neurons (see a review by Nag[14]). Later, astroglial cells were named as housekeeping cells with somewhat more active supportive roles (such as buffering and siphoning of potassium after excessive firing of neuronal action potentials). Regulation of brain interstitial potassium concentration is especially important after the enhanced potassium outflow from neurons during epileptic seizures.[15]

Conditions like epilepsy also cause an increased need for energy, and energy metabolism (lactate production) is one of the major functions of astrocytes.[16] Among the many functions of astrocytes are: control of CNS blood circulation, extracellular ion homeostasis, and release of energy substrates, growth factors, and transmitters for neurons. All these functions make the astrocytes able to actively modulate the dynamics of neurons by regulating and organizing local or distant (extrasynaptic) synaptic activity, excitability, transmission, and plasticity at the cellular and system levels.[17–21]

During CNS development, astrocytes influence neurogenesis and synapse formation (reviewed in Refs. 9,22). In the adult brain, astrocytes continue to transfer information via flow of molecules through their gap junctions and by releasing gliotransmitters (D-serine, ATP, GABA, and glutamate) and other neuroactive substances.[23] Neurons respond to these signals from astrocytes, and astrocytes in turn respond to neuronal activity via activation of their own ion channels, neurotransmitter receptors, and transporters especially in their various types of processes and endfeet residing in the close proximity of neurons.[24–26]

Astrocytes are known to play an important role in many neurological disorders and neurodegenerative diseases. Recent studies have indicated the link between astrocytes and many of these neurological disorders, such as migraine, epilepsy, stroke, inflammatory diseases, Alzheimer's disease (AD), Parkinson's disease, and amyotrophic lateral sclerosis (ALS), through pathways that include inflammation, oxidative stress, cell signaling, necrosis, and apoptosis.[27,28] Astrocytes have also been shown to be cells regulating respiration[29] and motor nerves.[30] A better understanding of the cell–cell interactions will be useful in view of the development of new therapeutic approaches.[31] In addition to the direct neuron–astrocyte–neuron communication, astrocytes are essential in neurovascular coupling (together with endothelial cells, pericytes, and even microglia) by release of vasoactive substances such as nitric oxide and metabolites of arachidonic acid.[32–34] Under conditions such as hemorrhage, ischaemia, and tumors in the brain, the complexity of glial–neuronal–vasculature interactions increases the challenges set for the treatments to be provided.[35,36] Furthermore, astrocyte–oligodendrocyte–neuron interactions are becoming more important in searching for causes and treatments for many demyelinating CNS diseases.[37]

Astrocytes, in contrast to neurons, do not fire action potentials, due to the lack of sufficient number of voltage-sensitive Na^+ channels.[38] Nonetheless, experimental research in the last two decades has revealed that astrocytes can respond to neurotransmitters with Ca^{2+} elevations to generate feedback signals, which modulate synaptic transmission and excitability in neurons. During tripartite synaptic neurotransmission, astrocytes have the capacity to regulate both the neuronal pre- and postsynaptic compartments through the Ca^{2+}-dependent release of gliotransmitters.[39] The excitation of neurons caused by glutamate can be modulated by inhibition via glutamate uptake-induced release of GABA from astrocytes, as has been shown in the pyramidal neurons in the hippocampus.[40] This provides an additional modulatory mechanism on neural network activity and dynamics, making the tripartite synaptic neurotransmission more favorable over the bipartite synaptic transmission typically used in earlier biophysical models of neural networks.[41]

Increasing evidence from both experimental and computational studies points out how astrocytes enhance the computational power of neuronal networks and information-processing capabilities of the brain in previously unexpected ways.[42–45] In addition to direct effects by balancing the microenvironment in their immediate local networks, astrocytes most probably participate in higher cognitive functions related to learning and memory (see, e.g., Refs. 7,46). Based on our current understanding, astrocytes can no longer be considered as purely passive support cells in the brain. Instead, astrocytes should be considered as active processing elements, which communicate closely with neurons.

Despite the growing awareness of the importance of astrocytes, the exact molecular and cellular level mechanisms underlying astrocyte–neuron communication, and the consequences of these interactions on local network level dynamics, still remain largely unclear. Computational models describing neuronal–glial function can be used to obtain insights into the distinct contributions by different mechanisms, such as signal transduction pathways, membrane-bound receptors and ion channels, and release of neuro- and gliotransmitters (see, e.g., Refs. 1,47–51). Models can also help to compare the healthy and unhealthy functioning of cells and networks of cells in the brain. Computational models facilitate the analysis of the responses of neural systems to various stimuli and conditions that are otherwise difficult to obtain experimentally, in particular, the responses at the subcellular level. Though there is too little information about astrocytes' interactions with the neurons in the human brain to enable building detailed mathematical models, we here summarize the present state of the research from the modeler's point of view and also seek to bring some crucial points to the attention of the nonmathematically proficient reader.

2. MODELS AND SIMULATIONS OF CELLS AND NETWORKS

There are various strategies to choose when creating mathematical models of biological molecular level phenomena, tissue- and organ-specific cellular interactions, and especially the complex cell–cell signaling found in the cellular networks of the brain. According to Gerstner et al.,[52] modeling in neuroscience can be classified by using two different criteria: (i) the complexity of the model and (ii) the direction of workflow (see also Fig. 9.1).

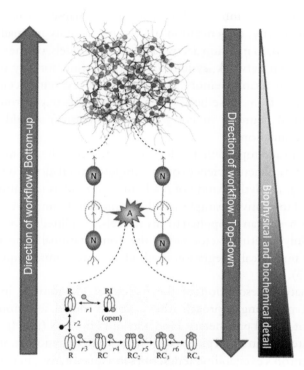

Figure 9.1 The effects of astrocyte–neuron interactions can be studied at different levels of organizational detail in the brain (here, molecular, cellular, and network levels are illustrated). At any level, the model can be either simplified conceptual or detailed biophysical, depending on scientific question and computational resources. Due to computational limitations, the higher level models (e.g., network) are typically described using simplified conceptual models with few details or mechanisms captured from the molecular level. The direction of workflow is different depending on whether astrocyte–neuron interactions are studied from "bottom-up" or "top-down." A denotes astrocyte, N neuron, R activated receptor, RC a closed state of a receptor, RI inactivated state of a receptor, and r reaction rate constant. (See the color plate.)

Complexity of a model can range from simplified conceptual model to detailed biophysical one. Direction of workflow in modeling can be from microscopic to macroscopic scales (bottom–up) or from behavioral and system functions to properties of components (top–down).[52] The so-called bottom–up models integrate information on a lower level (e.g., properties of ion channels, transmembrane receptors, and signaling pathways) to explain phenomena observed on a higher level (e.g., generation of action

potentials in neurons, functionality of tripartite synapse, formation of anatomical connectivity, or generation of rhythmic activity in small-scale networks of neurons, to mention a few). Top-down models, on the other hand,
start with known network, system, or cognitive functionality observed in
the brain (e.g., working memory, associative memory, reinforcement learning) and extract from these how system components (e.g., groups of neurons, individual neurons, or components of neurons) should behave to
achieve these functions.

Bottom-up and top-down models can be studied either by mathematical
theory (theoretical neuroscience) or by computer simulation (computational
neuroscience). Analytical study of mathematical models (equations) is usually restricted to relatively simple, small-size models. Computer simulations,
on the other hand, can be applied to any models, simplified as well as detailed
ones, the only limiting factor being the computational time to run the
models (for numerical integration methods to approximate equations, see,
e.g., Ref. 53).

Mathematical description of the biological phenomena involves the
selection of modeling approach (the so-called model selection problem)
and fitting of model parameters. Hodgkin and Huxley's (H–H) description
of action potentials in squid giant axon[54] is an electrophysiologically accurate
model of neuron's excitability. It describes in a quantitative, detailed manner
the ionic currents along the axon as the action potential propagates. An
approach originally presented by Hodgkin and Huxley, therefore, provides
a basis for today's standard simulator software to model neuronal excitability
(for discussion of the suitability of the approach in comparison to other
approaches to model neural functions, see, e.g., Refs. 52,53,55,56). In
H–H-type models, each patch (or compartment) of cell membrane is
described by its composition of ion channels and ion channel–receptor complexes. Many H–H-type models provide beautiful examples of the model
selection problem where the model is formulated based on experimental data
and the parameters of the model are estimated based on the data. Because
H–H-type models are computationally expensive due to their complexity,
several other approaches to model the excitability of a neuron have been
developed. The FitzHugh–Nagumo neuron oscillator model,[57] as well as
the biophysical Morris–Lecar neuron model,[58] is simplified when compared
to the H–H model, but still complex enough to be computationally intensive.
These models also require much parameter tuning. Some of the less computationally expensive neuron models are the different types of integrate-and-
fire-neuron models (IF neuron models; see, e.g., Ref. 59) as well as the

Izhikevich neuron model.[60] These are often more feasible for simulations of large neuronal networks, if not as biophysically detailed as are the H–H-type models. For description of different modeling approaches in neuroscience and review of simulation tools and software built based on these modeling approaches, see, for example, Sterratt et al.[53] and Brette et al.[55]

Part of the model selection problem is the question whether to use a deterministic or stochastic simulation approach. In neuronal modeling, the function of the neuron's components, including the cell membrane and voltage-dependent ion channels, is typically described using deterministic ordinary differential equations that always provide the same model output when repeating computer simulations with fixed model parameter values. The selection of simulation method heavily depends on the scientific questions that are sought to be asked with the model. In the less complex modeled systems, a deterministic model may be enough to accurately describe the phenomenon under study when only the phenomenological behavior of the system is studied instead of detailed mechanistic understanding of the phenomenon.

Noise plays an essential role in all biochemical and biophysical systems. It derives from many sources, especially from reaction events (i.e., protein conformation changes) that are discrete and occur at random times. Since stochasticity (fluctuations, noise) is an inherent part of all biological processes, it is often not enough to use deterministic approaches to describe these processes, and a stochastic model then more accurately describes the functionality of the system. It is well known, for example, that the behavior of cells (neurons), as well as the behavior of cellular structures, from lipid bilayer membranes, down to the protein level of voltage-dependent ion channels, is stochastic in nature.[61] Stochastic modeling approaches, based on probabilistically describing the transition rates of ion channels, ion channel–receptor complexes, and other entities in the cell membranes, have therefore gained a lot of interest due to their ability to produce more accurate results than the deterministic approaches. These Markov chain type of models are, however, relatively time consuming to simulate, and with bottom-up, detailed modeling other stochastic approaches may be more reasonable to use for computational reasons. Besides including in the model the noise introduced by neurons, it has recently been suggested that astrocytes could also be regarded as a possible internal source of noise in brain networks.[62] This opens up interesting new venues for neural modeling.

In our previous studies, we have developed new modeling and simulation approaches that take into account the stochasticity inherently present in the function of ion channels, receptors, and in the interactions between signal transduction molecules, in a computationally meaningful manner (see, e.g., Refs. 63,64). We have applied stochastic differential equations (SDEs; an SDE is a differential equation in which one or more of the terms of the mathematical equation are stochastic processes) and Brownian motion, which are also commonly used in the aerospace industry and in economics, to neural modeling. We have, for example, successfully applied SDEs to model the stochastic function of ion channels,[65] spontaneous activity in neurons,[64] and the variability of the neurotransmitter-induced astrocyte calcium signals.[49] Furthermore, we have also created models for the stochasticity of neuronal intracellular signal transduction molecules and signaling networks, when small numbers of molecules are involved.[63,66]

Noise might be the factor that keeps the system (e.g., a cell and a network of cells) spontaneously in a state of preparedness of action and tests the ability to function properly. As shown, for example, in microglia, astrocytes, human embryonic kidney cell, and processed lipoaspirate cells, cells may use a stochastic mechanism to generate local or global Ca^{2+} spikes.[67,68] These intracellular Ca^{2+} spikes occur both randomly and following specific stimuli (as seen in astrocytes; Refs. 69,70), and the variability of these glutamate- or serotonin-induced calcium signals in astrocytes can be more accurately mimicked by stochastic approaches[49] (see also Fig. 9.2). We have also shown that experimental data of IP3 receptor functioning (open and closed states and the open probability) can accurately be described using stochastic models,[71] the Gillespie stochastic simulation algorithm[72,73] being sufficient in describing the statistics of data. It may therefore be important to model the stochastic mechanisms present in both neuronal and glial cells to understand the transitions between different dynamic states of cell functions—in both healthy and dysfunctional cells. Noise can, for example, make the system more robust to external perturbations or might lead to a specific noise-induced bistability with oscillations.[74] In neurons, specifically in very small surface areas, subthreshold oscillations caused by stochastic behavior of ion channels may be important in determining the reliability and accuracy of action potential timing, as well as in coincidence detection and multiplication of inputs.

3. COMPUTATIONAL MODELS OF ASTROCYTE–NEURON INTERACTIONS IN INFORMATION PROCESSING

Experimental and computational evidence is increasingly supporting the active participation of glial cells, especially astrocytes, in information

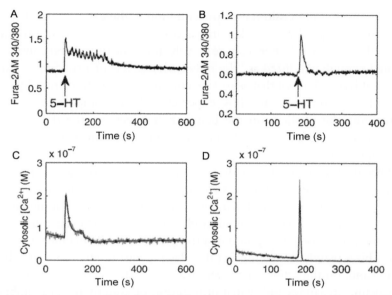

Figure 9.2 Generation of simulated graphs from *in vitro* experiments. Intracellular calcium signals were obtained in cultured rat cortical astrocytes following additions of serotonin (10 μM in A, 1 μM in B) and corresponding simulations using stochastic computational model (C, D). The model takes into account the major mechanisms known to be involved in calcium signaling in astrocytes, such as Ca^{2+} leak from/to extracellular matrix, capacitive Ca^{2+} entry from extracellular matrix, Ca^{2+} entry via ionotropic receptors, Ca^{2+} leak from intracellular stores, storage of Ca^{2+} to endoplasmic reticulum (ER) via sarco(endo)plasmic Ca^{2+} ATPase pumps, and Ca^{2+} release from ER mediated by inositol 1,4,5-trisphosphate. The model can further be used to explore additional transmitter-induced effects in astrocytes. *Figure is modified from Ref. 49.*

processing, and in structural and synaptic plasticity in the brain (for review, see, e.g., Ref. 7). The details of these proposed roles for astrocytes are still only partially known and can be debated (see, e.g., Refs. 75–78). It is evident that astrocytes can modulate short–term plasticity and, potentially, be integral partners in the induction and maintenance of long-term potentiation (LTP) and depression (LTD) at least in the developing brain. So far, however, mostly neuronal functions have been incorporated in computational models for both short- and long-term plasticity (see, e.g., extensive reviews in Refs. 79,80). Though more information is at present becoming available on astroglial calcium sources, gliotransmission, calcium signaling routes, and modulation of the synaptic structures,[39] this knowledge has not as yet been fully used in models of brain networks and associated simulations. Thus far, only regulation of potassium ion concentration in the extracellular space by astrocytes has typically been taken into account in neural models, as research data of the potassium buffering have already, for some time, been considered detailed enough and listed as "the only major effects of astrocytes." However, as it is now known that astrocytes communicate with neurons by multiple other ways and are thus expected to have important roles in maintaining the plasticity in the brain, their effects on the development of neural network structure and function should in the future be more carefully addressed.

The effects of astrocytes are diverse, and calcium signaling is increasingly accepted as the major factor in astrocyte–neuron interactions, as well as in the regulation of the multicellular neurovascular system in both the healthy and diseased brain.[81] Most previous studies do not model "excitability" (i.e., Ca^{2+} oscillations or Ca^{2+} spikes) in astrocytes in a similar manner as in neurons. This is mainly due to the fact that detailed quantitative data have been lacking and, on the other hand, there has not existed a complete picture of the calcium dynamics and spatial distribution of associated cell membrane ion channels in astrocytes. No concrete hypothesis on astroglia's effects on neuronal network function in the various areas in the brain has been proposed, other than the hypothesis about modulatory effects on the tripartite synapse.[82] Additionally, there is even less information about the regulation of astrocyte activity by neurons (but see Ref. 83). The astrocyte–neuron interactions are complex and may drastically vary in different brain structures. For example, Pirttimaki and Parri[84] saw no generation of NMDA receptor-mediated slow inward currents (SICs) in thalamocortical neurons following afferent stimulation and astrocytic intracellular Ca^{2+} elevations. This fact of variability in different regions in the brain together with computational limitations has most probably prevented serious large-scale modeling work on astrocyte–neuron interactions.

Below, we review recent attempts to model astrocyte–neuron interactions. Most of the models represent the bottom-up, detailed modeling approach, where the system behavior is built from microscopic components to produce the whole. The aim of these studies is to, step-by-step, show the regulatory effects of astrocytes on the dynamics of neural activity, first on a local scale in the brain (tripartite synapse and associated local areas) and, ultimately, on information processing on a larger scale in the brain.

Computational models of astrocyte–neuron interactions involve various types of models ranging from models of single tripartite synapse to models containing hundreds of neurons and their interactions with astrocytes. Most models are built according to the bottom-up approach and range from detailed and complex biophysical descriptions to simplified ones. The very early models of the tripartite synapse include the models by Nadkarni and Jung[47,48] and Volman et al.[85] The Nadkarni–Jung models make use of the Pinsky–Rinzel two-compartmental, reduced model for hippocampal pyramidal neurons[86] and the Li–Rinzel model for astrocytic Ca^{2+} dynamics.[87] The biophysical Li–Rinzel Ca^{2+} dynamics model describes Ca^{2+} dynamics in the cytosol of astrocytes through Ca^{2+} release from and uptake into the endoplasmic reticulum, mediated by the IP3 concentration, and it binds together synaptic neurotransmitter release and astrocyte Ca^{2+} dynamics. Calcium changes in the model occur in response to glutamate release from the presynaptic terminal of a neuron and the presynaptic feedback of the glutamate released by the astrocyte. These early models, although relatively simple, predicted that for a large production rate of IP3 and over-expression of metabotropic glutamate receptors in response to neuronal firing, the neuron can be switched to an oscillatory state through long-lasting stimulation, possibly mimicking an epileptic state.

The astrocyte–neuron model by Wade et al.[50] contains 20 mathematical equations for tripartite synapse function. The model employs two biophysically motivated models to describe the interactions between astrocytes and neurons in a tripartite synapse: the Volman gatekeeper model[85] and the Nadkarni–Jung model,[47,48] as well as the astrocyte-driven extrasynaptic NMDAR-mediated neuronal currents. Both Volman and Nadkarni–Jung models use the Li–Rinzel calcium dynamics model[87] as the basis of model building. The Volman gatekeeper model is a phenomenological model describing the dynamics of the astrocytic IP3 concentration, which is dependent on the internal dynamics of the astrocyte, as well as the concentration of neurotransmitter in the synaptic cleft. The model of Wade et al.[50] exhibits both amplitude and frequency modulation encoding of IP3 concentration, depending on how the model parameters are tuned. The resulting dynamics can also be a mixture of these two. The authors show that with this model, the extrasynaptic SIC originating from astrocytic glutamate release causes synchronized postsynaptic activity also in distant neurons. A further study from the same authors proposed a simple computational model for self-repair based on bidirectional interactions between astrocytes and neurons.[51]

The study of Amiri et al.[42] used a biologically inspired neuronal network model, which is constructed by connecting two Morris–Lecar neuron models (Morris–Lecar models are simplified descriptions of neuronal activity; see, e.g., Ref. 53), and a generalized and simplified mathematical model for the dynamics of the intracellular Ca^{2+} waves produced by astrocytes.[88] The study showed that an astrocyte increases the threshold value of synchronization (of neurons) and provides appropriate feedback control in regulating the neural activities. It is, therefore, suggested that astrocytes could have a key role in stabilizing neural activity. In other words, an astrocyte has the potential to desynchronize the communication between two coupled neurons, and, in turn, a malfunctioning astrocyte could initiate the hypersynchronous firing of neurons. Continuation studies from the same authors further demonstrate that astrocytes are able to change the threshold value of transition from synchronous to asynchronous behavior among neurons.[43,44] Results from the models are validated using recordings of population spikes in the hippocampus[44] and further studied using phase-plane analysis.[45]

The study of Reato et al.[89] used Izhikevich model of both excitatory and inhibitory neurons to construct a network of cells (Ref. 60; see also Ref. 53; note that the Izhikevich model neuron expresses the dynamics of excitability but lacks detailed biological mechanisms). The experimentally observed Ca^{2+} change in astrocytes, in response to neuronal activity, was modeled with linear equations. Computer simulations show that astrocytes can contribute to focal seizure-like ictal discharge (ID) generation by directly affecting the excitatory/inhibitory balance of the neuronal network. The threshold of ID generation was lowered when an excitatory feedback-loop between astrocytes and neurons was included. Simulation results replicate experimental results in a slice preparation of focal ID in entorhinal cortex. Additionally, a novel and efficient computational implementation of a spiking neuron–astrocyte network is reported in which neurons are modeled according to the Izhikevich formulation and the astrocyte–neuron interactions as tripartite synapses with nonlinear transistor-like models.[90]

Allam et al.[91] have presented a detailed glutamatergic synaptic model involving glutamate diffusion inside the synaptic cleft, neuronal AMPA, NMDA, and mGlu receptors, as well as glutamate uptake mediated by glial and neuronal transporters. Paired pulse stimulation results reveal that the effects of astrocytic glutamate uptake are more apparent when the input interspike interval is sufficiently long to allow the receptors to recover from desensitization. Results suggest an important functional role of astrocytes in spike timing-dependent processes and call for further studies of the molecular basis of certain neurological diseases specifically related to alterations in astrocytic glutamate uptake, such as epilepsy.

In addition to astrocyte–neuron interactions, a couple of important studies modeled the astrocyte–astrocyte interactions to understand the coupling strength between astrocytes and the capacity of astrocytes to spread Ca^{2+} waves across astrocyte networks (see, e.g., Refs. 92,93). Recently, Chander and Chakravarthy[1] have presented one of the first attempts to build a computational, biophysically detailed model of neuro–glio–vascular loop interaction.

4. TOWARD UNDERSTANDING BRAIN DISORDERS USING COMPUTATIONAL MODELS INVOLVING ASTROCYTE–NEURON INTERACTIONS

Recent studies indicate that astrocytes are involved in the genesis of increasing numbers of neurological and neurodegenerative disorders during the human lifespan. Some of these diseases are caused purely by dysfunctional astrocytes and are called astropathies (such as Alexander's disease and hepatic encephalopathy), some in their turn (epilepsy, AD) are caused by miscommunication between neurons and astrocytes, or even between neurons, astrocytes, other glia, and endothelial cells.[4,94] Astrocytes are suggested to play a critical role in the genetic pathologies in neurodevelopmental diseases such as Rett syndrome and fragile X mental retardation, as well as idiopathic (genetic) epilepsy.[10,95] Neurological and psychiatric disorders, such as various mood disorders and schizophrenia appearing in adolescence, may be partly brought on by developmental disruption of brain circuitry caused by functional deficiencies in astrocytes.

When the brain circuitry develops, spatial and temporal interaction patterns develop between the various cell types of the brain. The structure and efficacy of these multicellular networks depend on correct cell-to-cell signaling with release and uptake of trophic factors, transmitters, and ions. Intercellular signaling, formation of syncytia, and synchronization of activity via gap junctions are especially important in regulating local and distant neuronal and glial activity.[96,97] Calcium, potassium, chloride, and small metabolites pass through the gap junctions and have been shown to modulate neuronal synaptic transmission. Interestingly, it is not only calcium waves that are able to modify neuronal transmitter releases, as it has been shown that Cl^- conductance within the astrocytic network may contribute to maintaining neuronal GABAergic synaptic transmission by regulating extracellular chloride.[98]

Calcium, as a ubiquitous intra- and intercellular signaling molecule creating calcium waves through astrocyte networks, has been the target and main interest when searching for therapy of the so far untreatable neurological and neurodegenerative diseases. Both spontaneous and stimulated astrocyte calcium waves have been shown to interact with neuronal function. Mathematical models have predicted nonlinear and saltatory waves and regenerative amplification with different behavior of calcium oscillations in different locations inside the cells.[99,100] Mathematical models of hippocampus have proposed that astrocytes modulate hippocampal LTP and actively transfer and store synaptic information.[41,101–103] Despite the results presented in these models, it is as yet not fully clarified how essential the calcium-mediated astrocyte–neuron communication in, for example, hippocampus is for the overall spontaneous or evoked synaptic transmission.[75] Usually, glial–neuronal interactions are local, dynamic, and quickly terminated, but interesting long-term enhancement of astrocytic glutamate release, dependent on astrocytic intracellular calcium, has been shown in thalamocortical neurons. This interaction of nonsynaptic neuronal and astrocytic calcium, ionotropic NMDAR, and metabotropic glutamate receptors generates local neuronal firing for a long time after the original stimulus has been terminated.[104] Involvement of astrocytes creates interesting possibilities for therapies for patients with disabilities in learning and memory formation. Furthermore, it has been suggested that neuronal activity in hippocampus alters calcium ion dynamics in astrocytes and reflects then back to the formation of functional tripartite synapses.[105]

The interplay between neurons and glia, specifically due to the dynamic changes of cellular calcium and the release/uptake of various neurotransmitters (e.g., the glutamate/glutamine cycle) and activation of their respective receptors and transporters, is currently studied in connection to various neurological and neurodegenerative conditions.[106–108] In addition to genetic regulation, all diseases of the nervous system are also sensitive to short-term and long-term changes in the immediate cellular environment, as well as to changes caused by distant modulators carried by blood flow. In some diseases, disruption of the blood–brain barrier (BBB), because of errors in astrocyte–endothelial cell signaling, imbalances the brain interstitial environment and increases the vulnerability of neuronal networks for disease onset. New methodology is necessary for detailed characterization of the damage caused to the multiple cell types of the brain by rupture of the BBB.[109] Leakage through the BBB and more serious injuries caused by ruptured blood vessels (hemorrhages) cause swelling (edema) and reactive astrocytosis in the brain. Reactive astrocytosis can be seen as changes in astrocyte morphology and increase in the amount of glial fibrillary acidic protein (GFAP, an intermediate filament), as well as in imbalance in water and ion homeostasis (especially of potassium and calcium ions). Increased GFAP is detected in astrocytoma cells, as well as in dysfunctional astrocytes in Alexander's disease.[110] In addition to the increase in GFAP, reactive astrocytes often lose their stellate morphology and become hypertrophic with no extended processes or endfeet, and lack of gap junctions between the cells. These cells thus clearly lose their normal gap junction connectivity and are not able to function in potassium (or calcium or chloride) buffering and siphoning. This immediately disrupts the network structure normally seen in the brain.

Though most of the known neurodegenerative and neurological diseases were originally thought to be solely neuronal diseases, it is now clear that they contain a strong glial component. Demyelinating diseases (Multiple Sclerosis disease in CNS and peripheral myopathies in PNS), which were earlier mainly seen as lack of myelin secretion by dysfunctional oligodendrocytes or Schwann cells, now have astrocytes and neurons increasingly named as coculprits in the demyelination progress. In ALS, degeneration of cortical and spinal motoneurons is accompanied by impaired astrocytic clearance of extracellular glutamate and release of neurotrophic factors.[111,112] Below, we list some additional neurological and neurodegenerative diseases, which have been shown to be at least in part caused by imbalance in astroglial–neuronal signaling, specifically highlighting the role of calcium in the outset and progress of the disease.

4.1. Alzheimer's disease

AD is a slowly progressing neurodegenerative disease characterized by reduced brain volume, atrophy of the cholinergic basal forebrain, presence of neurons containing hyperphosphorylated tau protein tangles, and formation of amyloid–beta plaques, often surrounded by reactive astrocytes.[113] Simultaneously with these general morphological changes in the brain and neurons, clear changes are also detected in astrocyte morphology (hypertrophy) with increase in GFAP content, increase in chloride and potassium channel activity, and fluctuations in intracellular calcium concentrations.[114] AD can be either sporadic or familial, with large number of mutations in genes for amyloid precursor or presenilin 1 and 2 proteins. As the disease originates from different genes and different mutations in them, it can be expected that there is also a variety of changes to be detected at the cellular level. The main focus is, however, on the negative effects of accumulation of amyloid-beta protein fragments in the brain. Still, because of the heterogeneity of the astrocytes in different brain regions, and even inside the same local network, only a subset of astrocytes might show amyloid-beta-induced elevated intracellular calcium levels and thus show varying degrees of changes in the regulatory capabilities for environmental challenges. Amyloid-beta-treated astrocytes also respond with enhanced transient increase in intracellular calcium (release from intracellular stores, such as endoplasmic reticulum, and influx through L-type calcium channels) when exposed to neurotransmitters (such as glutamate or serotonin).[49] As both astrocytes and neurons are sensitive to any imbalance in the fine-tuned transmitter-mediated information transfer, the deregulated calcium homeostasis and, especially in neuronal synapses, the ryanodine-receptor-mediated calcium upregulation slowly destroys the vitally important balance in glia–neuron communication throughout the brain networks.[115]

Due to the loss of memory formation seen in patients with AD, special interest lays in the communication between astrocytes and neurons in hippocampus and related brain structures, and in the way that propagation of calcium waves in astrocyte syncytia is changed during advancing brain degeneration. In the cortex, amyloid-beta plaque formation both silences a fraction of neurons and renders other neurons hyperactive. Interestingly, hippocampal neuronal hyperactivity might prove to be a very early sign of developing AD.[116] Computational modeling of altered cellular mechanisms would further advance our understanding of the induction and progression of AD. As an example, Anastasio[117] presents a computational model of the contribution of estrogen to amyloid-beta regulation. The model illustrates how the various effects of estrogen could work together to reduce amyloid-beta levels, or prevent them from rising, in the presence of pathological triggers. As many other brain endogenous factors affect the release and accumulation of amyloid-beta fragments (such as secretase enzymes), it is necessary to collect and combine the information so far gathered from the multitude of involved factors and use it for an enhanced model to be used in therapeutic interventions and drug development.

4.2. Epilepsy

Epilepsy is a neurological disorder characterized by recurrent seizures that can arise focally at restricted areas and/or propagate throughout the whole brain. The seizures show a variety of manifestations, from absence and tonic–clonic (grand mal) seizures to status epilepticus.[118] Early on, epilepsy was seen as a disease of neurons, arising from injury–related excessive action potential firing. Increased depolarization in neighboring brain areas due to increased potassium ion release from neurons would then widen the damaged brain area. Recently, several possible mechanisms have been indicated in the process of epilepsy, and the role of astrocytes in the regulation of seizure activity has been discussed and modeled (for a review of models, see Ref. 119). Besides regulation of extracellular potassium concentrations, modification of calcium signaling pathways, calcium and sodium channels, and gap junction activity have been targeted as treatment possibilities.[118,120] Methods for tuning down the excitatory transmission (glutamate release or receptor activation) and enhancing the inhibition of neuronal networks via GABA have been used in drug development. Disturbances of glutamine metabolism and/or transport contribute to changes in glutamatergic or GABAergic transmission associated with different pathological conditions of the brain.[106] Drugs developed toward controlling the neurotransmission by diseased reactive glia, especially astrocytes and microglia, will provide new lines of therapy to treat the hyperexcitability and inflammation detected in epilepsy, especially in those patients whose seizures do not respond to existing medications.[121] The heterogeneity of astrocytes and their emerging association with immune and inflammatory systems will be a further challenge in drug development.

The seriously prolonged seizure activity, status epilepticus, is a life-threatening condition, causing acidosis, delayed neuronal death, and, later, cognitive decline. Unremitting epileptiform activity leads to elevated astrocytic Ca^{2+} signaling and release of the excitotoxic transmitter glutamate. It has been found that attenuation of these glial Ca^{2+} signals by selective NMDAR antagonists was neuroprotective.[122] To balance and stabilize the astrocytic calcium signaling and propagation of calcium waves may be one useful therapeutic target, though it is still unclear if intervention with astrocyte calcium waves would prove to be detrimental or beneficial for neuronal networks in general.

Computational modeling provides a useful tool to cost-effectively explore various mechanisms possibly responsible for epilepsy. As an example, propagation of epileptic-like seizures and the effects of excitatory and inhibitory drives can already be easily explored in localized networks by computational means. A computational study of Reato et al.[89] showed that astrocytes can contribute to focal seizure-like ID generation by directly affecting the excitatory/inhibitory balance of the neuronal network. Hall and Kuhlmann[123] recently coupled excitatory and inhibitory single neuron models with approximated cerebral cortex connectivity to explore alternative mechanisms responsible for seizure-like activity propagation and compared the obtained simulated results with animal-slice-based models. Recent theoretical, systematic work on how the structure of a network affects its functionality will provide further tools for understanding structure–dynamics relationships and activity propagation in the brain.[124]

4.3. Pain and migraine

Interactions between peripheral impulse input and CNS pain mechanisms, facilitation and inhibition, all define the intensity of experiencing pain. Pain can be either acute or chronic with different causes and mechanisms, and therefore open for varying treatments. Chronic pain involves more CNS factors and can be treated by neuromodulatory agents. Acute and peripheral pain usually responds to anti-inflammatory drugs and opioids. Neuropathic pain and migraine (categorized as a pain disorder) have recently been classified as being caused by dysfunctional ion channels and thus belong to neurological channelopathies caused by mutations in one or more ion channel genes (similar to epilepsy, myotonia, or ataxia). These mutations can be present in either neurons or glia. Of the many types of calcium channels, mutated CaV2.1 channels (voltage-activated P/Q-type Ca^{2+} channels) have been shown to activate at more hyperpolarizing potentials than the wild-type channels and lead to a gain-of-function in synaptic transmission, which might alter the excitatory/inhibitory balance of synaptic transmission. Because of this, the cortical neurons would be in a persistent state of hyperexcitability causing increased release of glutamate and increased spreading depression. This suggested mechanism might initiate the attacks of migraine with aura, as well as various other forms of pain. Drugs that are at present commonly used for limiting hyperexcitability of the nervous system include, besides calcium channel regulators, also sodium channel blocking agents, NMDAR antagonists, and opioids.[125–128] For some time, it has been questioned whether astrocytes surrounding the neuronal synapses together with other glia are involved in regulating pain responses at the level of the spinal cord or cortex.[129] Activation of spinal cord astrocytes has been shown to contribute to nerve injury, inflammation-induced persistent pain, and chronic opioid-induced antinociceptive tolerance.[130] As astrocytes are a multitasking cell type, and calcium is a major player in intracellular and intercellular signaling, it would be expected that astrocytic calcium waves would participate in coordinating the neuronal excitability also in various types of pain syndromes.

There has so far been only little interest in creating computational models of migraine or any of the pain syndromes. In one study by DaSilva et al.[131] with human subjects with chronic migraine, transcranial direct current stimulation was directed through brain regions associated with pain perception and modulation. The study showed an improvement in the stimulated group over the sham treated. The computational model predicted electric current flow not only in the main areas associated with the migraine pathophysiology (primary motor cortex) but also in a number of other areas (insula, cingulate cortex, thalamus, and brainstem regions). It would be interesting to follow up this model by adding details of the functional and structural differences found in each of the affected brain areas and build a model predicting the most sensitive areas for onset (and treatment) of pain or migraine, as well as which of the cell types, astrocytes or neurons, are most prominent in originating the pathological phenomena. Furthermore, the differential roles for specific transmitters and the underlying calcium metabolism (together with changes of concentrations of other ions) would be an informative addition to models of human behavior in various nervous system diseases.

5. CONCLUSIONS

There is little doubt that astrocytes are dynamic players in the brain. Experimental research in the last three decades has revealed that astrocytes can respond to neurotransmitters with Ca^{2+} elevations and generate feedback signals to neurons in order to modulate synaptic transmission, neuronal excitability, and to an extent also plasticity. These results have changed our basic understanding of brain function and provided various new perspectives for how astrocytes can participate in information processing in the healthy brain, and also when the nervous system is faced with acute injury or chronic disease. Recent computational and theoretical studies support and clarify the mechanisms of the complex exchange of information between the astroglia and neurons. All of the currently known mechanisms by which astroglia contribute to synaptic information processing, and additionally mediate other types of chemical (humoral) information exchange in the brain, have not as yet been fully modeled. More detailed knowledge about the biophysical and biochemical mechanisms will in future dictate how the effects of astroglia can be incorporated in modeling neural activities.

The current trend is to use multiscale models to explain dynamical changes and mechanisms contributing to changes in network activities. Multiscale models will include fine details of all physical organizational levels (molecular networks, cells, local neural networks, long-range networks, systems). The key mechanisms are described in minute detail and others with less detail. Methodology to create multiscale models is currently under intensive development. Equally important is the development of validation strategies for comparison of various computational models (for more detailed discussion, see, Ref. 80). Better mathematical and computational tools have to be developed in order to provide easy and user-friendly evaluation and comparison. As can be seen from the presentation of previous modeling attempts, many new models are built on top of previous models, with some further parameter tuning based on experimental data. Often the validation against existing similar models is too tedious to do and, consequently, is not done. To increase the usability of models, future computational neuroscience research should pay more attention to the question of validation—both to the models and methods to analyze the models.

The involvement of astroglia in the development and plasticity in the brain is yet to be unveiled with a tight integration of multiscale computational models and experimental observations. A deeper understanding of altered astrocyte–neuron interactions is important in order to develop therapeutic strategies that can be used to modify the activity of glial cells, reduce their possible neurotoxic effects, and enhance their neuroprotective action. Furthermore, the computational models must account not only for the activities of astrocytes but also for the relationships astrocytes and astrocytic signaling networks have with other glial cells, the neurovascular unit and BBB, and, finally, the large number of different neuronal types in different CNS regions. Additional models should be developed to clarify how the peripheral nervous system and CNS interact and participate in encoding sensory and other types of information.

We are only beginning to have an idea about the diversity, connectivity, and function of astrocytic networks and how they can participate in the dynamical behavior of brain networks. It is important to study the single astrocyte and its local interactions with nearby neuronal cells, as this provides us with mechanistic understanding of the basis of the relationships astrocytes have with neuronal cells. Studies should also take into account the developmental variability, heterogeneity, of astrocytes and other glial cells (see, e.g., the role of microglia on developing retinal networks[3]). In addition to the single cell level, the effects of astrocytes should also be explored in a larger scale, as astrocytes are known to distribute ions and small molecules via their gap junctions, and this siphoning of material will permit the regulation of larger network areas in the vicinity of active astrocytes. The effects of the close communication of astrocytes and neurons in networks will increase the understanding of the information processing, learning, and plasticity in the brain, and in the nervous system as a whole. Combining data from experimental research with mathematical models and computer simulations enhances our knowledge of the exciting world of neural networks.

ACKNOWLEDGMENT

The support from the Foundation of Tampere University of Technology for M.-L. L. is acknowledged.

REFERENCES

1. Chander BS, Chakravarthy VS. A computational model of neuro-glio-vascular loop interactions. *PLoS One.* 2012;7(11):e48802.
2. Clarke LE, Barres BA. Emerging roles of astrocytes in neural circuit development. *Nat Rev Neurosci.* 2013;14(5):311–321.
3. Schafer DP, Lehrman EK, Kautzman AG, et al. Microglia sculpt postnatal neural circuits in an activity and complement-dependent manner. *Neuron.* 2012;74(4):691–705.
4. Verkhratsky A, Parpura V. Recent advances in (patho)physiology of astroglia. *Acta Pharmacol Sin.* 2010;31(9):1044–1054.
5. Hertz L, Xu J, Song D, Yan E, Gu L, Peng L. Astrocytic and neuronal accumulation of elevated extracellular K(+) with a 2/3 K(+)/Na(+) flux ratio-consequences for energy metabolism, osmolarity and higher brain function. *Front Comput Neurosci.* 2013;7:114.
6. Rose CR, Karus C. Two sides of the same coin: sodium homeostasis and signaling in astrocytes under physiological and pathophysiological conditions. *Glia.* 2013;61(8):1191–1205.
7. Min R, Santello M, Nevian T. The computational power of astrocyte mediated synaptic plasticity. *Front Comput Neurosci.* 2012;6:93.
8. Halassa MM, Haydon PG. Integrated brain circuits: astrocytic networks modulate neuronal activity and behavior. *Annu Rev Physiol.* 2010;72:335–355.
9. Kanski R, van Strien ME, van Tijn P, Hol EM. A star is born: new insights into the mechanism of astrogenesis. *Cell Mol Life Sci.* 2014;71(3):433–447.
10. Molofsky AV, Krencik R, Ullian EM, et al. Astrocytes and disease: a neurodevelopmental perspective. *Genes Dev.* 2012;26(9):891–907.
11. Chaboub LS, Deneen B. Developmental origins of astrocyte heterogeneity: the final frontier of CNS development. *Dev Neurosci.* 2012;34(5):379–388.
12. Oberheim NA, Takano T, Han X, et al. Uniquely hominid features of adult human astrocytes. *J Neurosci.* 2009;29(10):3276–3287.
13. Ransom BR, Ransom CB. Astrocytes: multitalented stars of the central nervous system. *Methods Mol Biol.* 2012;814:3–7.

14. Nag S. Morphology and properties of astrocytes. *Methods Mol Biol.* 2011;686:69–100.
15. Steinhäuser C, Seifert G, Bedner P. Astrocyte dysfunction in temporal lobe epilepsy: K^+ channels and gap junction coupling. *Glia.* 2012;60(8):1192–1202.
16. Barros LF. Metabolic signaling by lactate in the brain. *Trends Neurosci.* 2013;36(7):396–404.
17. Arcuino G, Lin JHC, Takano T, et al. Intercellular calcium signaling mediated by point-source release of ATP. *Proc Natl Acad Sci USA.* 2002;99:9840–9845.
18. Gordleeva SY, Stasenko SV, Semyanov AV, Dityatev AE, Kazantsev VB. Bi-directional astrocytic regulation of neuronal activity within a network. *Front Comput Neurosci.* 2012;6:92.
19. Parpura V, Haydon P. Physiological astrocytic calcium levels stimulate glutamate release to modulate adjacent neurons. *Proc Natl Acad Sci USA.* 2000;97:8629–8634.
20. Pérez-Alvarez A, Araque A. Astrocyte-neuron interaction at tripartite synapses. *Curr Drug Targets.* 2013;14(11):1220–1224.
21. Porter JT, McCarthy KD. Hippocampal astrocytes in situ respond to glutamate released from synaptic terminals. *J Neurosci.* 1996;16:5073–5081.
22. Barker AJ, Ullian EM. New roles for astrocytes in developing synaptic circuits. *Commun Integr.* 2008;1(2):207–211.
23. Van Horn MR, Sild M, Ruthazer ES. D-serine as a gliotransmitter and its roles in brain development and disease. *Front Cell Neurosci.* 2013;7:39.
24. Lalo U, Pankratov Y, Parpura V, Verkhratsky A. Ionotropic receptors in neuronal-astroglial signalling: what is the role of "excitable" molecules in non-excitable cells. *Biochim Biophys Acta.* 2011;1813(5):992–1002.
25. Stipursky J, Spohr TC, Sousa VO, Gomes FC. Neuron-astroglial interactions in cell-fate commitment and maturation in the central nervous system. *Neurochem Res.* 2012;37(11):2402–2418.
26. Theodosis DT, Poulain DA, Oliet SH. Activity-dependent structural and functional plasticity of astrocyte-neuron interactions. *Physiol Rev.* 2008;88(3):983–1008.
27. Agulhon C, Sun MY, Murphy T, Myers T, Lauderdale K, Fiacco TA. Calcium signaling and gliotransmission in normal vs. reactive astrocytes. *Front Pharmacol.* 2012;3:139.
28. Rossi D, Martorana F, Brambilla L. Implications of gliotransmission for the pharmacotherapy of CNS disorders. *CNS Drugs.* 2011;25(8):641–658.
29. Caravagna C, Soliz J, Seaborn T. Brain-derived neurotrophic factor interacts with astrocytes and neurons to control respiration. *Eur J Neurosci.* 2013;38(9):3261–3269.
30. Christensen RK, Petersen AV, Perrier JF. How do glial cells contribute to motor control? *Curr Pharm Des.* 2013;19(24):4385–4399.
31. Takahashi N, Sakurai T. Roles of glial cells in schizophrenia: possible targets for therapeutic approaches. *Neurobiol Dis.* 2013;53:49–60.
32. Ghosh S, Basu A. Calcium signaling in cerebral vasoregulation. *Adv Exp Med Biol.* 2012;740:833–858.
33. Jakovcevic D, Harder DR. Role of astrocytes in matching blood flow to neuronal activity. *Curr Top Dev Biol.* 2007;79:75–97.
34. Koehler RC, Roman RJ, Harder DR. Astrocytes and the regulation of cerebral blood flow. *Trends Neurosci.* 2009;32(3):160–169.
35. Stobart JL, Anderson CM. Multifunctional role of astrocytes as gatekeepers of neuronal energy supply. *Front Cell Neurosci.* 2013;7:38.
36. Stobart JL, Lu L, Anderson HD, Mori H, Anderson CM. Astrocyte-induced cortical vasodilation is mediated by D-serine and endothelial nitric oxide synthase. *Proc Natl Acad Sci USA.* 2013;110(8):3149–3154.
37. Amaral AI, Meisingset TW, Kotter MR, Sonnewald U. Metabolic aspects of neuron-oligodendrocyte-astrocyte interactions. *Front Endocrinol.* 2013;4:54.

38. Bordeay A, Sontheimer H. Electrophysiological properties of human astrocytic tumor cells in situ: enigma of spriking glial cells. *J Neurophysiol*. 1998;79:2782–2793.
39. Zorec R, Araque A, Carmignoto G, Haydon PG, Verkhratsky A, Parpura V. Astroglial excitability and gliotransmission: an appraisal of Ca^{2+} as a signalling route. *ASN Neuro*. 2012;4(2):e00080.
40. Héja L, Nyitrai G, Kékesi O, et al. Astrocytes convert network excitation to tonic inhibition of neurons. *BMC Biol*. 2012;10:26.
41. Perea G, Navarrete M, Araque A. Tripartite synapses: astrocytes process and control synaptic information. *Trends Neurosci*. 2009;32(8):421–431.
42. Amiri M, Montaseri G, Bahrami F. On the role of astrocytes in synchronization of two coupled neurons: a mathematical perspective. *Biol Cybern*. 2011;105:153–166.
43. Amiri M, Bahrami F, Janahmadi M. Functional contributions of astrocytes in synchronization of a neuronal network model. *J Theor Biol*. 2012;292:60–70.
44. Amiri M, Hosseinmardi N, Bahrami F, Janahmadi M. Astrocyte-neuron interaction as a mechanism responsible for generation of neural synchrony: a study based on modeling and experiments. *J Comput Neurosci*. 2013;34(3):489–504.
45. Amiri M, Montaseri G, Bahrami F. A phase plane analysis of neuron-astrocyte interactions. *Neural Netw*. 2013;44:157–165.
46. Han X, Chen M, Wang F, et al. Forebrain engraftment by human glial progenitor cells enhances synaptic plasticity and learning in adult mice. *Cell Stem Cell*. 2013;12(3):342–353.
47. Nadkarni S, Jung P. Dressed neurons: modeling neural-glial interactions. *Phys Biol*. 2004;1:35–41.
48. Nadkarni S, Jung P. Modeling synaptic transmission of the tripartite synapse. *Phys Biol*. 2007;4:1–9.
49. Toivari E, Manninen T, Nahata KK, Jalonen TO, Linne M-L. Effects of transmitters and amyloid-beta peptide on calcium signals in rat cortical astrocytes: fura-2AM measurements and stochastic model simulations. *PLoS One*. 2011;6(3):e17914.
50. Wade J, McDaid L, Harkin J, Crunelli V, Kelso S. Bidirectional coupling between astrocytes and neurons mediates learning and dynamic coordination in the brain: a multiple modeling approach. *PLoS One*. 2011;6(12):e29445.
51. Wade J, McDaid L, Harkin J, Crunelli V, Kelso S. Self-repair in a bidirectionally coupled astrocyte-neuron (AN) system based on retrograde signaling. *Front Comp Neurosci*. 2012;6:76.
52. Gerstner W, Sprekeler H, Deco G. Theory and simulation in neuroscience. *Science*. 2012;338:60–65.
53. Sterratt D, Graham B, Gillies A, Willshaw D. *Principles of Computational Modeling in Neuroscience*. Cambridge, UK: Cambridge University Press; 2011:390.
54. Hodgkin AL, Huxley AF. A quantitative description of membrane current and its application to conduction and excitation in nerve. *J Physiol*. 1952;117(4):500–544.
55. Brette R, Rudolph M, Carnevale T, et al. Simulation of networks of spiking neurons: a review of tools and strategies. *J Comput Neurosci*. 2007;23(3):349–398.
56. De Schutter E. Why are computational neuroscience and systems biology so separate? *PLoS Comput Biol*. 2008;4(5):e1000078.
57. FitzHugh R. Mathematical models of threshold phenomena in the nerve membrane. *Bull Math Biophys*. 1955;17:257–278.
58. Morris C, Lecar H. Voltage oscillations in the barnacle giant muscle fiber. *Biophys J*. 1981;35(1):193–213.
59. Brunel N, van Rossum MC. Lapicque's 1907 paper: from frogs to integrate-and-fire. *Biol Cybern*. 2007;97:337–339.
60. Izhikevich EM. Simple model of spiking neurons. *IEEE Trans Neural Netw*. 2003;14:1569–1572.

61. Huang Y, Rüdiger S, Shuai J. Channel-based Langevin approach for the stochastic Hodgkin-Huxley neuron. *Phys Rev E.* 2013;87(1):012716.
62. Liu Y, Li C. Stochastic resonance in feedforward-loop neuronal network motifs in astrocyte field. *J Theor Biol.* 2013;335:265–275.
63. Manninen M, Linne M-L, Ruohonen K. Developing Itô stochastic differential equation models for neuronal signal transduction pathways. *Comput Biol Chem.* 2006;30:280–291.
64. Saarinen A, Linne M-L, Yli-Harja O. Stochastic differential equation model for cerebellar granule cell excitability. *PLoS Comput Biol.* 2008;4(2):e1000004.
65. Saarinen A, Linne M-L, Yli-Harja O. Modeling single neuron behavior using stochastic differential equations. *Neurocomputing.* 2006;69:1091–1096.
66. Intosalmi J, Manninen T, Ruohonen K, Linne M-L. Computational study of noise in a large signal transduction network. *BMC Bioinforma.* 2011;12:252.
67. Skupin A, Kettenmann H, Winkler U, et al. How does intracellular Ca^{2+} oscillate: by chance or by the clock? *Biophys J.* 2008;94(6):2404–2411.
68. Skupin A, Kettenmann H, Falcke M. Calcium signals driven by single channel noise. *PLoS Comput Biol.* 2010;6(8).
69. Jalonen TO, Margraf RR, Wielt DB, Charniga CJ, Linne M-L, Kimelberg HK. Serotonin induces inward potassium and calcium currents in rat cortical astrocytes. *Brain Res.* 1997;758(1–2):69–82.
70. Kimelberg HK, Cai Z, Rastogi P, et al. Transmitter-induced calcium responses differ in astrocytes acutely isolated from rat brain and in culture. *J Neurochem.* 1997;68(3):1088–1098.
71. Hituri K, Linne M-L. Comparison of models for IP3 receptor kinetics using stochastic simulations. *PLoS One.* 2013;8(4).
72. Gillespie DT. A general method for numerical simulating the stochastic time evolution of coupled chemical reactions. *J Comp Phys.* 1976;22:403–434.
73. Gillespie DT. Exact stochastic simulation of coupled chemical reactions. *J Phys Chem.* 1977;81(25):2340–2361.
74. Samoilov M, Plyasunov S, Arkin AP. Stochastic amplification and signaling in enzymatic futile cycles through noise-induced bistability with oscillations. *Proc Natl Acad Sci USA.* 2005;102(7):2310–2315.
75. Agulhon C, Fiacco TA, McCarthy KD. Hippocampal short- and long-term plasticity are not modulated by astrocyte Ca2+ signaling. *Science.* 2010;327:1250–1254.
76. Min R, Nevian T. Astrocyte signaling controls spike timing-dependent depression at neocortical synapses. *Nat Neurosci.* 2012;15:746–753.
77. Navarrete M, Perea G, Fernandez de Sevilla D, et al. Astrocytes mediate in vivo cholinergic-induced synaptic plasticity. *PLoS Biol.* 2012;10(2):e1001259.
78. Rodriguez-Moreno A, Gonzalez-Rueda A, Banerjee A, Upton AL, Craig MT, Paulsen O. Presynaptic self-depression at developing neocortical synapses. *Neuron.* 2013;77(1):35–42.
79. Graupner M, Brunel N. Mechanisms of induction and maintenance of spike-timing dependent plasticity in biophysical synapse models. *Front Comput Neurosci.* 2010;4:136.
80. Manninen T, Hituri K, Hellgren-Kotaleski J, Blackwell KT, Linne M-L. Postsynaptic signal transduction models for long-term potentiation and depression. *Front Comput Neurosci.* 2010;4:152.
81. Attwell D, Buchan AM, Charpak S, Lauritzen M, MacVicar BA, Newman EA. Glial and neuronal control of brain blood flow. *Nature.* 2010;468(7321):232–243.
82. Araque A, Parpura V, Sanzgiri RP, Haydon PG. Tripartite synapses: glia, the unacknowledged partner. *Trends Neurosci.* 1999;22(5):208–215.
83. Habas A, Hahn J, Wang X, Margeta M. Neuronal activity regulates astrocytic Nrf2 signaling. *Proc Natl Acad Sci USA.* 2013;110:18291–18296.

84. Pirttimaki TM, Parri HR. Glutamatergic input-output properties of thalamic astro-
    cytes. *Neuroscience*. 2012;205:18–28.
85. Volman V, Ben-Jakob E, Levine H. The astrocyte as gatekeeper of synaptic information
    transfer. *Neural Comput*. 2007;19:303–326.
86. Pinsky PF, Rinzel J. Intrinsic and network rhytmogenisis in a reduced traub model for
    CA3 neurons. *J Comp Neurosci*. 1994;1:39–60.
87. Rinzel J, Li YX. Equations of IP3 receptor-mediated Ca^{2+} oscillations derived from a
    detailed kinetic model: a Hodgkin and Huxley like formalism. *J Theor Biol*.
    1994;166:461–473.
88. Postnov DE, Koreshkov RN, Brazhe NA, Brazhe AR, Sosnovtseva OV. Dynamical
    patterns of calcium signaling in a functional model of neuron-astrocyte networks.
    J Biol Phys. 2009;35:425–445.
89. Reato M, Cammarota M, Parra LC, Carmignoto G. Computational model of neuron-
    astrocyte interactions during focal seizure generation. *Front Comput Neurosci*. 2012;6:81.
90. Valenza G, Tedesco L, Lanata A, De Rossi D, Scilingo EP. Novel spiking neuron-
    astrocyte networks based on nonlinear transistor-like models of tripartite synapses. *Conf
    Proc IEEE Eng Med Biol Soc*. 2013:6559–6562.
91. Allam SL, Ghaderi VS, Bouteiller J-MC, et al. A computational model to investigate
    astrocytic glutamate uptake influence on synaptic transmission. *Front Comput Neurosci*.
    2012;6:70.
92. Ceelen PW, Lockridge A, Newman EA. Electrical coupling between glial cells in the
    rat retina. *Glia*. 2001;35(1):1–13.
93. Goldberg M, De Pitta M, Volman V, Berry B, Ben-Jacob E. Nonlinear gap junctions
    enable long-distance propagation of pulsating calcium waves in astrocyte networks.
    PLoS Comput Biol. 2010;6(8):e1000909.
94. Krencik R, Ullian EM. A cellular star atlas: using astrocytes from human pluripotent
    stem cells for disease studies. *Front Cell Neurosci*. 2013;7:25.
95. Leung HT, Ring H. Epilepsy in four genetically determined syndromes of intellectual
    disability. *J Intellect Disabil Res*. 2013;57(1):3–20.
96. Escartin C, Rouach N. Astroglial networking contributes to neurometabolic coupling.
    Front Neuroenergetics. 2013;5:4.
97. Giaume C, Leybaert L, Naus CC, Sáez JC. Connexin and pannexin hemichannels in
    brain glial cells: properties, pharmacology, and roles. *Front Pharmacol*. 2013;4:88.
98. Egawa K, Yamada J, Furukawa T, Yanagawa Y, Fukuda A. Cl⁻ homeodynamics in gap
    junction coupled astrocytic networks on activation of GABAergic synapses. *J Physiol*.
    2013;591(Pt. 16):3901–3917.
99. Lavrentovich M, Hemkin S. A mathematical model of spontaneous calcium(II) oscil-
    lations in astrocytes. *J Theor Biol*. 2008;251(4):553–560.
100. Roth BJ, Yagodin SV, Holtzclaw L, Russell JT. A mathematical model of agonist-
    induced propagation of calcium waves in astrocytes. *Cell Calcium*. 1995;17(1):53–64.
101. Perea G, Araque A. Astrocytes potentiate transmitter release at single hippocampal syn-
    apses. *Science*. 2007;317(5841):1083–1086.
102. Tewari SG, Majumdar KK. A mathematical model of the tripartite synapse: astrocyte-
    induced synaptic plasticity. *J Biol Phys*. 2012;38(3):465–496.
103. Tewari S, Majumdar K. A mathematical model for astrocytes mediated LTP at single
    hippocampal synapses. *J Comput Neurosci*. 2012;33(2):341–370.
104. Pirttimaki TM, Hall SD, Parri HR. Sustained neuronal activity generated by glial plas-
    ticity. *J Neurosci*. 2011;31(21):7637–7647.
105. Tanaka M, Shih PY, Gomi H, et al. Astrocytic Ca2+ signals are required for the func-
    tional integrity of tripartite synapses. *Mol Brain*. 2013;6:6.
106. Albrecht J, Sidoryk-Węgrzynowicz M, Zielińska M, Aschner M. Roles of glutamine in
    neurotransmission. *Neuron Glia Biol*. 2010;6(4):263–276.

107. Parpura V, Verkhratsky A. Astrocytes revisited: concise historic outlook on glutamate homeostasis and signaling. *Croat Med J.* 2012;53(6):518–528.
108. Verkhratsky A, Rodríguez JJ, Parpura V. Astroglia in neurological diseases. *Future Neurol.* 2013;8(2):149–158.
109. Xue Q, Liu Y, Qi H, et al. A novel brain neurovascular unit model with neurons, astrocytes and microvascular endothelial cells of rat. *Int J Biol Sci.* 2013;9(2):174–189.
110. Sosunov AA, Guilfoyle E, Wu X, McKhann 2nd GM, Goldman JE. Phenotypic conversions of "protoplasmic" to "reactive" astrocytes in Alexander disease. *J Neurosci.* 2013;33(17):7439–7450.
111. Lasiene J, Yamanaka K. Glial cells in amyotrophic lateral sclerosis. *Neurol Res Int.* 2011;2011:718987.
112. Quinlan KA. Links between electrophysiological and molecular pathology of amyotrophic lateral sclerosis. *Integr Comp Biol.* 2011;51(6):913–925.
113. Grothe M, Heinsen H, Teipel S. Longitudinal measures of cholinergic forebrain atrophy in the transition from healthy aging to Alzheimer's disease. *Neurobiol Aging.* 2013;34(4):1210–1220.
114. Jalonen TO, Charniga CJ, Wielt DB. Beta-amyloid peptide-induced morphological changes coincide with increased K^+ and Cl^- channel activity in rat cortical astrocytes. *Brain Res.* 1997;746(1–2):85–97.
115. Chakroborty S, Briggs C, Miller MB, et al. Stabilizing ER Ca^{2+} channel function as an early preventative strategy for Alzheimer's disease. *PLoS One.* 2012;7(12):e52056.
116. Busche MA, Chen X, Henning HA, et al. Critical role of soluble amyloid-β for early hippocampal hyperactivity in a mouse model of Alzheimer's disease. *Proc Natl Acad Sci USA.* 2012;109(22):8740–8745.
117. Anastasio TJ. Exploring the contribution of estrogen to amyloid-beta regulation: a novel multifactorial computational modeling approach. *Front Pharmacol.* 2013;4:16.
118. Seifert G, Steinhäuser C. Neuron-astrocyte signaling and epilepsy. *Exp Neurol.* 2013;244:4–10.
119. Volman V, Bazhenov M, Sejnowski TJ. Divide and conquer: functional segregation of synaptic inputs by astrocytic microdomains could alleviate paroxysmal activity following brain trauma. *PLoS Comput Biol.* 2013;9(1):e1002856.
120. Bedner P, Steinhäuser C. Altered Kir and gap junction channels in temporal lobe epilepsy. *Neurochem Int.* 2013;63(7):682–687.
121. Devinsky O, Vezzani A, Najjar S, De Lanerolle NC, Rogawski MA. Glia and epilepsy: excitability and inflammation. *Trends Neurosci.* 2013;36(3):174–184.
122. Ding S, Fellin T, Zhu Y, et al. Enhanced astrocytic Ca2 + signals contribute to neuronal excitotoxicity after status epilepticus. *J Neurosci.* 2007;27(40):10674–10684.
123. Hall D, Kuhlmann L. Mechanisms of seizure propagation in 2-dimensional centre-surround recurrent networks. *PLoS One.* 2013;8(8):e71369.
124. Mäki-Marttunen T, Acimovic J, Ruohonen K, Linne M-L. Structure-dynamics relationships in bursting neuronal networks revealed using a prediction framework. *PLoS One.* 2013;8(7):e69373.
125. Ren K. Emerging role of astroglia in pain hypersensitivity. *Jpn Dent Sci Rev.* 2010;46(1):86.
126. Tuchmann M, Barrett JA, Donevan S, Hedberg TG, Taylor CP. Central sensitization and Ca(V)α2δ ligands in chronic pain syndromes: pathologic processes and pharmacologic effect. *J Pain.* 2010;11(12):1241–1249.
127. Uchitel OD, Inchauspe CG, Urbano FJ, Di Guilmi MN. CaV2.1 voltage activated calcium channels and synaptic transmission in familial hemiplegic migraine pathogenesis. *J Physiol Paris.* 2012;106(1–2):12–22.
128. Vranken JH. Elucidation of pathophysiology and treatment of neuropathic pain. *Cent Nerv Syst Agents Med Chem.* 2012;12(4):304–314.

129. Ji RR, Berta T, Nedergaard M. Glia and pain: is chronic pain a gliopathy? *Pain.* 2013;154:S10–S28.
130. Berta T, Xu ZZ, Ji RR. Tissue plasminogen activator contributes to morphine tolerance and induces mechanical allodynia via astrocytic IL-1β and ERK signaling in the spinal cord of mice. *Neuroscience.* 2013;247:376–385.
131. DaSilva AF, Mendonca ME, Zaghi S, et al. tDCS-induced analgesia and electrical fields in pain-related neural networks in chronic migraine. *Headache.* 2012;52:1283–1295.

CHAPTER TEN

Dynamic Metabolic Control of an Ion Channel

Bertil Hille[*], **Eamonn Dickson**[*], **Martin Kruse**[*], **Bjoern Falkenburger**[†]

[*]Department of Physiology and Biophysics, University of Washington, Seattle, Washington, USA
[†]Department of Neurology, RWTH Aachen University, Aachen, Germany

Contents

Abstract

G-protein-coupled receptors mediate responses to external stimuli in various cell types. We are interested in the modulation of KCNQ2/3 potassium channels by the G_q-coupled M_1 muscarinic (acetylcholine) receptor (M_1R). Here, we describe development of a mathematical model that incorporates all known steps along the M_1R signaling cascade and accurately reproduces the macroscopic behavior we observe when KCNQ2/3 currents are inhibited following M_1R activation. G_q protein-coupled receptors of the plasma membrane activate phospholipase C (PLC) which cleaves the minor plasma membrane lipid phosphatidylinositol 4,5-bisphosphate (PI(4,5)P_2) into the second messengers diacylglycerol and inositol 1,4,5-trisphosphate, leading to calcium release, protein kinase C (PKC) activation, and PI(4,5)P_2 depletion. Combining optical and electrical techniques with knowledge of relative abundance of each signaling component has allowed us to develop a kinetic model and determine that (i) M_1R activation and M_1R/Gβ interaction are fast; (ii) Gα_q/Gβ separation and Gα_q/PLC interaction have intermediate time constants; (iii) the amount of activated PLC limits the rate of KCNQ2/3 suppression; (iv) weak PLC activation can elicit robust calcium signals without net PI(4,5)P_2 depletion or KCNQ2/3 channel inhibition; and (v) depletion of PI(4,5)P_2, and not calcium/CaM or PKC-mediated phosphorylation, closes KCNQ2/3 potassium channels, thereby increasing neuronal excitability.

This chapter concerns receptor modulation of potassium channels of the KCNQ family. By modulation, we mean changes of the properties of ion channels due to receptor-initiated biochemical signaling within the cell. Modulation is typically much slower than the very rapid action of membrane voltage on channel voltage sensors or of extracellular neurotransmitters on the fast ligand-gated receptor channels of synapses. Ongoing modulatory signals continually alter the firing decisions and the input–output relations of neurons and nerve circuits. They underlie major changes of mental state and are the target of most of the major drugs of psychiatry.

The classical work of Hodgkin and Huxley[1] introduced neurobiologists to highly deterministic computation of stereotyped electrical responses of neurons from kinetic models of ion-channel gating. With the recognition of ion-channel modulation, it became clear that channel gating does not have fixed stable properties. Instead the Hodgkin–Huxley equations must be continuously changing to reflect the dynamic impact from slower modulatory signals. Thus, full simulations of neuronal excitability would need to incorporate the kinetics of modulatory signals that act through receptors on intracellular enzymes and metabolism. Neurobiologists frequently call the underlying receptors, metabotropic receptors. Most of them are G-protein coupled. We describe the history and background of receptor-mediated ion-channel modulation and then outline our work toward a fuller mechanistic and kinetic analysis of one example.

1. BACKGROUND AND HISTORY OF CHANNEL MODULATION AND KCNQ CURRENT

Since the late 1800s, physiologists have noted that hormones and plant extracts change the heart rate, contraction of the gut and arteries, and secretion from glands. During the 1960s and 1970s, it was recognized that cell membranes have ion channels and G-protein-coupled receptors and that many physiological actions of hormones and neurotransmitters are accounted for by modulation of the channels via receptor activation. Probably, the pioneering example was augmentation of cardiac calcium currents by beta-adrenergic stimulation of the heart.[2] Eventually, that augmentation was explained by cyclic AMP-dependent phosphorylation of the calcium channel itself.[3] Table 10.1 shows additional, very early examples of ion-channel modulation through G-protein-coupled receptors.[4–10] Judging from the pace of current research, we can suggest that every type of ion channel has several kinds of modulation that change its gating properties in response to physiological inputs. Such modulation, prominent in neurons, is also present in all other types of cells of the body, muscle, epithelia, etc.

Brown and Adams (1980) discovered that cholinergic agonists such as acetylcholine acting on muscarinic receptors turn off a novel K^+ channel in frog sympathetic neurons.[4] Regulation of this channel is the focus of this chapter. Because of the muscarinic regulation, they called the current I_M. Subsequent work showed that the receptor is the M_1 muscarinic receptor and that many other receptors that couple to the G-protein now called G_q regulate the M-current channel as well as several Ca^{2+} channels in various neurons.[11,12] Already partially open at resting potentials, this channel regulates the excitability of neurons. When the K^+ current is suppressed in a graded way by receptor action, the cell depolarizes and becomes more excitable. The channel was cloned and the principal subunits given the gene name $KCNQ$, a family with five gene subtypes. A heterotetramer of KCNQ2 and KCNQ3 subunits forms the classical neuronal M-current channel. Regulation of KCNQ channels by G_q-coupled receptors had an

Table 10.1 Early examples of ion-channel modulation

Current	Modulator
I_{Ca} heart↑	Adrenaline (cAMP)[2]
I_M sympathetic neuron↓	mACh (several messages)[4]
I_{Ca} DRG neuron↓	Adrenaline (several messages)[5]
I_S aplysia↓	5-HT (cAMP)[6]
I_{GIRK} heart↑	mACh $(G_{i/o})$[7,8] mACh $(G\beta\gamma)$[9]
I_{AHP} ↓	Adrenaline (cAMP)[10]

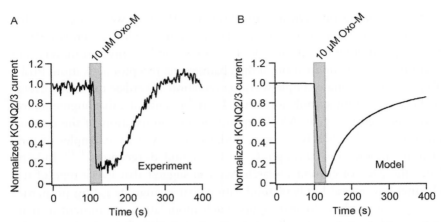

Figure 10.1 Muscarinic modulation of KCNQ2/3 current and simulation by the model. (A) KCNQ potassium current recorded under whole-cell patch clamp. A tsA-201 cell has been transfected with KCNQ2 and KCNQ3 subunits and the M_1 muscarinic receptor so that it expresses the M-current ion channel. When the muscarinic agonist oxotremorine-M is applied to the cell, the KCNQ current is nearly fully suppressed. It recovers slowly when agonist is removed. This example is from a single cell. Each cell is slightly different from the next. (B) The output of our kinetic model mimics this behavior.[7,13,15–21] The model simulates the average of our observations and is not optimized to match the cell in panel (A). (See the color plate.)

unexpected mechanism. Activation of G_q canonically activates the enzyme phospholipase C (PLC), which cleaves the minor plasma membrane phospholipid phosphatidylinositol 4,5-bisphosphate ($PI(4,5)P_2$). The well-known products of this cleavage are the two second messengers diacylglycerol (DAG) and inositol 1,4,5-trisphosphate (IP_3). However, rather than depending on the signaling of these two messengers, the strong muscarinic modulation of the channel is primarily due to the depletion of the lipid $PI(4,5)P_2$.[13,14] The channel requires plasma membrane $PI(4,5)P_2$ to function and turns off when that phospholipid is depleted (Fig. 10.1A). A dependence of some ion channels and transporters on $PI(4,5)P_2$ was first proposed by Donald Hilgemann.[8,22] All members of the KCNQ family need this membrane lipid to function.

Here, we describe our modeling of the muscarinic modulation of KCNQ channels. The experiments and development of the model have been published in a series of papers.[13,15–20,23] In conceptual terms, the model and this chapter deal with the following modular components: Agonist action turns on the PLC enzyme, the lipid is cleaved, and the channel turns off. In parallel, the messenger products IP_3 and DAG initiate signaling of

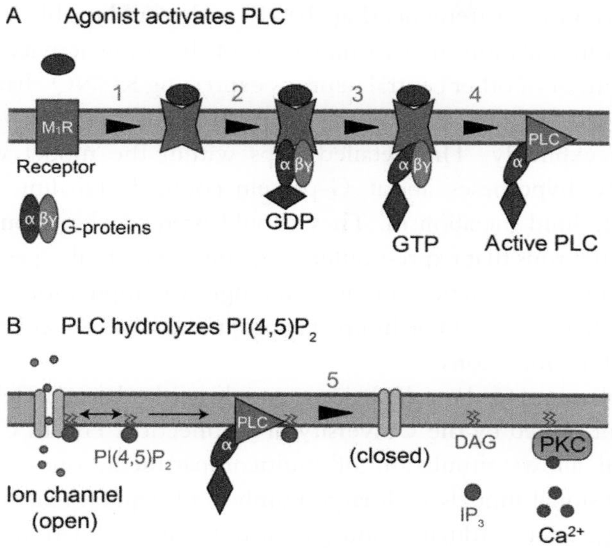

A   Agonist activates PLC

B   PLC hydrolyzes PI(4,5)P$_2$

Figure 10.2 Schematic cartoon of early steps in the signaling pathway to channel modulation. (A) The G$_q$-coupled muscarinic receptor binds the agonist ligand and activates the G-protein. The G-protein turns on the enzyme PLC. (B) Active PLC cleaves membrane PI(4,5)P$_2$, the KCNQ channel is inhibited as it loses PI(4,5)P$_2$, and several second messengers are generated. *Modified from Ref. 17.* (See the color plate.)

their own, and during recovery PLC is turned off again and lipid synthesis restores the pools of PI(4,5)P$_2$. Most of the underlying signaling steps represent enzymatic modifications of membrane lipids and G-proteins. The individual steps are discussed in detail below. Many steps in the onset of agonist action are summarized in Fig. 10.2, which will serve as a point of reference to guide the discussion. While developing the model, we tried to measure the time course of all the intermediate steps using optical and electrical methods.[15–20,23,24] The measurements helped to refine the assumptions of the model, and the model helped to specify further measurements.

2. MODELING APPROACH

2.1. Goals and programming environment

Our goals were to understand and describe the time course of suppression and recovery of KCNQ channels during addition and removal of agonists like acetylcholine (Fig. 10.1A). The overall model output would have to simulate current measurements (Fig. 10.1B) and would enable inclusion of ion-channel modulation in any simulations of the firing activity of sympathetic neurons or of other central neurons expressing KCNQ channels. Ideally, on a cell biological level, every identifiable enzymatic step would be represented explicitly. The detailed steps within the model would test and describe hypotheses about G-protein-coupled signaling and phosphoinositide lipid metabolism. They should predict what would happen in different neurons that express different quantities and subtypes of the signaling enzymes. In practice for neurobiological computations, the many internal mechanisms could be lumped appropriately into fewer components to allow faster simulations.

For implementing the calculations, we chose the Virtual Cell software environment hosted by the University of Connecticut Health Center.[25,26] Virtual Cell allows simulation of multicompartment, one-, two-, and three-dimensional models with large numbers of explicit reaction mechanisms that generate ordinary- and partial-differential equations. The user places reactants, enzymes, and products in different compartments using a graphical interface and specifies the rate laws. Typically, the user interface is on a personal computer and the actual calculations are done at high speed on the hosting mainframe. The calculation engine generates the differential equations and integrates them in time in a manner transparent to the user. The system handles multiphase models that, for example, combine membrane surface compartments and cell volume compartments. Thus, the transformation from surface units to volume units in hydrolysis of membrane $PI(4,5)P_2$ to yield membrane DAG and cytoplasmic IP_3 is handled automatically. Finished models can be posted in a public library for public use and simulated results exported in various formats including graphs, spreadsheet data (comma separated value .csv), image files (GIF, NRRD), or movies (Quicktime or Animated GIFs). Equations from Virtual Cell can be exported to MATLAB or as SBML files. A public version of our latest model is available and can be run by logging into Virtual Cell and loading the biomodel: FalkenburgerDicksonHille2013 from shared: Hillelab.

2.2. Tools to obtain quantitative kinetic data

To enable quantitative kinetic modeling, we needed experimental techniques that measured the speed, concentrations, and buffering capacity of many signaling steps in living tsA-201 cells. Overexpressed fluorescent

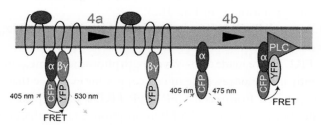

Figure 10.3 Expanded illustration of step 4 from Fig. 10.2, showing two examples of the use of real-time FRET. Schematic illustration of the overexpressed fluorescent reporter proteins used as real-time FRET monitors to track $G\alpha_q$ dissociation from $G\beta1$ (4a) and subsequent activation of PLC (4b). In step 4a, $G\beta$ is the FRET acceptor and in 4b, PLC is the FRET acceptor. (See the color plate.)

reporter proteins were used as real-time monitors to track dynamic inter-molecular or intramolecular interactions along the activated muscarinic receptor signaling cascade.[16–19,23] Transfecting CFP- and YFP-labeled signaling proteins into cells allowed the progression of the muscarinic signal through each intermediate signaling step (Fig. 10.2) to be measured via Förster resonance energy transfer (FRET). The fundamental concept for FRET is that blue light normally evokes blue fluorescence from an isolated CFP, but when a YFP-labeled probe is in the close vicinity, energy can be transferred from CFP (donor) to YFP (acceptor), which then emits yellow light. FRET is a nonlinear measure of the proximity of donor molecules to acceptor molecules. One example is shown in Fig. 10.3 (step 4a), where two subunits of the G-protein have donor and acceptor fluorescent proteins. When they are together, the fluorescence is at longer wavelength because of FRET, and when signaling steps move them apart, the fluorescence changes to shorter wavelength. The amount of FRET is often quantified simply as the ratio of yellow emission YFP_C to blue emission CFP_C (YFP_C/CFP_C) during blue excitation.[27–31] We call this the FRET ratio (FRETr). Thus, from early in the signaling pathway, the following donor and acceptor pairs were constructed (numbered as in Fig. 10.2A):

(1) CFP and YFP attached at two different positions on the M_1 muscarinic receptor protein reported the kinetics of conformational changes within the receptor upon ligand binding.

(2) CFP attached to the receptor and YFP attached to a G-protein subunit reported the changing distance or orientation between receptor and G-protein.

(3) CFP attached to $G\alpha_q$ and YFP to $G\beta1$ reported the kinetics of G-protein separation as in Fig. 10.3.

(4) $G\alpha_q$–CFP and PLC–YFP reported the time course of $G\alpha_q/PLC\beta1$ interaction.

To measure FRETr, we made two-wavelength photometric measurements on single cells with an epifluorescence microscope. Our evidence that the quantitative data gathered from each CFP–YFP FRET pair represented actual energy transfer between CFP and YFP fluorophores, rather than some other optical change is that: (i) during perfusion of agonist, CFP_C and YFP_C values typically changed in opposite directions with identical time courses; (ii) fluorescence changes were reversible following removal of agonist; and (iii) when strong 500 nm illumination was used to bleach the YFP fluorophore (known as acceptor photobleaching), CFP_C increased and FRETr fell to near zero. We measured FRETr at different sampling intervals depending on the kinetics of the reaction in question. For example, the time constant for M_1 receptor activation is very fast, requiring sampling at 20 Hz, whereas the time constant for $PI(4,5)P_2$ depletion is slow, allowing sampling at 0.25 Hz.

Quantitative aspects of signaling depend on the absolute amounts of signaling molecules present. Therefore, transfection of fluorescently tagged signaling molecules, although extremely informative in terms of the kinetic information, also can perturb the system. The new proteins can reach much higher densities than the endogenous proteins and may alter the steady-state and kinetic properties of signaling. For instance, transfecting G-proteins increases the percentage of G-protein-bound receptors, and transfecting PLC accelerates G-protein deactivation (see below). Thus, amounts of endogenous and overexpressed signaling components needed to be measured. Conversely, changing the amounts of signaling molecules in this way also is an interesting intervention that further probes the signaling system. To quantify the density of overexpressed fluorescent molecules, the fluorescence intensities of cells were compared (i) to fluorescent beads, (ii) to solutions of recombinant fluorescent proteins, and (iii) to levels of transfected and endogenous proteins measured using Western blot analysis.[17–19,23] These three independent assessments of protein expression levels indicated that proteins overexpressed enough to give good optical signals have densities around 3000 μm^{-2} at the plasma membrane, whereas endogenous proteins have typically only $<10\ \mu m^{-2}$ (see Table 10.2).

As stated earlier, a major goal of the model is to describe the time course and amplitude of suppression and recovery of KCNQ channels following M_1R activation, as in Fig. 10.1. KCNQ2/3 channels require $PI(4,5)P_2$

Table 10.2 Some of the measured initial conditions and rates

Parameter	Value (units)	Method/rationale
Initial conditions		
R (endogenous)	1 μm^{-2}	From Western blot[18]
R (overexpressed)	500 μm^{-2}	From fluorescence[17]
G-protein (endogenous)	40 μm^{-2}	To fit concentration–response curve of current[16]
G-protein (overexpressed)	3000 μm^{-2}	From fluorescence[17]
PLCβ1 (endogenous)	3 μm^{-2}	From Western blot[17]
PLCβ1 (overexpressed)	3000 μm^{-2}	From fluorescence[17]
(Free) PI(4,5)P$_2$	5000 μm^{-2}	From distribution of PH domains[17,32,33]
(Free) PI(4)P	4000 μm^{-2}	Ratio to PI(4,5)P$_2$[15]
Bound PI(4,5)P$_2$	10,000 μm^{-2}	Response of IP$_3$ probe to 200 s Oxo-M[18]
KCNQ2/3 channels	4 μm^{-2}	From whole-cell current, open probability, and single-channel conductance[23]
Rate constants		
Receptor ligand unbinding (L1)	5.5 s^{-1}	From M$_1$R–Y–C FRET recovery
Receptor–G-protein binding (G1)	2.7×10^{-3} s^{-1}	From M$_1$R/Gβ FRET onset
Receptor–G-protein unbinding (G2)	0.68 s^{-1}	From M$_1$R/Gβ FRET recovery
G-protein nucleotide exchange (NX_RLG)	0.65 s^{-1}	From Gα/Gβ FRET onset, fits Gα/PLC FRET onset
GTPase1	0.026 s^{-1}	From Gα/Gβ FRET recovery
PLC dissociation	0.71 s^{-1}	From Gα–PLC FRET recovery
PI 4-kinase (rest)	0.00078 s^{-1}	KCNQ2/3 current recovery after Oxo-M[23]
PI 5-kinase (rest)	0.06 s^{-1}	KCNQ2/3 current recovery after VSP[23]
IP$_3$		From LIBRAvIII
IP$_3$ase	0.08 s^{-1}	To reproduce the duration of LIBRAvIII and Fura-4F responses[19]
DAG		From FRET between C1 domain–Caax
DAGase	0.05 s^{-1}	From C1–Caax FRET decay[19]

as a positive cofactor for function and can serve as effective monitors of free PI(4,5)P$_2$ at the plasma membrane. To measure KCNQ2/3 currents, whole-cell, or perforated patch–clamp electrophysiological recordings are made while activating the M$_1$ receptor. The information obtained includes: (i) the time taken from the initial decay to total current inhibition informs us about the rate of PLC activation, PI(4,5)P$_2$ hydrolysis, and the KCNQ2/3 channel affinity for PI(4,5)P$_2$; (ii) the delay between agonist application and onset of current inhibition should reflect the kinetics of intermediate steps reported by the individual FRET indicators discussed above; and (iii) the recovery of KCNQ2/3 current indicates the time taken for both the PI 4-kinase and PI(4)P 5-kinase enzymes to synthesize monophosphorylated PI(4)P and diphosphorylated PI(4,5)P$_2$ from the precursor phosphatidylinositol, PI, respectively.

Zebrafish (*Danio rerio*) express a useful voltage-sensitive lipid phosphatase enzyme (Dr-VSP[34]). When activated by large membrane depolarizations, this enzyme quickly dephosphorylates PI(4,5)P$_2$ on the 5 position to produce PI(4)P. Transfected into cells, Dr-VSP confirms the fundamental requirement of KCNQ2/3 channels for PI(4,5)P$_2$, and also reveals (i) the residence time of PI(4,5)P$_2$ on channel subunits; (ii) a cooperative effect of PI(4,5)P$_2$ for activation of KCNQ channels; and (iii) as discussed later, the time constant for resynthesis of PI(4,5)P$_2$ by PI(4)P 5-kinase.

We monitored four signals downstream from cleavage of PI(4,5)P$_2$: production of IP$_3$ and DAG, and stimulation of calcium release and PKC (Fig. 10.2B). To measure IP$_3$ production, we used membrane-targeted LIBRAvIII, a FRET reporter based on the ligand-binding domain of the rat IP$_3$ receptor type III.[35] The kinetics of the LIBRAvIII signal informed us about the rate of IP$_3$ production by PLC and the rate of the IP$_3$ degradation by IP$_3$ 5-phosphatase, dephosphorylating IP$_3$ to IP$_2$. Calibrating LIBRAvIII by dialyzing different concentrations of IP$_3$ into the cell through a patch pipette allowed us to estimate the absolute quantities of IP$_3$ produced during receptor activation. Calcium release was monitored using the low-affinity calcium indicator Fura-4F. To study the production of DAG, we developed a FRET assay modified from an existing strategy based on migration of the DAG-binding C1 domain of PKCγ. The time constant of the increase in FRETr represents the rate of DAG production, whereas the subsequent decay constrains the rates of the enzymes responsible for degradation of DAG. The activation of PKC by DAG was monitored using a FRET C-kinase activity reporter (CKAR).[36,37] CKAR provides a real-time readout of the phosphorylation state of PKC substrates.

Our model incorporates the density of proteins, kinetic interactions, and any buffering of small molecules by binding to proteins. These parameters were used to fit the experimental observations and then were scaled back to endogenous levels when we were simulating physiological measurements. In particular, the model contains all the buffering effects of any expressed translocation- and FRET-indicator molecules. Concentrations of indicators, such as for DAG, IP_3, and calcium, were scaled back to zero for physiological simulations shown in the figures.

3. MODELING RECEPTOR AND G-PROTEIN ACTIVATION

3.1. General considerations

In their classical study of ion channels of the squid giant axon, Hodgkin and Huxley[1] looked at a distal event (ionic current) and presumed internal, "hidden" gating states to best fit their data. They did not assume that their four "gating particles" had a molecular substrate, but when, decades later, ion channels were found to indeed contain four voltage sensors, their work appeared almost prophetic. We believe that two factors were instrumental for their successful predictions. One is their aim to simulate not a single-current trace, but currents over wide ranges of voltages, pulse durations, and ion concentrations. The second is that currents of giant axons are readily measured with high accuracy and good signal-to-noise ratio. This allows accurate fitting of activation and deactivation kinetics at many voltages.

Similarly, in our modeling of the G-protein-coupled cascade, we used measurements of kinetics over a wide range of agonist concentrations to probe the system we study. However, this system is different. We did not have to assume "hidden" states because much is known about the intermediate molecules of G-protein-coupled signaling. Our aim was not to "best fit the data," but to incorporate known intermediate states that represent actual measurable molecular steps of the signaling cascade. We wanted to see how all the known steps worked together to produce the macroscopic channel modulation.

Another difference from the situation of Hodgkin and Huxley is that responses of small cells are more variable than those of giant axons. Hence, individual recordings are less reproducible and dose–response curves more "noisy." Further, many of the optical measurements we use have lower signal-to-noise ratio than whole-cell current. Taken together, these factors make it more difficult to tweak intermediate steps by fitting the exact shape of a downstream curve.

3.2. Ligand, receptor, and G-protein: A ternary complex

We now consider the earliest steps of the receptor signaling cascade, comprising steps 1 and 2 in Fig. 10.2A. At the same time, we introduce the notation of the model, shown in Fig. 10.4A. Signaling begins with binding of an activating ligand (L) to the receptor (R) to form the LR complex, step 1, and receptor interaction with G-proteins, step 2. Modeling the binding step has a long tradition. Historically, researchers had used radioligands to study steady-state receptor occupancy and the kinetics of ligand dissociation from the receptor (see an overview in Ref. 39). Careful analysis of such experiments gave evidence for high-affinity and low-affinity binding states.

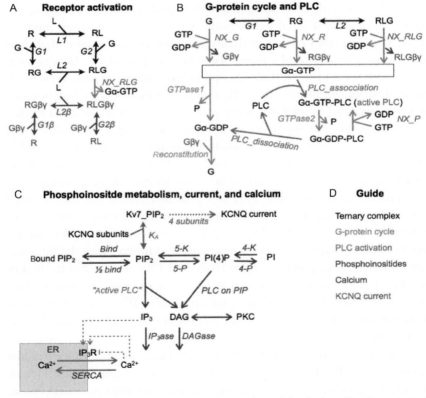

Figure 10.4 The formal reaction steps in our model of the signaling cascade from G_q-coupled receptors to channel modulation, including the signal shutdown and lipid recovery paths.[17,19,23,38] The steps, written in standard chemical kinetic notation, are either first-order or second-order reactions as appropriate. (See the color plate.)

G-protein-bound receptors (RG) are thought to underlie the high-affinity binding state. Receptors not bound to G-proteins underlie the low-affinity binding state. Manipulations that dissociate G-proteins from receptors (e.g., guanylyl imidodiphosphate, a nonhydrolyzable analogue of GTP) abolish the high-affinity binding and leave only the low-affinity binding. Ligands dissociate faster from the low-affinity binding state LR than from the high-affinity binding state LRG. Consequently, the time course of ligand dissociation has a fast component and a slow component in real cells, and ligand dissociation is faster in cells where G-protein binding to receptors is inhibited. The basic laws of chemical equilibrium (microscopic reversibility) require a reciprocity: "G-protein binding to receptors facilitates ligand binding" is equivalent to "ligand binding facilitates G-protein binding." Thus, the two levels of ligand affinity provide an explanation why ligand binding to the receptor increases G-protein binding to the receptor.

This mutual interaction of receptors, ligands, and G-proteins has been formalized in the "ternary complex model" (Fig. 10.4A).[40] Such interaction is critical for understanding the behavior at the level of receptors, and we believe that any model describing G-protein signaling should include it. It is particularly important when explaining behavior for different membrane densities or concentrations of receptors, ligand, and G-proteins.

Fluorescent protein sequences inserted into the sequence of receptors report the time course of conformational changes within receptors in response to agonist application.[16,29] In our work, the rate of ligand unbinding was indicated by the time course of FRET decrease after removing the ligand, and the very fast rate of ligand binding was then inferred using the known dissociation constant. Another important parameter in the ternary complex model is the cooperativity factor α. It signifies the extent with which G-protein binding facilitates ligand binding and vice versa. In radioligand binding studies, it is determined from the affinity ratios of the high- and low-affinity binding states. We have used $\alpha = 100$, which means that ligand binding increases G-protein binding by a factor of 100.

Newer measurements with smaller fluorophores and faster measurement techniques (spectroscopy) revealed that what is observed as activation and deactivation kinetics on a whole-cell level is in fact the convolution of individual receptors rapidly transitioning among multiple microstates. External effects such as binding of G-proteins or ligands preferentially stabilize one of these states. This model is called the "conformational selection" model.[41,42] There are parallels with single-channel recordings where individual ion channels rapidly fluctuate between open and closed states and voltage or ligands affect the probabilities of going from one state to the other, whole-cell current being a smoothly graded summation of many single-channel currents each with binary and stochastic behavior. Most models of neuronal behavior do not need to include such fast timescales to represent ion channels and receptors. For our purpose, it has worked very well to simulate receptor activation as a slow, smooth process and current as a graded phenomenon. Yet, the binary and probabilistic effects may become more relevant when simulating smaller structures with fewer molecules, such as individual synaptic spines or synaptic boutons.

Some authors and models discriminate between inactive receptors R and activated receptors R^*. These models are "extended ternary complex models".[41,43] Most agree that activation is a fast first-order reaction. We have not included such states because they multiply the number of species and reactions to consider and introduce free parameters for which we have no restraining data. Our model contains R, RL, RG, and RLG (Fig. 10.4A); others have all of these plus R^*, R^*L, R^*G, and R^*LG. Some models assume that the amounts of R^* without L and G bound are small, so they have R, RL, RG, RLG, and R^*LG. We think that this is problematic since if one R^* state is considered, we feel that all R^* states must be included to not miss transitions that may become significant under some conditions.

3.3. G-protein activation

G-proteins are heterotrimers composed of $G\alpha$, $G\beta$, and $G\gamma$ subunits. Each subunit is encoded by a family of related genes. It is primarily the $G\alpha$ subunit that determines which receptors a G-protein trimer binds to. M_1 muscarinic receptors bind G-proteins containing $G\alpha_q$; others bind G-proteins containing $G\alpha_i$ or $G\alpha_o$, etc. Some receptors can bind several classes of G-proteins. There appear to be fewer constraints in the selection of $G\beta$ and $G\gamma$ subunits for the G-protein heterotrimer. In our experiments, we have used $G\beta1$ and $G\gamma2$ along with $G\alpha_q$. As determined from FRET between the M_1 receptor and $G\beta$ subunits, it takes about 0.5 s for the effects of addition of agonist to reach the G-protein (Fig. 10.5A).

The $G\alpha$ subunit can bind a guanine nucleotide, either GDP or GTP.[44] At rest, it has GDP bound. G-proteins become activated by the exchange of GDP for GTP (step 3, Fig. 10.2A, and reactions NX in the model, Fig. 10.4B). This exchange is facilitated by ligand-bound receptors (LR) and by other guanine nucleotide exchange factors. Specifically, binding of the G-protein to the receptor as the ternary complex LGR facilitates

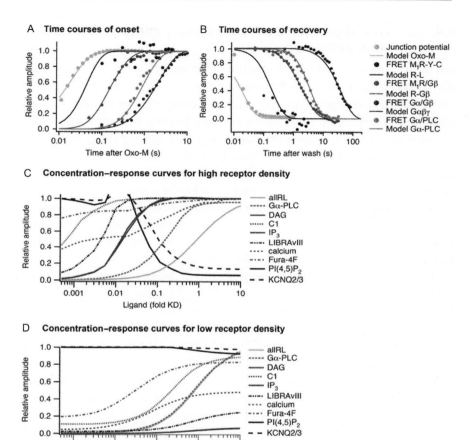

Figure 10.5 Outputs of the model. Overview of normalized time courses for onset and recovery of components of the G-protein signaling pathway after addition (A) and removal (B) of a near-saturating concentration of agonist. Note logarithmic time axes. Ligand concentration–response curves for high (C) and low (D) receptor densities. High density is 500 receptors per μm^2 and low density is 1.25 receptors per μm^2. The concentration axis is in reduced units of multiples of the agonist equilibrium dissociation constant. *Modified from Ref. 19*. (See the color plate.)

unbinding of GDP from the G-protein. This is the rate-limiting step for nucleotide exchange.[45] The GTP-bound form, $G\alpha_q$-GTP, is the active form. The classical view holds that GTP binding leads to dissociation of the G-protein heterotrimer into the $G\alpha$-GTP subunit and the $G\beta\gamma$ subunit. Both can activate downstream targets. Well-known examples are the activation of PLC by $G\alpha_q$-GTP and the activation of K_{ir} channels by $G\beta\gamma$.

3.4. Unresolved problems with the ternary complex model and G-protein activation

For many steps in the early signaling pathway, it is not clear whether they actually represent full protein association and dissociation events or rather local conformational changes within a preformed complex. For instance, ligand binding to the receptor could either facilitate binding of G-proteins to the receptor or lead to a conformational change in a preexisting receptor–G-protein complex. Our FRET measurements do not differentiate between these two possibilities. Computationally, a conformational change would be implemented by a first-order reaction whereas association would be a second-order reaction. Thus, they are computationally different and may lead to different predictions when changing G-proteins. Similarly, it is unclear whether G-proteins fully dissociate upon nucleotide exchange or whether this step is also sometimes just a conformational change.

During agonist application, we observed a ligand-induced decrease in FRET between $G\alpha_q$–CFP and $G\beta\gamma$–YFP[16] and therefore have implemented the classical view of G-proteins binding to receptor and G-protein dissociation (as in step 4a of Fig. 10.3). In another system, a ligand-induced increase of FRET was seen between $G\alpha_i$–CFP and $G\beta\gamma$–YFP that can be explained only by a conformational change.[27] Indeed, the rapid cycling of receptor-induced nucleotide exchange at $G\alpha$, and PLC-activated GTPase activity, can plausibly occur in a receptor–$G\alpha$–PLC complex. FRET measurements by others have indicated that the scaffolding protein AKAP79 makes an even larger molecular complex containing the M_1 receptor and the regulated KCNQ channel,[46] which might further contain G-proteins and PLC. Yet additional, elegant studies with immobilized $G\alpha$ have provided evidence for the classical view and kept this issue controversial.[47]

Several interesting features are missing from our model and this presentation. We just list them without further discussion. They include: analysis and kinetic modeling of activation of K_{ir} channels by $G\beta\gamma$[48,49]; the phenomenon of inverse agonism[43]; possible receptor dimerization[50]; activity of G-proteins at intracellular membranes[51,52]; and receptor phosphorylation, arrestin binding, and internalization.[53] We also call attention to useful modeling papers from the groups of Linderman, Simon, and Ross.[43,54,55]

4. ACTIVATION OF PLC, MODULATION OF CHANNELS, AND DEACTIVATION OF G-PROTEINS

The preceding paragraphs describe kinetic steps generating the activated G-protein, $G\alpha_q$–GTP. Within about 2–3 s of application of a high concentration of agonist, $G\alpha_q$-GTP has bound to and activated PLC (step 4B, Fig. 10.3), initiating the breakdown of PI(4,5)P$_2$ lipid (Fig. 10.5A). PLC begins to deplete PI(4,5)P$_2$ and simultaneously generates the DAG and IP$_3$ second messengers (Figs. 10.2B and 10.4C). There is an excess of PLC in the cell to catalyze this reaction. Even when receptor occupancy is only a few percent ("all RL" in Fig. 10.5C) and only a small percentage of PLC molecules are activated ("Gα–PLC" in Fig. 10.5C), there can be significant production of IP$_3$ and DAG (see below). However, the response is slow, and weak activation of PLC does not deplete the PI(4,5)P$_2$ pool or decrease KCNQ current much since PI(4,5)P$_2$ resynthesis can keep up. Stronger activation of PLC at higher ligand concentrations allows the cell to reach its full response within <10 s, including PI(4,5)P$_2$ depletion and KCNQ current suppression. Thus, the advantages of a high available PLC activity are rapid responses and signaling by lipid depletion.

We have finally come to modulation of the KCNQ2/3 ion channel itself. As free PI(4,5)P$_2$ falls (step 5, Fig. 10.2B), lipid molecules dissociate from the channel subunits as well as from the other postulated PI(4,5)P$_2$-buffering proteins. Both reactions are assumed to be in high-speed equilibrium with the pool of free lipid. Therefore, the bound pools fall, tracking the free pool. The tetrameric channel has two KCNQ2 and two KCNQ3 subunits. As in recent crystal structures of K$_{ir}$2.2 and Girk2 channels,[56,57] each subunit is assumed to bind one activating lipid, which helps the channel fold into an active conformation. We assume therefore that four lipid molecules are needed for channel activity (Fig. 10.4C). However, since KCNQ2 subunits have a higher affinity for PI(4,5)P$_2$ than KCNQ3, we can simplify the modulation model to consider only the two low-affinity sites.[38] The channel activity is taken as the square of the binding to a single site.

The activity of $G\alpha_q$-GTP is terminated by hydrolysis of the bound GTP. The Gα subunit has a slow GTPase activity (GTPase1 in Fig. 10.4B). GTPase activity is accelerated by GTPase-activating proteins (GAPs). In our measurements with overexpression of receptors and G-proteins for FRET, G-proteins outnumbered endogenous GAPs, and G-protein deactivation was very slow. This allowed us to calculate the endogenous GTPase activity from this slow G-protein reassociation. The resulting rate constant was in accord with values obtained by *in vitro* GTPase activity measurements.[45]

When $G\alpha_q$-GTP binds to PLC, the PLC molecule acts as a GAP for $G\alpha_q$ and thus eventually turns off its own activator.[58] (In the model, GTPase2 $\gg$ GTPase1, Fig. 10.4B) This may appear counter-intuitive at first. However, the resulting free Gα-GDPs bind again to the receptor–ligand complex and become rapidly reactivated. Accordingly, we see sustained (re-) binding of $G\alpha_q$ to PLC in our FRET measurements and not a transient response. By speeding cycling of GTP and GDP at $G\alpha_q$, the GAP activity of PLC speeds the deactivation of the agonist response when the ligand is removed. If the amounts of PLC and G-proteins are about equal, G-protein activation is terminated quickly (Fig. 10.5B, "$G\alpha$–PLC").[16] If PLC did not act as a GAP and G-proteins relied on their own GTPase activity for deactivation, G-proteins would remain active for a long time after ligand removal and brief responses would be impossible. Given that the PLC-inhibited KCNQ channels enhance neuronal excitability, prolonged PLC activity could be dangerous for the organism.

5. PLC MESSENGERS

Activated PLC generates DAG and IP$_3$ as it cleaves PI(4,5)P$_2$ (step 5, Fig. 10.2B). These messengers also affect ion-channel activities. DAG modulates some ion channels, such as TRPC3, 6, and 7, directly by binding to them,[59,60] and it activates PKC, which can phosphorylate ion-channel subunits. By binding to IP$_3$ receptors, IP$_3$ releases Ca^{2+} into the cytoplasm from internal stores. Some ion channels are modulated by direct calcium binding[61,62]; others are modulated through calcium–effector proteins like calmodulin.[63,64] The KCNQ2/3 potassium channel has been suggested to be regulated by PI(4,5)P$_2$ depletion, PKC activation, and calcium/calmodulin.[13,65,66] We have modeled the production of the calcium and PKC signals because they follow stoichiometrically from the hydrolysis of PI(4,5)P$_2$, but so far have not included their modulatory actions on channels in the model.

5.1. Production of DAG and activation of PKC

We determined the kinetics of DAG production upon receptor activation using our C1-domain FRET reporter and an Oxo-M concentration–response protocol (Fig. 10.5C). The DAG response amplitude was graded with increasing concentration of Oxo-M up to 0.1 μM; higher concentrations up to 10 μM prolonged the DAG signal but did not increase it further. The saturation of the peak FRETr response at 0.1 μM Oxo-M reflected saturation of our reporting system. This can also explain the longer duration of the reporter response (i.e., longer time spent at or above saturation) with 10 versus 0.1 μM agonist. Thus, we expect production of DAG to increase further and to increase faster above 0.1 μM Oxo-M.

How much DAG is necessary to activate PKC? PKC activity was measured using CKARs that responded with a decrease in FRET to phosphorylation of their PKC substrate peptide.[37] The phosphate is rapidly removed by endogenous phosphatases, and thus CKARs are suitable to monitor dynamic changes in PKC activity in intact cells. Two CKAR probes were tested, one being localized in the cytoplasm and one close to the cytoplasmic leaflet of the plasma membrane. Both responded to agonist application. The amplitudes of the CKAR responses were independent of the agonist concentration, meaning that even a low concentration of agonist was sufficient for saturating activation of PKC-induced phosphorylation.

5.2. IP$_3$ and Ca^{2+} signals

Even the lowest agonist concentration (1 nM Oxo-M) evoked cytoplasmic Ca^{2+} signals that were readily detected with the low–affinity fluorescent calcium indicator Fura-4F (Fig. 10.5C). Although the time to half-maximum responses became significantly shorter with increasing agonist concentration, the peak amplitude of the Ca^{2+} signal was only weakly dependent on the agonist concentration over many orders of magnitude.[19] Similar to our experience with production of DAG and activation of PKC, the experiments and the model show that very little PI(4,5)P$_2$ needs to be hydrolyzed to make enough IP$_3$ to open IP$_3$ receptor channels. Nevertheless, with stronger PLC activation, there was faster IP$_3$ production and the apparent threshold for initiation of the calcium response was reached sooner. Several factors give calcium release a somewhat regenerative quality. Both IP$_3$ receptors (in our model) and ryanodine receptors (not implemented yet) would participate in a positive feedback due to calcium-induced calcium release, and in addition, the IP$_3$ receptor response to IP$_3$ obeys a power law that steepens the response as IP$_3$ rises. Cytoplasmic calcium, when it rises to micromolar concentrations, also acts as an inhibitor of calcium release. In our model, we adopted an old mathematical description of these events derived from work with IP$_3$ receptors incorporated into lipid bilayers (Fig. 10.4C, lower right; Refs. 67,68). This description was convenient because it was the only one that represented the release process as fully defined continuous functions of calcium and IP$_3$ concentrations over the very broad range encountered in the cell. Eventually, it should be supplanted by continuous equations based on newer direct patch-clamp studies on IP$_3$ receptors in native membranes when appropriate equations become available.[69–72] Our model also included a simple representation of calcium clearance from the cytoplasm.

5.3. Concentration–response relations and spare receptors

We have been considering the concentration–response relations in Fig. 10.5C. They are the output of the model corresponding to graded activation of overexpressed M_1 receptors that have an assumed surface density of ~500–1000 μm^{-2}. The x-axis is ligand concentration expressed in reduced units of the dissociation constant for ligand. Thus, by definition, at 1.0 on the axis, receptor occupancy is 50%. The line labeled "allRL" is the effective binding curve for ligand. We have emphasized that with only a few percent occupancy, there already is strong production of DAG, IP_3, and Ca^{2+} signals. In the vocabulary of classical pharmacology, there are "spare receptors" or a "receptor reserve" for these responses, and the agonist Oxo-M is a "full agonist" for all responses. With overexpressed receptors and much endogenous PLC, it is easy to saturate all measureable outputs of the system. Comparing the computed responses with those of a sympathetic neuron shows that the neuron also behaves like this system with high receptor density.

The model allows us to vary parameters. For example, a clear change can be seen when comparing these high receptor densities (Fig. 10.5C) with much lower receptor densities (Fig. 10.5D). The mid-point of the ligand-binding curve normalized to the dissociation constant remains the same, but because the activation of PLC is much less at any agonist concentration, much more agonist is needed to have an effect. All response curves move to the right on the concentration axis. Further, because there are fewer receptors than G-proteins and PLC enzymes, it is not possible to get full production of DAG or IP_3, hydrolysis of $PI(4,5)P_2$, or inhibition of KCNQ2/3 current. The agonist has become a "partial agonist" for most responses simply by changing receptor densities. The cell line we study expresses endogenous purinergic receptors ($P2Y_2Rs$) that couple to the G-protein $G\alpha_q$. These purinergic receptors have a low surface density we estimate at about 1 μm^{-2}. Their responses to a saturating concentration of UTP as an agonist resemble the predictions for Fig. 10.5D, a nearly full-sized calcium response, but markedly slowed, and the same for the DAG probe and PKC probe responses, yet no net depletion of $PI(4,5)P_2$ and no inhibition of KCNQ2/3 channels.[18,19] UTP is a partial agonist, and all of the results are well imitated just by adjusting receptor density without assuming other special properties for purinergic receptors. If we overexpress exogenous $P2Y_2$ receptors in this cell line, the cells respond to saturating UTP just as they responded to saturating muscarinic agonist. With a high density of receptors, UTP too becomes a full agonist.

6. PHOSPHOINOSITIDE METABOLISM AND COMPARTMENTS

6.1. The surface density of PI(4,5)P$_2$

How much PI(4,5)P$_2$ is there in the cell? Earlier, we and others had estimated the pool of free PI(4,5)P$_2$ to be ~5000 μm^{-2} of PM, which was sufficient to generate 5 μM IP$_3$ in the cytoplasm upon full hydrolysis by PLC. However, then our calibration experiments with IP$_3$ and LIBRAvIII showed that the LIBRA probe's response to a saturating concentration of Oxo-M was more comparable to the infusion of 10 μM IP$_3$ via the patch pipette. Therefore, needing more PI(4,5)P$_2$, we postulated an additional reserve membrane pool of ~10,000 μm^{-2} PI(4,5)P$_2$ molecules reversibly bound to proteins in addition to the free pool. This bound pool would be in rapid dynamic equilibrium with the free pool, in agreement with the findings of Golebiewska et al. that two-thirds of PI(4,5)P$_2$ is reversibly bound to proteins with millisecond or submillisecond residence times.[73] The total of 15,000 μm^{-2} also brings us in better alignment with Hilgemann's estimate that total amounts of PI(4,5)P$_2$ may be larger than 20,000 μm^{-2}.[73,74]

6.2. Lipid pools

Our kinetic model contains three interconverting pools of membrane phosphoinositides (phosphatidylinositol PI, PI(4)P, and PI(4,5)P$_2$), which the model places within a single-membrane kinetic compartment (Fig. 10.4C). Such a model is probably anatomically incorrect. Synthesis of PI is typically said to take place at the ER or at recently discovered ER–derived mobile platforms, and it may reach the Golgi and PM by multiple mechanisms.[75–77] Since PI is abundant compared with PI(4)P and PI(4,5)P$_2$ and we have no kinetic measurements relating to PI synthesis, we represented PI as a constant pool of 140,000 molecules μm^{-2} of PM surface based on estimates from [^{3}H]inositol equilibrium experiments.[78] The single pool of PI(4)P, the precursor for PI(4,5)P$_2$, is represented as being synthesized from PI by the kinase "4-K" in a single reaction. Probably, it would be more realistic to represent PI(4)P synthesis not by a single reaction, but rather by several 4-kinases in different compartments,[79,80] and transfer of PI(4)P from other cellular compartments to the PM.[75] Currently, the model lumps these steps into one "4-K" for lack of better information.

6.3. Relative rates of the lipid kinases

After removal of agonist, it takes several hundred seconds for the $PI(4,5)P_2$ and hence KCNQ current to recover (Fig. 10.1A). The activation of PLC depletes two lipid pools, $PI(4)P$ and $PI(4,5)P_2$, so two phosphorylation steps are needed for this slow restoration of the $PI(4,5)P_2$ pool from PI (Fig. 10.4C). To describe recovery in the model, we needed to know the relative rates of the 4-kinase and the 5-kinase reactions. We did this using overexpressed voltage-sensitive lipid phosphatase, Dr-VSP, already introduced.[81] Activation of this enzyme on the plasma membrane converts the $PI(4,5)P_2$ pool to $PI(4)P$ in 1–2 s, suppressing KCNQ current. After Dr-VSP is turned off again, the $PI(4,5)P_2$ and KCNQ2/3 current recover. This time, the recovery requires only one step, catalyzed by the endogenous $PI(4)P$ 5-kinase(s). The recovery is about 20 times faster than the two-step recovery after PLC. Thus, we made the "5-K" rate constant 20 times faster than the "4-K" rate constant. The "4-K" would be the rate-limiting step in recovery after PLC. Values for the reverse 4- and 5-phosphatase ("4-P" and "5-P") reactions were assigned to keep $PI(4)P$ and $PI(4,5)P_2$ values stable at rest.

6.4. Stimulated phosphoinositide kinases

There is growing evidence that the rates of both the "4-K" and "5-K" are speeded up during receptor activation. We find the need for "4-K" acceleration following M_1 receptor activation based on experiments with the LIBRAvIII IP_3 probe. Prolonged (~200 s) stimulation of the M_1 receptor evokes continuous IP_3 production, even though $PI(4,5)P_2$ levels fall dramatically within 20 s. This continued $PI(4,5)P_2$ synthesis fails if type III PI 4-kinase activity is blocked pharmacologically.[18] In the model, IP_3 molecules with a mean lifetime of only 12.5 s need to be replaced continuously, but the chosen resting rates for the "4-K" and "5-K" are too slow to support the observed IP_3 production. In order to reconcile the kinetic model with the experimental data, the rate of the "4-K" needs to be transiently elevated during receptor activation. Acceleration of phosphoinositide synthesis has been proposed previously in studies estimating IP_3 production and showing a failure of bradykinin and purinergic agonists to induce net $PI(4,5)P_2$ depletion in superior cervical ganglia neurons.[64,64,78,82,83] Similar to the "4-K," there is evidence that the $PI(4)P$ 5-kinase(s) (5-K) could also be regulated, potentially through Rho family kinases.[84,85] We also found it necessary to increase the rate for "5-K" transiently during receptor activation.

An acceleration of the rate of the PI 4-kinase by a factor of 7.5 and of the PI(4)P 5-kinase by a factor of 10 upon PLC activation reproduced the time course of IP_3 production as observed with LIBRAvIII. Both these rate constants are in close agreement to values published by others.[78]

The biochemical mechanism(s) for stimulation of "4-K" or "5-K" remain unknown. The model does provide some interesting temporal constraints for the acceleration and subsequent slowing of $PI(4,5)P_2$ synthesis during receptor activation. For instance, the stimulation of synthesis cannot occur too soon (<1 s), as this would result in a transient increase in $PI(4,5)P_2$ that we did not observe experimentally. Equally, if stimulation is delayed by longer than a few seconds, steady-state $PI(4,5)P_2$ levels are not reached within 20 s, which again contradicts our experimental findings. Next, if the acceleration of $PI(4,5)P_2$ synthesis decays too quickly after removal of agonist, $PI(4,5)P_2$ levels would continue to decrease, which does not concur with the experimental data. Given this information, the model predicts that the acceleration of the kinase should develop with a time constant of 1 s or longer and decline with a time constant of 10 s or longer after receptor activation. Such predictions suggest that the molecular mechanism responsible for kinase acceleration is more likely to involve G-protein $G\alpha$ and $G\beta\gamma$ subunits, rather than calcium which had been previously been suggested to underlie the acceleration of $PI(4,5)P_2$ synthesis.[86]

6.5. Compartments

As stated earlier, the species at the plasma membrane are modeled with the assumption of uniform distribution of all membrane reactants. This neglects possible nanodomains with clustered localization of specific components like $PI(4,5)P_2$, G-protein-coupled receptors, or enzymes. In addition, barriers for free diffusion of species like $PI(4,5)P_2$ in the plasma membrane as postulated by some are not included in our model.[87]

The activity of PLC turns off once its activating $G\alpha_q$-GTP decreases in concentration, and then levels of $PI(4,5)P_2$ increase again. In the cell, the resynthesis occurs by phosphorylation of PI to PI(4)P by a 4-kinase and of PI(4)P by a 5-kinase partly at the plasma membrane, but again probably at other membranes as well. In the model, these different processes are condensed into few reactions leading to a simulated homeostasis of $PI(4,5)P_2$ levels in the cell. Thus, the rate constants for these reactions are based on experimentally measured changes in $PI(4,5)P_2$ levels, which result from several reactions taking place inside the cell. The model simulates the summed effect of all reactions without "knowing" about the specific parameters of each compartment's reaction and transport processes. Deconstructing simplified reactions, such as PI(4)P synthesis by "4K," will nevertheless be important as it includes additional parameters that may play modulatory roles and aid in our understanding of PI(4,5)P$_2$ synthesis.

A kinetic determination of transport processes of reactants from one compartment to another would allow for a truly compartmental simulation of phospholipid metabolism and its effects on ion-channel activity in a cell. For this, suitable reporter systems for the individual reactions and processes are needed. In addition, concentrations for each species in these reactions will have to be determined for inclusion in a model. Recent advances in the development of new fluorescent probes for phosphoinositides and new microscopy techniques allowing for detection of signals below classical optical resolution will be of great help in monitoring dynamic changes in phosphoinositide levels, their organization, and localizations. Lastly, knowledge about the volumes, geometry, topography, and specific compartment juxtaposition inside a cell will be necessary to construct a complete three-dimensional model. Expressing fluorescent markers for specific compartments in a cell and creating a three-dimensional image via confocal laser microscopy is a new challenge.

Introducing compartments, transport systems, and dimensionality to our model would allow the study of a variety of interesting questions, with some examples given here: How does impairment of exo- or endocytosis change phosphoinositide levels and, eventually, ion-channel activity? If distinct "domains" of PI(4,5)P$_2$ metabolism do exist within a compartment like the plasma membrane, how would such subcompartmentalization be organized to mimic the whole cellular response we observe experimentally? For very large cells like cardiac myocytes or highly polarized cells, how do signals spread in them? We have so far conducted our studies in tsA-201 cells, which are small in comparison to ventricular myocytes and most neurons with dendrites. In such cells, transport and diffusion processes, and their restrictions, are presumably of much higher importance for the response to a stimulus like receptor activation as the distances for a signal to spread are greater. In addition, how does a signal develop if the activating stimulus is polarized and not uniformly present over the entire cell surface? This situation is of great physiological importance as many cells express surface proteins like receptors in a polarized manner.[88,89] Upon presentation of an activating ligand, the signaling cascade would start at this specific site and spread out from there.

6.6. Closing words

The model that we made is generalizable to many cells that use $G\alpha_q$-coupled receptors and to other ion channels and cell processes regulated by $PI(4,5)P_2$ depletion and signaling from IP_3 and DAG. The model is modular so that the G-protein activation part can be extracted for description of other classes of G-protein signaling. Although we have focused on a single signaling pathway, the methods we describe are quite generally applicable to other signaling problems. Their application forces a certain discipline because in the end you must write down explicit steps in making a model. Each step expresses a hypothesis that the model tests. The comparison with observations forces mechanistic refinement or abandoning an idea. The result is verification of cell biological and biochemical concepts.

REFERENCES

1. Hodgkin AL, Huxley AF. A quantitative description of membrane current and its application to conduction and excitation in nerve. *J Physiol*. 1952;117(4):500–544.
2. Reuter H. The dependence of slow inward current in Purkinje fibres on the extracellular calcium-concentration. *J Physiol*. 1967;192(2):479–492.
3. Fuller MD, Emrick MA, Sadilek M, Scheuer T, Catterall WA. Molecular mechanism of calcium channel regulation in the fight-or-flight response. *Sci Signal*. 2010; 28:3(141).
4. Brown DA, Adams PR. Muscarinic suppression of a novel voltage-sensitive K^+ current in a vertebrate neurone. *Nature*. 1980;283(5748):673–676.
5. Dunlap K, Fischbach GD. Neurotransmitters decrease the calcium conductance activated by depolarization of embryonic chick sensory neurones. *J Physiol*. 1981;317:519–535.
6. Siegelbaum SA, Camardo JS, Kandel ER. Serotonin and cyclic AMP close single K^+ channels in Aplysia sensory neurones. *Nature*. 1982;299(5882):413–417.
7. Pfaffinger PJ, Martin JM, Hunter DD, Nathanson NM, Hille B. GTP-binding proteins couple cardiac muscarinic receptors to a K channel. *Nature*. 1985;317(6037):536–538.
8. Breitwieser GE, Szabo G. Uncoupling of cardiac muscarinic and beta-adrenergic receptors from ion channels by a guanine nucleotide analogue. *Nature*. 1985;317(6037):538–540.
9. Logothetis DE, Kurachi Y, Galper J, Neer EJ, Clapham DE. The beta gamma subunits of GTP-binding proteins activate the muscarinic K^+ channel in heart. *Nature*. 1987;325(6102):321–326.
10. Madison DV, Nicoll RA. Actions of noradrenaline recorded intracellularly in rat hippocampal CA1 pyramidal neurones, in vitro. *J Physiol*. 1986;372:221–244.
11. Hille B, Beech DJ, Bernheim L, Mathie A, Shapiro MS, Wollmuth LP. Multiple G-protein-coupled pathways inhibit N-type Ca channels of neurons. *Life Sci*. 1995;56(11–12):989–992.
12. Delmas P, Crest M, Brown DA. Functional organization of PLC signaling microdomains in neurons. *Trends Neurosci*. 2004;27(1):41–47.
13. Suh BC, Hille B. Recovery from muscarinic modulation of M current channels requires phosphatidylinositol 4,5-bisphosphate synthesis. *Neuron*. 2002;35(3):507–520.
14. Zhang H, Craciun LC, Mirshahi T, et al. PIP$_2$ activates KCNQ channels, and its hydrolysis underlies receptor-mediated inhibition of M currents. *Neuron*. 2003;37(6):963–975.

15. Horowitz LF, Hirdes W, Suh BC, Hilgemann DW, Mackie K, Hille B. Phospholipase C in living cells: activation, inhibition, Ca^{2+} requirement, and regulation of M current. *J Gen Physiol*. 2005;126(3):243–262.

16. Jensen JB, Lyssand JS, Hague C, Hille B. Fluorescence changes reveal kinetic steps of muscarinic receptor-mediated modulation of phosphoinositides and Kv7.2/7.3 K^+ channels. *J Gen Physiol*. 2009;133(4):347–359.

17. Falkenburger BH, Jensen JB, Hille B. Kinetics of M_1 muscarinic receptor and G protein signaling to phospholipase C in living cells. *J Gen Physiol*. 2010;135(2):81–97.

18. Dickson EJ, Falkenburger BH, Hille B. Quantitative properties and receptor reserve of the IP_3 and calcium branch of G_q-coupled receptor signaling. *J Gen Physiol*. 2013;141(5):521–535.

19. Falkenburger BH, Dickson EJ, Hille B. Quantitative properties and receptor reserve of the DAG and PKC branch of G_q-coupled receptor signaling. *J Gen Physiol*. 2013;141(5):537–555.

20. Suh BC, Horowitz LF, Hirdes W, Mackie K, Hille B. Regulation of KCNQ2/KCNQ3 current by G protein cycling: the kinetics of receptor-mediated signaling by G_q. *J Gen Physiol*. 2004;123(6):663–683.

21. Falkenburger BH, Jensen JB, Dickson EJ, Suh BC, Hille B. Phosphoinositides: lipid regulators of membrane proteins. *J Gen Physiol*. 2010;588(Pt. 17):3179–3185.

22. Hilgemann DW, Ball R. Regulation of cardiac Na^+, $Ca2^+$ exchange and KATP potassium channels by PIP2. *Science*. 1996;273(5277):956–959.

23. Falkenburger BH, Jensen JB, Hille B. Kinetics of PIP_2 metabolism and KCNQ2/3 channel regulation studied with a voltage-sensitive phosphatase in living cells. *J Gen Physiol*. 2010;135(2):99–114.

24. Suh BC, Hille B. Does diacylglycerol regulate KCNQ channels? *Pflugers Arch*. 2006;453(3):293–301.

25. Moraru II, Schaff JC, Slepchenko BM, et al. Virtual Cell modelling and simulation software environment. *IET Syst Biol*. 2008;2(5):352–362.

26. Slepchenko BM, Loew LM. Use of virtual cell in studies of cellular dynamics. *Int Rev Cell Mol Biol*. 2010;283:1–56.

27. Bünemann M, Frank M, Lohse MJ. Gi protein activation in intact cells involves subunit rearrangement rather than dissociation. *Proc Natl Acad Sci USA*. 2003;100(26): 16077–16082.

28. Lohse MJ, Vilardaga JP, Bünemann M. Direct optical recording of intrinsic efficacy at a G protein-coupled receptor. *Life Sci*. 2003;74(2–3):397–404.

29. Vilardaga JP, Bünemann M, Krasel C, Castro M, Lohse MJ. Measurement of the millisecond activation switch of G protein-coupled receptors in living cells. *Nat Biotechnol*. 2003;21(7):807–812.

30. Frank M, Thumer L, Lohse MJ, Bünemann M. G Protein activation without subunit dissociation depends on a $G_{\alpha i}$-specific region. *J Biol Chem*. 2005;280(26):24584–24590.

31. Hein P, Frank M, Hoffmann C, Lohse MJ, Bünemann M. Dynamics of receptor/G protein coupling in living cells. *EMBO J*. 2005;24:4106–4114.

32. McLaughlin S, Murray D. Plasma membrane phosphoinositide organization by protein electrostatics. *Nature*. 2005;438(7068):605–611.

33. McLaughlin S, Wang J, Gambhir A, Murray D. PIP_2 and proteins: interactions, organization, and information flow. *Annu Rev Biophys Biomol Struct*. 2002;31:151–175.

34. Murata Y, Iwasaki H, Sasaki M, Inaba K, Okamura Y. Phosphoinositide phosphatase activity coupled to an intrinsic voltage sensor. *Nature*. 2005;435(7046):1239–1243.

35. Tanimura A, Morita T, Nezu A, Shitara A, Hashimoto N, Tojyo Y. Use of fluorescence resonance energy transfer-based biosensors for the quantitative analysis of inositol 1,4,5-trisphosphate dynamics in calcium oscillations. *J Biol Chem*. 2009;284(13):8910–8917.

36. Hoshi N, Langeberg LK, Gould CM, Newton AC, Scott JD. Interaction with AKAP79 modifies the cellular pharmacology of PKC. *Mol Cell.* 2010;37(4):541–550.
37. Violin JD, Zhang J, Tsien RY, Newton AC. A genetically encoded fluorescent reporter reveals oscillatory phosphorylation by protein kinase C. *J Cell Biol.* 2003;161(5): 899–909.
38. Hernandez CC, Falkenburger B, Shapiro MS. Affinity for phosphatidylinositol 4,5-bisphosphate determines muscarinic agonist sensitivity of Kv7 K^+ channels. *J Gen Physiol* 2009;134(5):437–448.
39. Limbird LE. *Introduction to Receptor Theory Cell Surface Receptors: A Short Course on Theory & Methods.* New York: Springer; 2005, 1–28.
40. De Lean A, Stadel JM, Lefkowitz RJ. A ternary complex model explains the agonist-specific binding properties of the adenylate cyclase-coupled β-adrenergic receptor. *J Biol Chem.* 1980;255(15):7108–7117.
41. Gether U, Kobilka BK. G protein-coupled receptors. II. Mechanism of agonist activation. *J Biol Chem.* 1998;273(29):17979–17982.
42. Rajagopal S, Rajagopal K, Lefkowitz RJ. Teaching old receptors new tricks: biasing seven-transmembrane receptors. *Nat Rev Drug Discov.* 2010;9(5):373–386.
43. Kinzer-Ursem TL, Linderman JJ. Both ligand- and cell-specific parameters control ligand agonism in a kinetic model of G protein-coupled receptor signaling. *PLoS Comput Biol.* 2007;3(1):e6.
44. Gilman AG. G proteins: transducers of receptor-generated signals. *Annu Rev Biochem.* 1987;56:615–649.
45. Mukhopadhyay S, Ross EM. Rapid GTP binding and hydrolysis by G_q promoted by receptor and GTPase-activating proteins. *Proc Natl Acad Sci USA.* 1999;96(17): 9539–9544.
46. Zhang J, Bal M, Bierbower S, Zaika O, Shapiro MS. AKAP79/150 signal complexes in G-protein modulation of neuronal ion channels. *J Neurosci.* 2011;31(19):7199–7211.
47. Digby GJ, Lober RM, Sethi PR, Lambert NA. Some G protein heterotrimers physically dissociate in living cells. *Proc Natl Acad Sci USA.* 2006;103(47):17789–17794.
48. Raveh A, Riven I, Reuveny E. Elucidation of the gating of the GIRK channel using a spectroscopic approach. *J Physiol.* 2009;587(Pt. 22):5331–5335.
49. Benians A, Leaney JL, Milligan G, Tinker A. The dynamics of formation and action of the ternary complex revealed in living cells using a G-protein-gated K^+ channel as a biosensor. *J Biol Chem.* 2003;278(12):10851–10858.
50. Hern JA, Baig AH, Mashanov GI, et al. Formation and dissociation of M_1 muscarinic receptor dimers seen by total internal reflection fluorescence imaging of single molecules. *Proc Natl Acad Sci USA.* 2010;107(6):2693–2698.
51. Saini DK, Kalyanaraman V, Chisari M, Gautam N. A family of G protein βγ subunits translocate reversibly from the plasma membrane to endomembranes on receptor activation. *J Biol Chem.* 2007;282(33):24099–24108.
52. O'Neill PR, Karunarathne WK, Kalyanaraman V, Silvius JR, Gautam N. G-protein signaling leverages subunit-dependent membrane affinity to differentially control betagamma translocation to intracellular membranes. *Proc Natl Acad Sci USA.* 2012;109(51):E3568–E3577.
53. Gainetdinov RR, Premont RT, Bohn LM, Lefkowitz RJ, Caron MG. Desensitization of G protein-coupled receptors and neuronal functions. *Annu Rev Neurosci.* 2004;27:107–144.
54. Turcotte M, Tang W, Ross EM. Coordinate regulation of G protein signaling via dynamic interactions of receptor and GAP. *PLoS Comput Biol.* 2008;4(8):e1000148.
55. Yi TM, Kitano H, Simon MI. A quantitative characterization of the yeast heterotrimeric G protein cycle. *Proc Natl Acad Sci USA.* 2003;100(19):10764–10769.
56. Hansen SB, Tao X, MacKinnon R. Structural basis of PIP_2 activation of the classical inward rectifier K^+ channel Kir2.2. *Nature.* 2011;477(7365):495–498.

57. Whorton MR, MacKinnon R. Crystal structure of the mammalian GIRK2 K^+ channel and gating regulation by G proteins, PIP_2, and sodium. *Cell.* 2011;147(1):199–208.

58. Berstein G, Blank JL, Smrcka AV, et al. Reconstitution of agonist-stimulated phosphatidylinositol 4,5-bisphosphate hydrolysis using purified m1 muscarinic receptor, $G_{q/11}$, and phospholipase C-β 1. *J Biol Chem.* 1992;267(12):8081–8088.

59. Vazquez G, Tano JY, Smedlund K. On the potential role of source and species of diacylglycerol in phospholipase-dependent regulation of TRPC3 channels. *Channels.* 2010;4(3):232–240.

60. Imai Y, Itsuki K, Okamura Y, Inoue R, Mori MX. A self-limiting regulation of vasoconstrictor-activated TRPC3/C6/C7 channels coupled to $PI(4,5)P_2$-diacylglycerol signalling. *J Physiol.* 2012;590(Pt. 5):1101–1119.

61. Horrigan FT, Aldrich RW. Coupling between voltage sensor activation, Ca^{2+} binding and channel opening in large conductance (BK) potassium channels. *J Gen Physiol.* 2002;120(3):267–305.

62. Adelman JP, Maylie J, Sah P. Small-conductance Ca^{2+}-activated K^+ channels: form and function. *Annu Rev Physiol.* 2012;74:245–269.

63. Gamper N, Li Y, Shapiro MS. Structural requirements for differential sensitivity of KCNQ K^+ channels to modulation by Ca^{2+}/calmodulin. *Mol Biol Cell.* 2005;16(8):3538–3551.

64. Zaika O, Tolstykh GP, Jaffe DB, Shapiro MS. Inositol triphosphate-mediated Ca^{2+} signals direct purinergic P2Y receptor regulation of neuronal ion channels. *J Neurosci.* 2007;27(33):8914–8926.

65. Shapiro MS, Roche JP, Kaftan EJ, Cruzblanca H, Mackie K, Hille B. Reconstitution of muscarinic modulation of the KCNQ2/KCNQ3 K^+ channels that underlie the neuronal M current. *J Neurosci.* 2000;20(5):1710–1721.

66. Brown DA, Hughes SA, Marsh SJ, Tinker A. Regulation of M(K_V7.2/7.3) channels in neurons by PIP_2 and products of PIP_2 hydrolysis: significance for receptor-mediated inhibition. *J Physiol.* 2007;582(Pt. 3):917–925.

67. De Young GW, Keizer J. A single-pool inositol 1,4,5-trisphosphate-receptor-based model for agonist-stimulated oscillations in Ca^{2+} concentration. *Proc Natl Acad Sci USA.* 1992;89(20):9895–9899.

68. Li YX, Rinzel J. Equations for InsP$_3$ receptor-mediated $[Ca^{2+}]_i$ oscillations derived from a detailed kinetic model: a Hodgkin-Huxley like formalism. *J Theor Biol.* 1994;166(4):461–473.

69. Foskett JK, White C, Cheung KH, Mak DO. Inositol trisphosphate receptor Ca^{2+} release channels. *Physiol Rev.* 2007;87(2):593–658.

70. Mak DO, Pearson JE, Loong KP, Datta S, Fernandez-Mongil M, Foskett JK. Rapid ligand-regulated gating kinetics of single inositol 1,4,5-trisphosphate receptor Ca^{2+} release channels. *EMBO Rep.* 2007;8(11):1044–1051.

71. Gin E, Falcke M, Wagner LE, Yule DI, Sneyd J. Markov chain Monte Carlo fitting of single-channel data from inositol trisphosphate receptors. *J Theor Biol.* 2009;257(3): 460–474.

72. Ullah G, Daniel Mak DO, Pearson JE. A data-driven model of a modal gated ion channel: the inositol 1,4,5-trisphosphate receptor in insect Sf9 cells. *J Gen Physiol.* 2012;140(2):159–173.

73. Golebiewska U, Nyako M, Woturski W, Zaitseva I, McLaughlin S. Diffusion coefficient of fluorescent phosphatidylinositol 4,5-bisphosphate in the plasma membrane of cells. *Mol Biol Cell.* 2008;19(4):1663–1669.

74. Hilgemann DW. Local PIP_2 signals: when, where, and how? *Pflugers Arch.* 2007;455(1):55–67.

75. Szentpetery Z, Varnai P, Balla T. Acute manipulation of Golgi phosphoinositides to assess their importance in cellular trafficking and signaling. *Proc Natl Acad Sci USA.* 2010;107(18):8225–8230.

76. Kim YJ, Guzman-Hernandez ML, Balla T. A highly dynamic ER-derived phosphatidylinositol-synthesizing organelle supplies phosphoinositides to cellular membranes. *Dev Cell.* 2011;21(5):813–824.

77. Kim S, Kedan A, Marom M, et al. The phosphatidylinositol-transfer protein Nir2 binds phosphatidic acid and positively regulates phosphoinositide signalling. *EMBO Rep.* 2013;14(10):891–899.

78. Xu C, Watras J, Loew LM. Kinetic analysis of receptor-activated phosphoinositide turnover. *J Cell Biol.* 2003;161(4):779–791.

79. Nakatsu F, Baskin JM, Chung J, et al. PtdIns4P synthesis by PI4KIIIα at the plasma membrane and its impact on plasma membrane identity. *J Cell Biol.* 2012;199(6):1003–1016.

80. Balla A, Balla T. Phosphatidylinositol 4-kinases: old enzymes with emerging functions. *Trends Cell Biol.* 2006;16(7):351–361.

81. Iwasaki H, Murata Y, Kim Y, et al. A voltage-sensing phosphatase, Ci-VSP, which shares sequence identity with PTEN, dephosphorylates phosphatidylinositol 4,5-bisphosphate. *Proc Natl Acad Sci USA.* 2008;105(23):7970–7975.

82. Cunningham E, Thomas GM, Ball A, Hiles I, Cockcroft S. Phosphatidylinositol transfer protein dictates the rate of inositol trisphosphate production by promoting the synthesis of PIP_2. *Curr Biol.* 1995;5(7):775–783.

83. Willars GB, Nahorski SR, Challiss RA. Differential regulation of muscarinic acetylcholine receptor-sensitive polyphosphoinositide pools and consequences for signaling in human neuroblastoma cells. *J Biol Chem.* 1998;273(9):5037–5046.

84. Mao YS, Yin HL. Regulation of the actin cytoskeleton by phosphatidylinositol 4-phosphate 5 kinases. *Pflugers Arch.* 2007;455:5–18.

85. Oude Weernink PA, Schmidt M, Jakobs KH. Regulation and cellular roles of phosphoinositide 5-kinases. *Eur J Pharmacol.* 2004;500(1–3):87–99.

86. Gamper N, Reznikov V, Yamada Y, Yang J, Shapiro MS. Phosphatidylinositol 4,5-bisphosphate signals underlie receptor-specific $G_{q/11}$-mediated modulation of N-type Ca^{2+} channels. *J Neurosci.* 2004;24(48):10980–10992.

87. Golebiewska U, Kay JG, Masters T, et al. Evidence for a fence that impedes the diffusion of phosphatidylinositol 4,5-bisphosphate out of the forming phagosomes of macrophages. *Mol Biol Cell.* 2011;22(18):3498–3507.

88. Shin DM, Luo X, Wilkie TM, et al. Polarized expression of G protein-coupled receptors and an all-or-none discharge of Ca^{2+} pools at initiation sites of $[Ca^{2+}]_i$ waves in polarized exocrine cells. *J Biol Chem.* 2001;276(47):44146–44156.

89. Karunarathne WK, Giri L, Kalyanaraman V, Gautam N. Optically triggering spatiotemporally confined GPCR activity in a cell and programming neurite initiation and extension. *Proc Natl Acad Sci USA.* 2013;110(17):E1565–E1574.

CHAPTER ELEVEN

Modeling Molecular Pathways of Neuronal Ischemia

Zachary H. Taxin[*], Samuel A. Neymotin[*], Ashutosh Mohan[*], Peter Lipton[†], William W. Lytton[*]

[*]Department of Physiology & Pharmacology, SUNY Downstate, New York, USA
[†]Department of Neuroscience, University of Wisconsin, Madison, Wisconsin, USA

Contents

Abstract

Neuronal ischemia, the consequence of a stroke (cerebrovascular accident), is a condition of reduced delivery of nutrients to brain neurons. The brain consumes more energy per gram of tissue than any other organ, making continuous blood flow critical. Loss of nutrients, most critically glucose and O_2, triggers a large number of interacting molecular pathways in neurons and astrocytes. The dynamics of these pathways take place over multiple temporal scales and occur in multiple interacting cytosolic and organelle compartments: in mitochondria, endoplasmic reticulum, and nucleus. The complexity of these relationships suggests the use of computer simulation to understand the

Progress in Molecular Biology and Translational Science, Volume 123
ISSN 1877-1173
http://dx.doi.org/10.1016/B978-0-12-397897-4.00014-0

interplay between pathways leading to reversible or irreversible damage, the forms of damage, and interventions that could reduce damage at different stages of stroke. We describe a number of models and simulation methods that can be used to further our understanding of ischemia.

Neuronal ischemia is a condition of transient or chronic low blood flow to the brain. Stroke (also called cerebrovascular accident, CVA) is a consequence of ischemia. In the United States, stroke is the fourth leading cause of death and is a major cause of disability and functional cognitive impairment, particularly in aging populations.[1]

The brain consumes more energy per gram of tissue than any other organ. Hence, ample cerebral blood flow and its autoregulation are critical to neuronal functioning. Loss of blood flow for even a few minutes triggers irreversible damage, only partially offset by neuroprotective mechanisms. Recently, much research has focused on elucidating these mechanisms for control of intrinsic neuroprotective pathways in order to help design therapeutic interventions. The complexity of the many different cellular and subcellular signaling pathways involved makes computer modeling an important tool for studying ischemia.

A stroke can be subdivided into a central area of severe ischemia, the ischemic core, and a surrounding area of damaged tissue, the ischemic penumbra. This penumbral area is considered the best target for recovery, as its cells still retain some viability. Therefore, in discussing the patterns of damage at the cellular and molecular level, we must explore the consequences of ischemia in both areas, considering also the damaging effects of cell death at the core through influence on the penumbra via local diffusion of released toxic cell contents.

Several of the models to be presented discuss the effects of *preconditioning*. Preconditioning is a therapeutic approach in which a pretreatment is used in order to prepare the brain for subsequent stroke, by initiating the varieties of intrinsic neuroprotective proteins and pathways. The most obvious preconditioning paradigm uses prior mild ischemia to trigger these changes, thereby providing partial protection against the effects of the more major subsequent event. However, this is not an approach that can be used clinically, due to the difficulty of titrating an appropriately minor ischemia, to avoid triggering the irreversible processes that would themselves produce an immediate stroke. Other preconditioning techniques utilize pharmacotherapeutic methods to activate these pathways.

1. CRITICAL INITIATOR EVENTS IN ISCHEMIC PATHWAYS

Ischemia triggers dysfunction in several pathways, with multiple final outcomes which include multiple paths to death or to recovery. The pathways to death include (1) apoptosis: programmed cell death which permits cells to fold inward in a controlled manner; (2) necrosis: uncontrolled cell collapse with membrane disruption releasing multiple toxic cell contents; (3) necroptosis: an intermediate condition where apoptosis is begun but cannot complete due to further disruption. Although the endgames differ, all ischemic events start with a similar set of initial cellular events that progress in a characteristic sequence, triggered after a 1- to 3-min period of metabolite reduction. However, following this initial sequence, ischemia sets in motion an extremely diverse set of enzymatic, control-cascade, mechanical, proteomic, and genomic changes that occur across many temporal scales, and which interact across milliseconds to days in various combinations. Some ischemic insults will cause immediate cell death via cytotoxicity and necrosis, whereas other levels of insult engage programmed cell death over several days. Still others will leave cells in various stages of prolonged attempted recovery. These many pathways suggest the possibility of many different points of possible therapeutic intervention at different temporal stages and in different parts of the umbra/penumbra.

All cells require glucose and O_2 to run oxidative phosphorylation in the mitochondria for production of adenosine triphosphate (ATP), the major energy currency within cells. Without these metabolites, cellular respiration will cease, rapidly diminishing available ATP. For this reason, some cell types maintain alternative short-term energy stores, allowing compensation for transient reductions in oxygen delivery to tissue. In particular, muscle and liver maintain significant stores of glycogen which serve to keep glucose levels from falling during exercise when metabolite levels are reduced. Creatine phosphate is another store of energy, good for up to 5 s of cellular activity. Unlike other tissues in the body, neurons and glia do not have a substantial energy reservoir. Glycogen is found in small amounts in glia, but is almost absent in neurons. Creatine phosphate exists in most brain tissue cells, but only provides brief respite from metabolic lack. Therefore, within 1–2 min of oxygen and glucose deprivation that occurs in complete ischemia, ATP levels in brain tissue fall to roughly 20% of their control levels.

Among the myriad homeostatic roles of ATP is the energy for the Na^+/K^+-ATPase pump that maintains the transmembrane ionic gradients for these two major ions, thereby maintaining resting membrane potential. In neurons, this pump must be tightly regulated to ensure appropriate conditions for firing action potentials. When the Na^+/K^+-ATPase fails, intracellular Na^+ increases dramatically. Functional consequences of this shift in equilibrium include: loss of signaling control, changes in cell volume, and swelling of cell organelles. Ionic dysregulation also causes anoxic membrane depolarizations, whereby large number of cells begin to depolarize significantly over long periods of time. These depolarizations cause further damage.

Loss of O_2 and reduced cellular respiration also results in the formation of reactive oxygen species (ROS), which are devastating to the cell. ROS are highly reactive species which damage both proteins and nucleic acids. ROS are generated during normal oxidative phosphorylation. But, in the healthy cell, ROS are then passed on to a set of conversion methods which transform them into water-soluble species that can be readily eliminated by the cell. As the mitochondrial machinery for respiration is interrupted, the cell begins to accumulate these intermediate ROS. Additional ROS are produced by generation of free fatty acids from mitochondrial membranes due to upregulation of oxygenase proteins (e.g., cyclo-oxygenase-2 or lipoxygenase). In addition to further damaging transmembrane pumps and other homeostatic control proteins, ROS are particularly damaging to DNA. Damage to DNA is a key initiating factor in control of the p53 protein, which is directly responsible for cell fate commitment leading to cell death via apoptosis. ROS production is particularly important in the penumbra and can increase markedly after reperfusion, reducing the efficacy of treatments that restore blood flow.

As respiration fails, increased intracellular Na^+ and consequent depolarization leads to glutamate being released into surrounding tissue. Extracellular glutamate, as an excitatory neurotransmitter, then opens synaptic channels causing influx of Ca^{++}, Na^+, and K^+, and further contributing to intracellular ion dysregulation, a phenomenon known as excitotoxicity. Ca^{++} is a major second messenger used in many pathways. A pathological increase in Ca^{++} is particularly deleterious, leading to further loss of homeostasis and disruption of multiple Ca^{++}-dependent pathways. Increased Ca^{++} levels also serve as a signal to initiate cell death pathways,[2] such that inhibitors of calpain (a Ca^{++}-activated protease) can markedly reduce cell death in *in vivo* animal models.[3] Ca^{++} rise is further augmented by release from intracellular stores in endoplasmic reticulum (ER) and mito-chondia through Ca^{++}-induced Ca^{++} release (CICR), as well as due to reversal of membrane Na^+–Ca^{++} ion exchange pumps.[4] In addition to its role in regulating cellular subprocesses (e.g., calmodulin, calbindin, CaMKII, etc.,), Ca^{++} is involved in mitochondrial and nuclear regulation. Although it remains somewhat unclear to what extent normal Ca^{++} regu-latory processes are localized to these separate organelles, it is clear that Ca^{++} does pass readily across compartments, for example, being taken up by mito-chondria from the surrounding cytosol.

In summary, the loss of ATP during ischemic insult is the prime mover for a series of interrelated signaling cascades with multiple important conse-quences that include:

- Ionic homeostasis dysregulation depolarizing resting membrane potential
- Generation of ROS, damaging proteins, DNA, and membranes
- Release of glutamate leading to excitotoxicity
- Increased intracellular Ca^{++}
- Loss of cell volume control
- Mitochondrial instability due to volume changes and ion flux

With the cell destabilized in these many ways, it may follow one of several paths that lead to cell death. Alternatively, following vascular reperfusion, the cell may return to a near-normal state. It is crucial to note that some changes are transient and reversible, while others represent major alterations in homeostatic cell processes that have inevitable negative consequences.

2. MODELS OF MOLECULAR PATHWAYS IN ISCHEMIA

There are several approaches used to model molecular pathways: sto-chastic modeling techniques at various levels of granularity, detailed mass-action kinetic models using systems of ordinary differential equations (ODEs), and reaction-scheme methods such as the Boolean network (BN) formalism.

This ordering of methods reflects decreasing modeling precision. The most detailed stochastic models, those that follow individual molecules using a Monte Carlo method,[5] require the most detailed parameterization, requir-ing the availability of large amounts of experimental detail, which may not always be available. Additionally, such models require substantial computer resources, making them impractical for very large problems such as a simu-lation of an entire neuron. At the other extreme, a BN provides only a very gross approximation of the underlying reality, capturing a sequence of molecules which turn other molecules on or off with no indication of rate or of other complexities of bimolecular reactions. Such schemes have the advantages of requiring only limited experimental data and running quickly as simulations. An additional aspect of molecular modeling is the movement of molecules by diffusion (reaction–diffusion techniques) or by transport across subcellular compartments and organelles.

Several models, using a variety of techniques, have been developed to assess molecular pathways in ischemic injury. We discuss several of them here.

2.1. Model of ionic dysregulation and cytotoxic edema

In the models of Dronne et al.[4,6], the key components of an evolving stroke can be grouped into 10 interacting submodels: (1) Metabolic reactions within the cell (O_2 lack, loss of ATP); (2) ionic dysregulation; (3) development of cytotoxic edema due to ionic disturbances and excitotoxic changes; (4) glutamate-dependent excitotoxicity; (5) spreading waves of glutamate depression; (6) nitric oxide (NO) synthesis; (7) tissue inflammation and local inflammatory response; (8) necrosis; (9) apoptosis; and (10) reperfusion-dependent tissue changes. Although the models interact, they can be evaluated independently as well.

We will focus here on one of their models,[4] a model of the development of cytotoxic edema—water accumulation within cells. Edema is caused by water following ion fluxes, and is one way that large-scale, tissue-wide changes can be linked to disturbances at the molecular scale. In addition to cytotoxic edema, stroke will cause other forms of edema, vasogenic and mechanical, that will also have profound effects.

In the Dronne model, intraneuronal space, intra-astrocytic space, and extracellular space were modeled as separate compartments. For simplification, it was assumed that the system was closed and that global volume remained constant. Five main species Na^+, K^+, Ca^{++}, Cl^-, and glutamate were considered in the model.

In this model, compartment volumes, represented as a fraction of a constant total volume, altered with species flux as water followed ionic gradients to maintain osmotic equilibrium. Volume changes were the consequence of time-dependent variations in species concentrations due to passage across membranes, modeled as occurring through the many voltage-gated channels, volume-activated channels, pumps, exchangers, and ligand-gated channels. However, open probabilities for channels were defined at steady state rather than using dynamic gating, because only consequences of long-term ischemic insult were being studied. Neuron and astrocyte compartments reached osmotic equilibrium following ionic disturbance. The volume of extracellular space was then calculated based on total volume. This is a clinically relevant measure because the ratio of the apparent diffusion coefficient of water can be measured using diffusion-weighted MRI scan, and is used as an indicator of the severity of acute stroke.[7]

A related change in the cellular microenvironment immediately after ischemic onset is a rise in extracellular K^+ due to activation of voltage- and Ca^{++}-dependent K^+ channels with depolarization as well as failure of the Na^+–K^+ pump. The other species which accumulated in the extracellular space was glutamate, released due to depolarizing potentials that triggered synaptic release, as well as failure of the glutamate carrier to uptake and remove glutamate from the extracellular space. Intracellularly, Ca^{++} increases of nearly 4 orders of magnitude were observed in the neuron compartment. These intracellular Ca^{++} dynamics were shown to be relatively slow, reaching their maxima nearly twice as slowly as changes in other ion concentrations. This slow dynamics was attributed to the Ca^{++} changes being primarily due to reversal of Na^+–Ca^{++} exchange leading to Ca^{++} influx. The process of Ca^{++} accumulation was increased further by NMDA receptor activation by extracellular glutamate, as well as Ca^{++}-induced Ca^{++} release from within the cell.

Having explored the natural history of ischemia in the model, the authors then looked at manipulation of individual channel conductances, in order to determine where modulation of channel activations might have protective effects in terms of reduction in edema. An increase in persistent Na^+ channel density was found to have a deleterious effect, promoting cytotoxic edema, suggesting that pharmacological blockade of this channel might be protective. By contrast, *increasing* conductance of the delayed rectifier K^+ channel was protective.

The model suggested the hypothesis of a causal sequence for development of edema, leading from pump failure to depolarization to pathological voltage- and Ca^{++}-dependent channel activation. These failures are then propagated by ionic imbalances across membranes compartments. Ion dysfunction cannot be fully compensated by non-ATP-based transmembrane ion exchangers or leak channels, which have little ability to stabilize fluxes after failure of energy-driven pumps.

As with all models, it is important to consider the results in the context of what the modelers left out. The model is primarily at the multicellular and tissue scale spatially and at the minutes-to-hours temporal scale. It therefore ignores neuronal network effects as well as more detailed intracellular molecular effects. Regional distribution of neurons and glia is not homogeneous throughout the brain, so the model might need to be modified to consider effects of ischemia in different brain areas. Of necessity, some types of channels and ions were omitted—in particular effects of pH and pCO_2, though acidity can have a large effect on cytotoxicity. The parameters used were optimized from a variety of sources to produce a steady state for the model. Not only were transient dynamics thereby omitted, but the approach also raises concerns as to whether typical steady-state values would still hold in the context of ischemic tissue, due to alterations in intrinsic ion–channel modulation by second messengers and phosphorylation state.

2.2. Model of astrocytic protection

Diekman *et al.* recently developed a model that explores how IP_3-mediated Ca^{++} release in astrocytes may protect against their involvement in cytotoxic edema and other local microenvironment damage.[8] This astrocyte model built on the classic ER CICR model of Magnus and Keizer.[9] The model looked at state variables for Ca^{++}, ATP, ADP, pH, NADH in the cytosolic, ER, and mitochondrial compartments. As in the prior model, ischemia caused the astrocytes to lose the ability to maintain electrochemical gradients due to loss of energy sources.

Ischemia was simulated by reducing glucose and O_2 inputs by a constant percentage at a given time. The two metabolites were tested individually and together: similar results were found with these three interventions—increased cell volume, depolarized membrane, and reductions in ATP, NADH, and Ca^{++}. Next, with both metabolites reduced jointly, IP_3 production was augmented, as can be done pharmacologically using an astrocyte-specific receptor. With a short period of ischemia, addition of IP_3 caused mitochondrial Ca^{++} to return to a higher level, improving ATP production. In another set of simulations, a mitochondrial K_{ATP} channel was included and also had a protective effect. These results suggested that loss of mitochondrial Ca^{++} homeostasis might be destabilizing, such that restoration of Ca^{++} levels could prevent an unfavorable outcome.

This work demonstrated a mechanism by which regulation of astrocyte Ca^{++} metabolism could have significant neuroprotective effects against loss of energy substrates in ischemia. However these protective mechanisms, stimulation of IP_3 and the K_{ATP}, would only work up to a certain point. Beyond that point Ca^{++} is too severely destabilized for recovery to occur. The model also suggested that metabolically impaired astrocytes could use high Ca^{++} to signal distress to downstream effectors of cell death.

2.3. Model of Ca^{++} dynamics and mitochondrial permeability

Alterations in mitochondrial permeability appear to be a common pathway leading to cell death in a number of pathological conditions, including ischemia. Experimental research has led to the identification of a permeability transition pore (PTP) as a cause of a pathological mitochondrial permeability transition which is related to increased intracellular Ca^{++}. Oster et al. explored a model of how intracellular Ca^{++} dynamics would produce changes in mitochondrial membranes which would lead to cell death.[10] This model was also built by extending the Magnus–Keizer model of CICR.[11]

Mitochondria use an electrochemical proton gradient to run the ATPase pump during oxidative phosphorylation, thereby generating ATP. This process has the secondary effect of sequestering free cytosolic Ca^{++} via a uniporter mechanism driven by the proton-based mitochondrial electrical gradient. ATP-dependent Ca^{++} pumps and Ca^{++}–Na^+ exchange also play a role and are also included in the model. The model added dynamic pH buffering by inorganic phosphates and a pH-dependent activation mechanism for the PTP to the basic model of CICR. Further channel dynamics were included, allowing for Ca^{++} uptake saturation.

The PTP was itself modeled as a three-state channel: closed, open low-conductance (PTP_l), and open high-conductance (PTP_h). Transitions from PTP_l to PTP_h required elevated mitochondrial Ca^{++} levels. Once the PTP_h state was achieved, it remained constitutively open. This led to cell death via extrusion of mitochondrial proteins such as Cytochrome C. Cytochrome C catalyzed feedforward activation of procaspases, the intrinsic cytosolic enzymes involved in the apoptosis pathway. The activation of the PTP_h state can therefore be considered an irreversible step towards cell death.

Changes in PTP_l state were modeled as dependent upon proton concentration in the mitochondria (H_M), where the opening rate and time constant of activation (τ_l) were also both dependent upon proton concentration using a variation on the Hodgkin–Huxley parameterization via steady state and time constants which are here dependent on H_M, and parameterized with amp_τ, and p_i:

$$\frac{dPTP_l}{dt} = \frac{[PTP_{l\infty}(H_M) - PTP_l]}{\tau_1(H_M)}$$

$$PTP_{l\infty}(H_M) = 0.5 \cdot \left(1 + \tan h\left(\frac{p_1 - H_M}{p_2}\right)\right)$$

$$\tau_1(H_M) = \frac{amp_\tau}{\cos h\left(\frac{H_M - p_3}{p_4}\right)} + p6$$

Ca^{++} flux through the PTP was assumed to be similar to flux through the uniporter.

Experiments have demonstrated that the cell's transition to an unhealthy state may require an extended period of time for full development, depending on the severity of damage. Therefore, modules were added to quantify PTP state transitions over long-term Ca^{++} fluctuations in the cell. This part of the model included a time–delay function in the equation for change in the PTP_h gate state. Similarly, a threshold equation was developed that took into account time-dependent Ca^{++} concentration changes within the mitochondria. Further equations modeled proton flux and proton buffering. Because these dynamics are poorly understood within mitochondria, a simple sigmoidal functional of acid buffering based on proton concentration within the mitochondria was used. Ca^{++} diffusion was modeled within the cytoplasm with no explicit cytoplasmic buffering. However, an effective diffusion constant (K_{eff}) was used based on the modification expected due to a fixed concentration of buffer (B_T) and a single dissociation constant (K_C): $D_{eff} = D_{ca} \cdot K_C / (B_T + K_C)$.

Simulations were performed by inducing pulses of Ca^{++} at varying intensities over varying time scales. This was intended to mimic aspects of oscillating Ca^{++} pulse waves that occur as a result of the CICR and increased cytosolic Ca^{++} that would result from an ischemic insult. While slow pulses (three pulses of 2 s each over 90 s interval) could be adequately buffered by the mitochondria, fast pulses (three pulses of 2 s each over 25 s) resulted in transient mitochondrial depolarizations that caused release of sequestered Ca^{++} into the cytoplasm. Slow and fast pulses also differed in how they affected mitochondrial pH. Slow pulses caused transient pH elevations that were compensated, whereas rapid pulses caused pH to exceed threshold levels for a long enough time that the PTP_l state was activated. Overall, Ca^{++} recovery via sequestration was found to occur faster than pH buffering.

2.4. Model of death by ROS

The generation of ROS is a powerful stimulus for cell death via damage to mitochondria and DNA. Low concentrations of ROS are produced during normal respiration. The cell eliminates these with scavenging enzymes and uncoupling proteins. As their names suggest, scavenger proteins search and absorb toxic species, while uncouplers physically disconnect elements of the electron transport chain to reduce ROS production. Excess ROS, exceeding the capacity of scavenging, is produced in the mitochondria when the electron transport chain fails.

The major electron donors to the oxidative chain are NADH and FADH$_2$, which are reduced through passage of the high-energy electron through the cytochromes. Heuett and Periwal developed a cell-scale model for production and regulation of ROS in response to metabolic dysfunction based on changes in NADH and FADH$_2$ concentration.[12] These were modeled as a function of glucose and oxygen availability. ATP concentration was also included and was dependent upon tricarboxylic acid cycle flux, F1–FO ATPase flux, and the efflux of ATP removal from the mitochondrium. The instantaneous rate of change for ROS concentration was given by the sum of its generation fluxes (from NADH and FADH$_2$) and its removal fluxes by scavenger and uncoupling proteins. Initial conditions included the concentration of NAD and FAD. Further modifications to the model took into account ROS-induced suspension of electron transport, and change in proton leak across the membrane. This allowed a way to test several mechanisms of pH change and Ca^{++} flux due to cellular damage.

2.5. Nitric oxide as a broadcast signal

Alam et al. explored the relationship between Ca^{++} concentrations and the feedback oscillation of apoptotic control proteins in the context of cellular stress.[13] Although cell stress can occur in a variety of ways, the most common is oxidative stress due to metabolic dysfunction. Once again, Ca^{++} is implicated, here increasing production of nitric oxide (NO) via its synthases. NO is a regulator of subcellular processes, and induction of NO has been demonstrated to be significant in the apoptotic decision.[14] A major downstream target of NO is Mdm2, which is a negative regulator of p53. p53 is a proapoptotic control protein and a primary regulator of cell fate decisions and changes in cell cycle activity. NO disrupts the inhibitory effect of Mdm2 by forming an Mdm2–NO complex that eliminates Mdm2 effects on p53. Mdm2 and p53 both have short half-lives, approximately

15–30 min. Therefore, levels of these proteins are constantly shifting, along with their transcription regulation. NO, a gas, diffuses extremely rapidly and can move directly through membrane, allowing it to have a volume effect.

Various levels of Ca^{++} concentration were tested for their effect on the oscillation dynamics of Mdm2 and p53. Levels of stability were identified which depended on the rate of Ca^{++} production: (1) a stable regime where Mdm2–p53 oscillation yielded apoptosis, (2) a partly stabilized regime with a damped oscillation. The effects of Ca^{++} oscillations were also tested. The results suggested that oscillating waves of diffusible Ca^{++} could induce widespread changes in p53 activity across the cell, and that independent Mdm2–p53 systems in different regions could be simultaneously activated by a sufficient Ca^{++} stimulus. The consequences of stochastic variation were then explored. High noise disrupted oscillations. With reduced noice, the multiple systems become synchronized, causing p53 oscillations to generate apoptosis. Stochastic noise could partially mask the effects of Ca^{++} on the Mdm2–p53 population when the system size was small. As the system increased, further coordination pushed the cell towards signal synchronization.

Taken together, these results showed how Ca^{++} levels could act as signal integrators for the downstream regulation of networks of cell cycle proteins. Such signaling would be highly dynamic and would involves feedback fluctuations and oscillations. An important stochastic element was suggested for the network regulation of apoptosis, with the level of noise directly correlated to system stability.

2.6. BN model of apoptotic protein regulation

In assessing complex system behavior, it can be helpful to distil the multifactorial elements such that stability and characteristics of the system under different conditions can be more readily modeled. The BN formalism provides a convenient way to model apoptotic pathways by using binary rules for pathway relationships.

The model of Mai et al. used BNs to explore the apoptosis network.[15] Both intrinsic and extrinsic pathways were represented: 40 nodes represented both extracellular mediators (TNF, tumor necrosis factor, which is proapoptotic; GF, growth factor—protective) and intracellular mediators (TRADD, Caspases, Akt, NFkB, p53, etc.,). The intrinsic pathway also included mitochondrial proteins in the Bcl-2 and Bax families. To simplify the model, some molecules with similar functions were combined in a single node. The model was divided into functional groups; pathways were generally grouped as being proapoptotic or prosurvival, although it was the behavior of the system with particular initial conditions and extrinsic influences that would determine how they interacted: proteins might have multiple functions depending on coactivations. Connections at p53 were given especial attention because of this protein's critical function as the final common pathway for cell fate. p53 received inputs from extrinsic and intrinsic apoptotic signals, as well as prosurvival signals from the Akt pathway.

The network was driven through activation of the extrinsic pathways which activate transmembrane domains leading to a sequence of feedforward reactions though the caspaces. Nodes are Boolean, either ON or OFF. To determine a node state, activating (A) and inactivating (H) inputs were compared at a given time-step:

$$S_i(t+1) = \begin{cases} \text{OFF} & \text{if } A_i(t) < H_i(t) \\ \text{ON} & \text{if } A_i(t) > H_i(t) \\ S_i(t) & \text{if } A_i(t) = H_i(t) \end{cases}$$

Current node state depends only on the inputs at the current iteration except when the S remains unchanged. Note that the network included both feedforward and feedback loops (Fig. 11.1).

At the beginning of each simulation, all internal nodes had an initial probability of 0.5 to be in the OFF or ON state. A network state could then be defined as a particular vector of node states. Each exploration of system fate assessed thousands out of the 2^{40} possible randomized initial network states in the context of particular inputs. This represented the concept that the network would be taken unaware by the sudden onset of a stroke with arrival of the extrinsic apoptotic signals. However, it neglects the likelihood that the network generally exists in some subspace of stable dynamical modes that are favored. Initial networks states that always led to cell death, for all inputs, were omitted from further analysis. After running the randomizations, the possible outputs were evaluated statistically. A *DNA damage event node* that remained turned on for more than five time-steps was considered lethal. If a certain number of time-steps occurred without activating that node that was considered survival. Survival in the context of an input could be defined by a unique set of initial network states. These states could then be tested for stability by perturbing the system by swapping one or two nodes and determining if the system still permitted survival. Network states which were stable for survival were explored in greater depth.

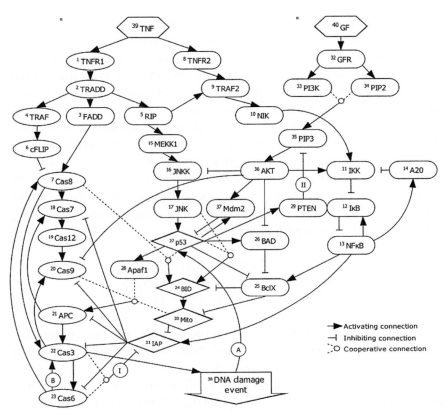

Figure 11.1 Schematic for BN model shows both feedforward and feedback loops from the extrinsic inputs at top to the potential for DNA damage at bottom. Note the sequence of caspaces down the left side which lead to the final *DNA damage event* at the bottom.[15]

Statistical analysis of different sets allowed determination of key nodes in initial states that were lethal in situations where no external apoptotic signals were given, or when protective growth factors were stimulated. These results were compared to damaged models, in which specific connections between internal nodes were severed (edge-ablations), simulating the potential for damage to intermediate signal proteins by ROS. An apoptosis ratio was defined as the number of initial states causing apoptosis divided by the total number of initial states. This ratio was then compared in 16 models with different edge-ablation combinations and across several different input signal combinations. For instance, the apoptosis ratios in a model with pathways blocked was compared for simulations with no external signal (i.e., steady state), growth factor alone, TNF alone, or both TNF and growth factor signals. These simulations also included signal interruptions for TNF and mitochondrial signals (represented as a single node).

Several conclusions emerged from this work. (1) The intrinsic state values are less important than extrinsic signaling in determining the cell fate. (2) A set of points in state space are identifiable which represent an irreversible pathway to apoptosis. In some cases, this fate could be predicted considerably ahead, indicating that the process could be a prolonged one that might be susceptible to external intervention. (3) Loss of a specific pathway could significantly affect probability of apoptosis. (4) Stability varied among survival network states. Greater concentrations of extrinsic (protective) growth factor increased cell stability. Overall, these results are consistent with experimental evidence, which has demonstrated surprising variability of apoptosis in ischemic preparations. Therapeutically, this approach will allow us to separate mechanisms with different pathways to their fate, with different mechanisms suggesting different approaches to intervention.

2.7. BN model of ROS effects and gene regulation

In the prior model, DNA damage was considered the final event that induces apoptosis. In a recent paper, Sridharan *et al.* (2012), modeling p53-mediated apoptosis in response to ROS-induced DNA damage, used gene regulation to explore mechanisms of ROS cytotoxicity.[21] This model also used the BN approach to localize points of pathway control. The modeling approach used here added the method of *fault identification*. Fault identification is an engineering method used to determine the vulnerability of electronics or other complex systems to failure of a single component.

Cells contain a variety of mechanisms for dealing with generation of ROS. Nrf2 is a transcription factor which upregulates antioxidant genes, producing enzymes to reduce ROS into metabolizable products, such as H_2O_2. Nrf2 is kept inactive in the cytosol by a Keap1–Nrf2 complex. Elevated ROS levels break up this complex, allowing Nrf2 to enter the nucleus and upregulate transcription, a long-term process that occurs over 6–24 h. Therefore, significant elevations in ROS requires a millisecond timescale to effect some cellular damage but a much longer time scale to push the cell into programmed death (if the initial cellular damage does not destroy the cell—necrosis). In the Sridharan model, a network map of Nrf2-related interaction proteins and genetic response elements was constructed as a BN.

The activity of the gene regulatory network was then explored in the context of the Akt pathway (which includes p53 and mdm2) for apoptosis activation. The model included a *Stuck-at-Faults* node state, which means that a particular pathway element's value remains constant regardless of input. This is akin to genetic mutations, whereby a protein remains constitutively ON or OFF, but might also occur in the setting of ROS damage to a protein.

The network was then simulated by adding an *external stress input signal* for 50 time-steps. This was intended to provide an adequately prolonged stimulus to activate apoptosis dynamics. When stress $= 0$, a network map demonstrated a single point attractor state for the entire network. However, when stress $= 1$, an attractor cycle (limit cycle) was established, producing activation oscillation.

By comparing multiple simulations over time, a set of fault locations was generated that produced aberrant behavior. The primary input was stress, and the primary output was the Bad/Bcl-2 ratio, where Bad (the Bcl-2-associated death promoter) and Bcl-2, pro- and antiapoptotic factors respectively, are critical in the pathway to apoptosis. Secondary outputs were also defined. Analysis of the network demonstrated the existence of *Homing sequences* as a set of system inputs that bring the network to a particular internal state, regardless of the initial state of genes within the network. Such a Homing sequence was applied to a particular set of initial states and then to the same set but with a fault added to see the effect of such a fault.

It was demonstrated that oxidative stress produced limit cycles where genes proceeded through cycles of upregulation and downregulation based on transcription factor changes. When combined with the Homing sequence algorithm, the network could identify likely fault locations for within the apoptotic network and associate a given fault with its likely outcome.

2.8. Kinetic model with Boolean macroparameters for gating of apoptosis signals

As noted above, Bcl-2 related proteins, including Bad, Bax (Bcl-2-associated X protein), Bak (Bcl-2 homologous antagonist/killer), and others of this family are critical in the process of programmed cell death. In order to extend the BN formalism to provide greater detail, ODEs or stochastic modeling can be used but analyzed in terms of Boolean logic or used in conjunction with BN. Such combined methods have a greater computational load, so generally reduce the scope of the number of interactions that can be explored. However, they provide greater depth and detail in understanding the dynamics of those pathways that are retained. Unlike BNs, ODE systems can incorporate experimentally determined rate kinetics, such as mRNA degradation rates and binding kinetics for protein–protein interactions.

A recent model by Bogda *et al.* set out to examine levels of control for proapoptotic effector proteins by using a combined ODE and Boolean approach.[16] Activity levels of Bcl-2 family proteins in response to DNA damage (ROS) and external growth factor (GF) signals were modeled in order to determine whether or not these proteins are effective integrators of upstream apoptotic signals transmitted by the Akt and p53 protein pathways. Although these two pathways are somewhat interdependent (i.e., they share common effector molecules), they are independent enough that they can be explored separately. When p53 and Akt are activated ($\{p53,Akt\} = \{1,1\}$), they initiate downstream effectors, resulting in increased accumulation of Bax. High concentration of Bax initiates apoptosis via release of Cytochrome C from mitochondria. Subsequent events involve feedforward and irreversible caspase activation in the common pathway to death.

Major sequences in the two pathways are as follows:

1. Dephosphorylation of the prosurvival Akt pathway leads to apoptosis:
   (a) Dephosphorylated Akt causes dephosphorylation (activation) of Bad; (b) dephosphorylated Bad is released from $Scaffold_{14-3-3}$ protein; (c) dephosphorylated Bad initiates Bad–Bcl-x_L binding; (d) Bad–Bcl-x_L binding causes dissociation of Bax–Bcl-x_L, thereby freeing Bax; (e) Bax causes Cytochrome C release; (f) Cytochrome C activates procaspase cleavage; and (g) Caspase cascades lead to apoptosis.

2. Changes in p53 dynamics lead to apoptosis:
   (a) p53 is activated in response to DNA damage (ROS); (b) activated p53 induces synthesis of its own inhibitors; (c) activated p53 also initiates cell cycle arrest and DNA repair; (d) at sufficiently high levels, p53 acts as a transcription factor for Bax and Bak; (e) Bax $\cdots$ —see 1e above.

The model used ODEs to represent species concentrations of relevant proteins and mRNA, including Bax_{mRNA}, Bax, Bcl-x_L, Bad, and Bad-related proteins. Activated procaspases and caspases were also modeled. As in many regulatory networks, bistability plays a critical role in activating and silencing a particular pathway. Bax concentration was identified as a bifurcation parameter: above a threshold concentration of Bax, caspase activation occurs in a feedforward manner and drives the cell irreversibly towards apoptosis. The threshold here was time dependent; that is, when Bax levels were sufficiently high over a period of time, the switch was activated. This condition was important because short threshold breaches of Bax occur from time to time and in nonapoptotic conditions, and do not normally cause apoptosis. Like other regulated cell decisions, apoptosis involves transcription of new protein. The rate of Bax_{mRNA} formation is modeled as a Hill equation, and is dependent upon activated p53.

The simulations' steady-state responses were analyzed for levels of signal necessary to initiate apoptosis, in the context of a background condition of {p53,Akt} represented by Boolean {0,0}, {1,0}, {0,1}, {1,1}, where 1 represented in each case a proapoptotic state (hence representing unphosphorylated Akt). The consequences of different conditions could then be described as OR or AND gates (OR requiring only activation of one of the two signals, AND requiring both). In these analyses, the Boolean represented initial low or high concentration of the factor; the simulation was then followed, based on its inputs, to determine the consequences once the steady state of the kinetic model was reached. When total levels of Bad were low, only the AND gate pathway led to apoptosis. In this case, both proapoptotic gates {p53,Akt} = {1,1} were required to synergistically activate apoptosis. By contrast, an apoptotic OR gate was found for Bad and Bcl-x_L activation. With high initial concentrations for Bad and Bcl-x_L, all cases except {0,0} generated enough Bax to push the cell towards apoptosis, demonstrating the OR gating. The {0,0} condition led to survival: a situation with low DNA damage and high growth factor. The model also includes the ability to transform an OR gate into an AND gate when Bcl-x_L levels rise, which is a physiologically important aspect of these feedback systems in real cells. Therefore, the relationship between Bax and Bcl-x_L concentrations can effectively determine how the cell responds to proapoptotic or prosurvival signals.

After testing several combinations of parameters and gates, the minimum duration of elevated signals necessary to trigger apoptosis was investigated. Simulation suggested that conditions that required coactivation of both pathways (AND gate), would not settle into an apoptotic state even after 10 h of simulated time. This was a surprising result, considering that Bax and caspase levels were both significantly elevated. After 11 h of simulated time, the system did settle into a state of constitutively high caspase expression, indicative of irreversible apoptosis. Therefore, there is a critical threshold period where proapoptotic and prosurvival signals compete and eventually one wins out over the other—this window would represent an opportunity for targeted therapy. Using the higher inputs which could lead to apoptosis via the OR gate, the critical duration was much lower, ~3 h.

Macroparameters were defined based on kinetically determined steady-state variables, which describe maximal and minimal concentrations of Bax, and Bax-related protein–protein interaction affinities. The benefit of this approach was to extend the use of Boolean logic by to these other system elements, allowing use of simple inequalities to define these as logic gates as well: for example $(Bax_{tot} + Bad_u > Bcl\text{-}x_L) \rightarrow$ apoptosis. Concentrations of prosurvival and proapoptotic proteins with macroparameters could then be simplified to determine the apoptotic effects of OR or AND gates for these variables. The analysis demonstrated that cells with low levels of free $Bcl\text{-}x_L$ will commit to apoptosis via either p53 activation or Akt withdrawal, with little resistance. In contrast, cells with high levels of free $Bcl\text{-}x_L$ require increased stimulation by both p53 and Akt to stimulate apoptosis. This model suggests several possible loci of therapeutic intervention, and implies that there is a time dependent, bistable threshold mechanism for pushing the cell towards apoptosis or survival.

2.9. Two-dimensional model of damage and stress response

Degracia *et al.* presented a model of cell damage and response to stress, based on a highly simplified two-dimensional model that lumped the details of molecular mechanisms that were explored in the prior models.[17] Detailed subcellular processes were omitted in order to develop a simplified, top-down understanding of cell fate decisions based on values for damage D and stress response S represented as two coupled ODEs responsive to an injury input I. The linkage between the injury stimulus and change in damage or stress was studied. Damage parameters increased with injury exponentially over time, whereas stress response parameters increased rapidly at first and then gradually declined. When injury was increased slowly from 0, all variables initially rose, leading to saturation of the stress response. However with continued increasing injury, damage is continuously compounded.

Graphically, an injury causes the $\{D,S\}$ point to deviate from the baseline stable attractor at the origin to a new location. The maximum deviation from the origin is described as point $\{D^*, S^*\}$, which is reached at time t_c. After this maximum deviation is reached, injury is dropped to 0 to represent the end of the insult. The state variables will return to $\{0,0\}$, which is interpreted in one of two ways: if $S^* > D^*$, then $\{0,0\}$ represents recovery;

whereas if $D^* > S^*$, $\{0,0\}$ represents cell death. The time it to reach $\{0,0\}$ is the recovery time (t_r) or time to death (t_d). When S^* and D^* were close in magnitude t_r or t_d was increased. This model suggested how the time dependence of recovery from a given magnitude of injury would differ based on the magnitude of competing damage/stress response.

Different response patterns were identified. A characteristic common to all patterns was the existence of a crossover value for injury (termed I_X), representing maximum sustainable injury. This value was dependent both upon the strength of the damage and on the cell's intrinsic strength of response based on the initial state. Cell responses could be classified into four categories, based on the mechanism of stability: (1) Monostable injury course: the manifold had a single attractor that determined cell response. For all I below a threshold value, the cell recovered. Above it, the cell died. (2) Bistable injury courses, type A: A lower and upper attractor were available drawing the system either towards a proapoptotic or prosurvival end result. Unlike the prior case, transition to cell death was abrupt and discontinuous. (3) Bistable injury courses, type B: Bistability was found over an entire range such that cell death could occur at any value of I. In this scenario, high values of S^* can be attained but only over a long period of time (high t_d). (4) Double bistable injury courses combine both types of bistability with a combined model. Due to bistability at the origin, a small injury can cause cell death.

These models can be used to help explain delayed death after injury, as t_d and t_r lengthen in situations where D^* is only slightly greater than S^*. During recovery from an injury, there is a point at which the damage has subsided, but the stress response remains elevated, due to the different time courses for these processes. This can be used to predict the effects of an ischemic preconditioning. By contrast, a situation in which the initial conditions involve preexisting damage can allow cell death even with small injury, as in the case with prior damage due to hypertension or diabetes mellitus. These results also suggest explanations for some of the difficulties involved in clinical translation due to the heterogeneity of the clinical conditions compared to the homogeneous $\{D^*, S^*\}$ utilized in animal experimentation.

2.10. Analyzing cell fate decisions with an ODE-based regulatory gene network model

Rapidly after ischemic insult, direct metabolic effects and changes in membrane potential dominate. After longer times, protein cascades take effect. Still later, for those cells that still survive, additional genetic factors kick in to further transform the cell through substantial new transcription and translation.

A model for cellular–level genetic factors involved in stroke was formulated by McDermott *et al.*[18] to assess how gene expression changes over time in response to stroke. A network of expressed proteins was defined as a set of gene modules (termed a cluster), each with a similar function. Information about how these modules interact was gleaned from a high–throughput analysis of transcription information from mouse stroke models, using mice that had been given a preconditioning stimulus of lipopolysaccharides, CpG–oligonucleotides, brief mild ischemia, saline (control), or sham surgery (control). Model parameters were defined based on levels of gene expression across many cells over time from 3 to 72 h after infarct. Probe sets were used to identify intensity of response in the various clusters (Fig. 11.2).

Cluster analyses were then used to determine hierarchical relationships. Inferred causative relationships were used to generate ODE models of steady-state behavior using an inference algorithm. The gene clusters with the highest downstream upregulation of effector proteins were used to generate an optimized network. Downstream expression of genes was modeled by changing regulatory inputs to look at differences in mean cluster expression over time. The observed results were then compared to results predicted from the optimal network, and against a consensus prediction from other nonoptimized network models.

A major result of the paper was identification of which gene clusters were upregulated by particular preconditioning stimuli. Mild ischemia strongly altered genes in cluster 5, which contains regulatory genes for metabolic processes and metabolic regulation. This suggested that cluster 5 might play a major role in neuroprotection following transient, mild ischemic insults. All of the preconditioning paradigms upregulated apoptosis and inflammatory gene clusters, and all led to significantly greater levels of protection compared to control. A neuroprotective gene for TNF-α was found in the inflammatory cluster, suggesting that mice deficient in this gene would lose some of the benefits of preconditioning.

2.11. Modeling evolution of the penumbra

Although the focus of this chapter is on cellular and subcellular pathways implicated in ischemia, these changes occur at the same time in different proportions in different cells depending on the location of a particular cell within the umbra or penumbra of a stroke. Effects at the single cell level are

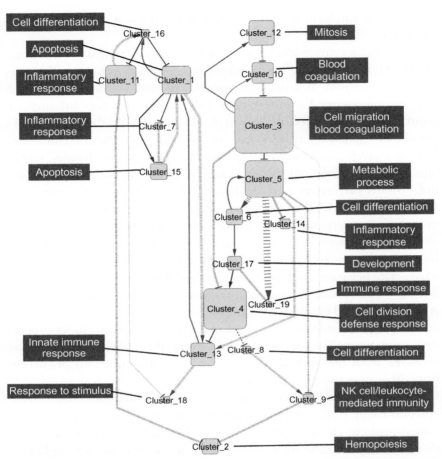

Figure 11.2 Multiple gene clusters involved in response to stroke. *Taken from fig. 8 of Ref. 18.* (See the color plate.)

also compounded at the higher scale, as toxic agents and ion shifts are spread between cells across the tissue. Ultimately, all damaging results of ischemia stem from loss of substrates for energy production. Nutrient delivery at the level of blood and of extracellular tissue is therefore highly relevant in creating the microenvironment within which the cell's molecular processes take place. Regional variation is an important feature in the brain as compared to other organs; variations in vascular patterns are common and the severity of ischemic damage can vary significantly based on whether or not a region has enough anastomotic connections to neighboring blood supplies. The affected stroke tissue is usually defined by examining regional differences in blood delivery. Assessing levels of nutrient flow is also clinically important in defining the severity of stroke. A variety of other factors, both environmental and genetic, can also contribute to morphological and molecular differences in the extent of penumbral regions.[14] Common to all penumbrae are characteristic edematous and inflammatory changes.

An early computational model linking molecular changes in the penumbral tissue with cerebral blood flow was published by Revett *et al.* in 1998.[19] This model focused on generation of the penumbra, at tissue scale, but incorporated cell-level modeling of basic metabolic processes. The model simulated depolarizations found in ischemic tissue that produce waves of pathological activation across the tissue, termed cortical spreading depression. In this model, cortex is represented as a hexagonal array of elements, each of which represents a single cell and contains variables for intracellular and extracellular ionic concentrations, levels of blood flow, and levels of metabolites. General impairment and intactness indices reflect levels of intact function of a given variable and of malfunction of the same variable, respectively. Extracellular diffusive K^+ was also simulated. Importantly, this model included discrete spatial zones. Varying parameters allowed exploration of effects such as relative damage depending on the ratio of core infarct to penumbra. The Revett model also predicted the relationship between cortical spreading depression and infarct size, and the balance between control of energy metabolism versus delivery of metabolites.

Duval *et al.* modeled cellular consequences of changes in cerebral blood flow.[20] Like Revett *et al.*, the authors model regions of brain tissue as a set of adjacent zones. Vascular regions are defined for specific zones. This method also allowed simulation of cellular regions with overlapping vasculature, an important factor in the brain's ischemic response. The Duval model used four primary parameters: cerebral blood flow, rate of oxygen extraction from blood, metabolic rate of oxygen use by tissue, and the apparent diffusion coefficient of water. A *survival delay variable* was also used to quantify the decay of tissue in the penumbra from functional, to salvageable, to necrotic. Results of running the model many times with varying initial conditions yielded information about patterns of morphological changes in penumbral regions. Severely ischemic penumbra tissue had altered cellular metabolism, and generally was not salvageable, whereas the edematous penumbra tissue appeared to retain greater recovery potential. This model clarified ways in which extracellular nutrient parameters might affect the generation of penumbral regions, and what types of edematous tissue might be salvageable.

The core of all ischemic infarcts becomes necrotic rapidly. It has been well documented that this necrotic core experiences changes different from the slower apoptotic and necrotic death that occurs in the surrounding tissue.[14] Necrotic cells initiate a series of local events, which will lead to cytotoxic edematous changes as noted in the prior section.

The model by Di Russo *et al.* was used to evaluate the contribution of inflammatory processes to the further development of edema in the penumbra.[22] The model uses proportional densities of healthy, dying, and dead cells, as well as leukocytic cells, to evaluate the diffusive effects of inflammatory chemokines and cytokines that react with local cells to modify the ischemic insult. Partial differential equations are used to describe how these molecular factors alter the densities of dead, dying and healthy cells over time. The change in necrotic tissue was considered to be the difference between the proportion of healthy cells that are dying and the proportion of necrotic cells that are being phagocytized at a given time-step. Results showed that the initial ratio of dying cell types (necrotic vs. apoptotic) was critical in determining brain response to inflammatory events, with a higher percentage of initial necrotic cells, causing a larger inflammatory response, which then augmented apoptosis in the surrounding region. Depending on the size of the infarct, the inflammatory response could have either negative or neuroprotective effects, related to the concept of ischemic preconditioning.

3. PITFALLS AND OUTLOOK

The most basic problem in all of these models is the necessity for omission of many biological parameters either because they are not known or because their inclusion makes the model unmanageable. Simplification is always a necessity in modeling but gives the risk of leaving out some critical factors. For example, none of the models include the more than 20 different channels and pumps that play a role in ionic fluctuation across the membrane, including some that have not yet been fully defined experimentally. In the case of intracellular organelles, the ER and mitochondria, there are still no experimental methods that allow activity across these internal membranes to be fully explored.

Some of the formalisms themselves necessarily limit the degree of verisimilitude. Clearly the BN formalism oversimplifies the relationships between state variables by defining them in terms of binary state switching without kinetics. BN modeling also does not account for transition states for variables, which may be important in ischemic pathways. However the reductionism of BN models is useful for exploration of large cascade models, in which many simulations are run to determine system dynamics.

Along similar lines, it has been noted in this chapter that ischemia is a multicellular, complex network phenomenon. Breaking the pathway components into chunks that can be adequately modeled is a necessary step given our current abilities in computing and our understanding of how these pathways interact. However, such a divisive and reductive approach may limit the usefulness of models of a system that is inherently multiscale, where the whole is greater than the sum of its parts due to emergent phenomenology. It has, in fact, been argued that the strength of modeling lies in the possibility of exploration across scales. Such cross-scale investigation is often difficult to do experimentally, where different techniques are required for explorations at different scales, different techniques that often are difficult or impossible to use in combination.

Another major difficulty, common to nearly all models of ischemia, is setting of parameters and initial values. Parameters are often be taken from models of healthy tissue and may be different than those found in ischemic tissue, where proteins are present in different phosphorylation states. Additionally, these models are typically chimeric with parameters gathered from different cell types from different species, at different animal ages. For example, the Magnus and Keizer model of ATP and Ca^{++} regulation, widely adopted for use in ischemia models, was based on pancreatic beta cells. It is now appreciated that cells vary enormously in their proteomic expression, genetic regulation, and intercellular communication, based on species type, tissue type, and even location within a seemingly homogeneous tissue.

Many of the models presented in this chapter do not deal at all with the spatial aspects of intracellular communication, making the assumption of a *well-mixed* solution so as to work with reactions without diffusion. However neurons and glia are among the most morphologically diverse and highly compartmentalized cells in the body: there are major differences between molecular machinery in spines, dendrites, soma, and axon. Full reaction–diffusion modeling of these factors is still rudimentary, made more difficult by the fact that different simulation techniques will need to be used for these different compartments. An additional factor that can still not be incorporated into models is changes in cell morphology due to plasticity or damage.

Computer modeling of ischemia can expand our understanding of the cellular and subcellular processes involved in ischemia, particularly in defining the ionic fluctuations of cytotoxicity, gene regulation and transcription, and cell fate decisions of cells under metabolic stress. Many questions still remain. What are the mechanisms by which cells switch between recovery and death in response to stressful stimuli? How do inflammatory factors affect apoptotic signaling in the penumbra? What is the role of cell-to-cell signaling in modulating the tissue response to ischemia? What genetic and environmental factors can predispose a patient to, or protect a patient from, ischemic damage? Several approaches to modeling have been used to tackle various aspects of ischemic pathways. A major challenge for the future will be figuring out how to effectively integrate these approaches to create multiscale models of ischemia incorporating information from the molecular to brain levels.

REFERENCES

1. *Center for Disease Control*, Leading causes of death, CDC fact sheet; 2011.
2. Saftenku EE, Friel DD. Chapter 26: combined computational and experimental approaches to understanding the CA^{2+} regulatory network in neurons. In: Islam S, ed. *Calcium Signaling*. New York, NY: Springer; 2012 Advances in Experimental Medicine and Biology. Vol. 740.
3. Bartus RT. The calpain hypothesis of neurodegeneration: evidence for a common cytotoxic pathway. *Neuroscientist*. 1997;3(5):314–327.
4. Dronne M-A, Boissel J-P, Grenier E. A mathematical model of ion movements in grey matter during a stroke. *J Theor Biol*. 2006;240(4):599–615.
5. Bartol Jr TM, Land BR, Salpeter EE, Salpeter MM. Monte carlo simulation of miniature endplate current generation in the vertebrate neuromuscular junction. *Biophys J*. 1991;59:1290–1307.
6. Dronne M-A, Boissel J-P, Grenier E, et al. Mathematical modelling of an ischemic stroke: an integrative approach. *Acta Biotheor*. 2004;52(4):255–272.
7. Verheul HB, Balazs R, Berkelbach van der Sprenkel JW, et al. Comparison of diffusion-weighted MRI with changes in cell volume in a rat model of brain injury. *NMR Biomed*. 1994;7(1–2):96–100.
8. Diekman CO, Fall CP, Lechleiter JD, Terman D. Modeling the neuroprotective role of enhanced astrocyte mitochondrial metabolism during stroke. *Biophys J*. 2013;104(8):1752–1763.
9. Magnus G, Keizer J. Model of beta-cell mitochondrial calcium handling and electrical activity. I. Cytoplasmic variables. *Am J physiol*. 1998;274(4 Pt 1):C1158–C1173.
10. Oster AM, Thomas B, Terman D, Fall CP. The low conductance mitochondrial permeability transition pore confers excitability and CICR wave propagation in a computational model. *J Theor Biol*. 2011;273(1):216–231.
11. Magnus G, Keizer J. Minimal model of beta-cell mitochondrial Ca^{2+} handling. *Am J Physiol*. 1997;273(2 Pt 1):C717–C733.
12. Heuett WJ, Periwal V. Autoregulation of free radicals via uncoupling protein control in pancreatic beta-cell mitochondria. *Biophys J*. 2010;98(2):207–217.
13. Alam MJ, Devi GR, Ravins, Ishrat R, Agarwal SM, Brojen Singh RK. Switching p53 states by calcium: dynamics and interaction of stress systems. *Mol BioSyst*. 2013;9(3):508–521.
14. Lipton P. Ischemic cell death in brain neurons. *Physiol Rev*. 1999;79(4):1431–1568.

15. Mai Z, Liu H. Boolean network-based analysis of the apoptosis network: irreversible apoptosis and stable surviving. *J Theor Biol.* 2009;259(4):760–769.
16. Bogda MN, Hat B, Kocha Czyk M, Lipniacki T. Levels of pro-apoptotic regulator Bad and anti-apoptotic regulator Bcl-xL determine the type of the apoptotic logic gate. *BMC Syst Biol.* 2013;7(1):67.
17. DeGracia DJ, Huang ZF, Huang S. A nonlinear dynamical theory of cell injury. *J Cereb Blood Flow Metab.* 2012;32(6):1000–1013.
18. McDermott JE, Jarman K, Taylor R, et al. Modeling dynamic regulatory processes in stroke. *PLoS Comput Biol.* 2012;8(10):e1002722.
19. Revett K, Ruppin E, Goodall S, Reggia JA. Spreading depression in focal ischemia: a computational study. *J Cereb Blood Flow Metab.* 1998;18(9):998–1007.
20. Duval V, Chabaud S, Girard P, Cucherat M, Hommel M, Boissel JP. Physiologically based model of acute ischemic stroke. *J Cereb Blood Flow Metab.* 2002;22(8):1010–1018.
21. Sridharan S, Layek R, Datta A, Venkatraj J. Boolean modeling and fault diagnosis in oxidative stress response. *BMC Genomics.* 2012;13(Suppl 6):S4.
22. Di Russo C, Lagaert J-B, Chapuisat G, Dronne M-A. A mathematical model of inflammation during ischemic stroke. In: *ESAIM: Proceedings*; 2010:15–33. EDP Sciences; Vol. 30.

CHAPTER TWELVE

Modeling Intracellular Signaling Underlying Striatal Function in Health and Disease

Anu G. Nair*, **Omar Gutierrez-Arenas***, **Olivia Eriksson†**,
Alexandra Jauhiainen‡, **Kim T. Blackwell§**, **Jeanette H. Kotaleski***,¶

*School of Computer Science and Communication, Royal Institute of Technology, Stockholm, Sweden
†Department of Numerical Analysis and Computer Science, Stockholm University, Stockholm, Sweden
‡Department of Medical Epidemiology and Biostatistics, Karolinska Institutet, Stockholm, Sweden
§Krasnow Institute for Advanced Study, George Mason University, Fairfax, VA, USA
¶Department of Neuroscience, Karolinska Institutet, Stockholm, Sweden

Contents

Abstract

Striatum, which is the input nucleus of the basal ganglia, integrates cortical and thalamic glutamatergic inputs with dopaminergic afferents from the substantia nigra pars compacta. The combination of dopamine and glutamate strongly modulates molecular and cellular properties of striatal neurons and the strength of corticostriatal synapses. These actions are performed via intracellular signaling networks, containing several intertwined feedback loops. Understanding the role of dopamine and other neuromodulators requires the development of quantitative dynamical models for describing the intracellular signaling, in order to provide precise unambiguous descriptions and quantitative predictions. Building such models requires integration of data from multiple data sources containing information regarding the molecular

Progress in Molecular Biology and Translational Science, Volume 123
ISSN 1877-1173
http://dx.doi.org/10.1016/B978-0-12-397897-4.00013-9

interactions, the strength of these interactions, and the subcellular localization of the molecules. Due to the uncertainty, variability, and sparseness of these data, parameter estimation techniques are critical for inferring or constraining the unknown parameters, and sensitivity analysis evaluates which parameters are most critical for a given observed macroscopic behavior. Here, we briefly review the modeling approaches and tools that have been used to investigate biochemical signaling in the striatum, along with some of the models built around striatum. We also suggest a future direction for the development of such models from the, now becoming abundant, high-throughput data.

1. INTRODUCTION

The basal ganglia, a group of phylogenetically conserved structures in vertebrates, are critical for the motivational and habitual control of motor and cognitive behaviors[1,2] in both health and disease. Several serious diseases, such as Parkinson's,[2] Huntington's,[3] schizophrenia,[4] addiction to psychostimulants,[5] and pharmacologically induced dyskinesia (e.g., LID),[6] are caused by degeneration or dysfunction in various basal ganglia structures. Despite the prevalence and severity of diseases affecting the basal ganglia, the development of drugs for these diseases and other central nervous system disorders suffers from the highest attrition rates in the pharmaceutical industry.[7,8] This is due to our lack of mechanistic understanding about the underlying system. The ultimate causes of these disorders are poorly understood, and most drugs are chosen based on symptom relief rather than on identified disease mechanisms. In order to design effective pharmacological interventions, the complexity of the nervous system requires an integrated understanding at the molecular, cellular, and neuronal network levels.

The striatum, the basal ganglia input stage, integrates cortical and thalamic inputs in functionally segregated pathways and loops.[9] The output of the striatum innervates the globus pallidus and substantia nigra pars reticulata, which control both the thalamus and additional targets in the brain stem,[10] see Fig. 12.1. The function of the basal ganglia circuitry is complicated by feedback loops between and within these structures. Further compounding this complexity is the nonlinear dynamic nature of cellular properties and synaptic connections. Dopaminergic afferents from the midbrain play a crucial role in modulating signaling through the basal ganglia, in particular through modification of molecular and cellular properties of striatal neurons as well as connection strength between neurons. The importance of dopamine is further highlighted by the number of pharmaceuticals which target dopamine receptors to treat basal ganglia disorders.[2,4,5]

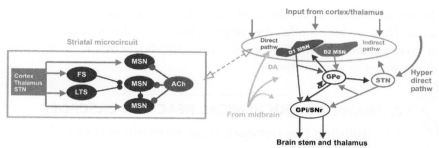

Figure 12.1 An overview of the basal ganglia. Left: A schematic representation of the striatal microcircuitry. The main neuron type in the striatum is the medium spiny neuron (MSN), constituting ∼95% of striatal neurons. MSNs are the projection neurons from the striatum and receive convergent excitatory glutamatergic input mainly from cortex and thalamus, inhibitory GABAergic input from neighboring MSNs and striatal fast-spiking (FS) and low-threshold spiking (LTS) neurons, cholinergic input from cholinergic inter-neurons (ACh), and dopaminergic input from substantia nigra pars compacta in the midbrain (SNc). Right: an illustration of the basal ganglia macrocircuitry, showing the glutamatergic and dopaminergic afferent inputs to striatum, as well as the projections via the direct, indirect, and hyperdirect pathways. Approximately half of the MSNs belong to the direct pathway; they carry dopamine type 1 receptors and project to GPi/SNr (globus pallidus interna/substantia nigra reticulata). The other MSNs express dopamine type 2 receptors and project to GPe (globus pallidus externa) before reaching basal ganglia output structures. These output structure neurons control the activity levels of (motor) programs in the brain stem and thalamus. (See the color plate.)

There is a large body of experimental data gathered over the years concerning information processing in the basal ganglia in health and disease. The size and complexity of these data make it difficult to synthesize a coherent picture. Development of quantitative models is needed for integrating the knowledge obtained from diverse experimental approaches. Such models further provide a compact and standardized means to represent current knowledge and constitute a tool for guiding experiments and generating predictions that can be tested experimentally. These models comprise not just a collection of nodes and edges within each scale but also the dynamics of these interactions which are an integral part of the information flow. Since most pharmaceuticals target transmembrane receptors and downstream signaling molecules, and due to the importance of dopamine in basal ganglia disorders, quantitative models of dopaminergic system along with the signaling pathways affected by dopamine have the potential for greatly enhancing our understanding of underlying mechanisms of pathological conditions.

In Section 2, we explain the procedures for developing quantitative models of intracellular signaling, with a focus on data sources, model building, parameter estimation, and sensitivity analysis. Then we review many of the published models addressing different aspects of the information flow in the striatum at the molecular level under normal and pathological conditions. These include models for synaptic plasticity in the principal neurons of striatum, known as medium spiny neurons (MSN), as well as the life cycle of dopamine in the striatum.

2. MODELING BIOCHEMICAL REACTION CASCADES

2.1. Building the network from multiple sources of information

Several types of information are required to build a quantitative mathematical model of intracellular signaling. First, proteins and other molecules that process and transmit information must be identified. Second, the pairwise interactions between these molecules must be determined. Third, the dynamics of these interactions must be accounted for. The specific proteins and their interactions represent the nodes and edges, respectively, of a chemical reaction network. The kinetics of these interactions transforms a static network into a dynamical one, in which the reaction rates define the behavior of the system. Most of the data that are used to define the interactions between signaling molecules comes from literature surveys of papers employing classical experimental studies addressing only a few interacting molecules at a time. These experiments could be pull-down assays where the physical interaction between two proteins is determined. There have also been several efforts to determine protein–protein interactions using high-throughput amenable techniques such as yeast two-hybrid.[11,12] A major challenge still remaining is to interpret these available high-throughput data to develop hypotheses on a systems level because the interactions detected in such screenings may lack functional mapping, that is, the actual effect of the interaction *in vivo*. However, the importance of high-throughput data generation is increasingly being acknowledged in formulating biological hypothesis.[13,14]

Similarly, information regarding strength and rate of molecule interactions, that is, kinetics and affinity of reactions, is mainly obtained from surveying papers addressing the interaction dynamics of two or three molecular species. There are no high-throughput techniques for generating these data yet. Classically, such kinetic experiments target a specific protein to identify its role in the system of interest.[15,16] Notably, there are large amounts of experimental data of this type for certain molecules, and very little such data for most of the other molecules. Thus, this data source is far from sufficient to identify all model parameters and also have some other limitations (see below). The process of model building can further be supplemented by specific *in vitro* experiments, particularly to identify the range of critical parameters.

An important part of the available interaction information has been organized into a few signaling pathways databases which have proved to be valuable. Databases like Reactome[17] and KEGG[18] contain static interaction information and can assist the building of neuron–specific reaction networks. On the other hand, model databases like BioModels,[19] ModelDB,[20] and DOQCS[21] contain published quantitative signaling models which are amenable to be reused, especially those that have been validated for signaling in neurons.

2.2. Modeling techniques and tools

Once the structure of the signaling network has been identified, as described in Section 2.1, this information is translated into a mathematical form suitable for rendering a quantitative dynamical description. Depending on the biological system being studied, the nature of the question, the assumptions made, and the available computational power, different simulation techniques can be used. In kinetic models of striatal signaling, two types of modeling strategies have been used, ordinary differential equations (ODE) and stochastic reaction modeling. In both cases, the rate of change of each of the system's species results from the contribution of individual reactions which produce and consume the species. Each of these reaction rates is expressed following the law of mass action, which states that the rate is proportional to the product of a kinetic constant and some power of the reactant concentrations. Whether the concentration change is implemented as deterministic or stochastic in a given time step is the main difference between these two modeling strategies. This in turn depends on a critical assumption regarding the number of molecules in the system.

In the deterministic implementation, the rate equations are modeled as a system of ODE with one ODE per species, which render the same solution for repeated simulations with a given initial condition. For most relevant cases, this system of ODEs does not have analytical solutions so that numerical integration utilizing one among several existing solvers is required to run simulations of the system dynamics. The ODE–based simulation is underlined by a fundamental assumption: the number of species' particles is high enough to ignore the stochasticity of the reaction events. However, when modeling cellular processes occurring in a very small volume like the dendritic spine, this assumption may not hold true for many molecular species.

In a system with very few number of particles, the effect of the stochasticity of the reaction events becomes more pronounced and the evolution of the system differs from what an ODE-based description renders not just because fluctuations around the mean are not considered but, more critically, because the mean dynamics can be qualitatively different.[22,23] This common situation then requires a stochastic simulation which, however, is more computationally demanding. In this approach, the rate constants and the reactant concentrations do not unequivocally determine the rate of change in each time step but define a probability for the reaction to occur which may be realized or not depending on the outcome of a random sampling.[24]

Another relevant issue is the simulation of spatial heterogeneities which result either from reactions occurring far faster than the diffusion of reactants and products or due to biological design, as in the case of compartmentalized signaling. This situation is the norm in intracellular signaling, and it is commonly addressed in simulations by meshing the space of interest into subvolumes which are considered homogeneous (or well-stirred) and exchange mass according to a first-order process.[25,26] It is worth mentioning that while the definition of subvolumes corresponding to functionally relevant compartments (e.g., a dendritic spine, cell membrane, PSD, etc.) is done at the level of the model building, as it entails biological knowledge,[26] the meshing to account for diffusion is rather a physical problem constrained by the diffusion properties.[27,28] Thus, both ODEs[29] and stochastic simulation approaches[26] have been used to simulate the reaction within each subvolume and the mass exchange between subvolumes.

There are more accurate stochastic and deterministic methods where no meshing is performed and heterogeneities are considered all over the space at far higher resolutions. On the stochastic side, there is single particle tracking,[30] and on the deterministic one, there are partial differential equations. These are quite computationally demanding.

2.3. Parameter estimation and sensitivity analysis

One persistent concern in these types of models is the estimation of all rate constants and total amounts of conserved moieties. A conserved moiety can be a molecule that can exist in several forms, such as phosphorylated or non-phosphorylated forms, but the summed quantity of all forms is conserved. Biochemical estimates of these parameters obtained in test tube experiments with isolated components are valuable.[31] However, this kind of data typically provides affinities and rarely provides estimates of the forward and backward reaction rates. Furthermore, the experimental conditions in a test tube do not match those inside the cell. In addition, most models are built under a considerable number of simplifying assumptions about the modular nature of real systems and the possibility to lump reaction steps or additional factors involved. Therefore, many intracellular signaling models are to different extents of a phenomenological character.[32] The models correspond to "gray" boxes where, despite their apparent straightforward interpretability, a variety of processes have been lumped together in each model reaction. This certainly obscures the use of parameters estimated from biochemical experiments with purified components in intracellular signaling models. A complementary approach of estimating parameters is to fit the model to sets of data obtained in experiments that can better represent the modeled system,[33] such as cell cultures, tissue slices, or whole organisms. This process of matching outputs from a more complex model to experimental data to find parameter estimates is sometimes referred to as calibration and the broader term reverse engineering appears frequently as well.

Within the field of modeling intracellular signaling in striatum and in fact in other systems,[34] most calibration methods are built on trial and error experience, where parameters are fitted by hand until the model output corresponds to what is seen in experiments. There is a lot to learn from the field of systems biology, where automatic parameter estimation is an increasingly used procedure.[35,36] Here, different local and global optimization techniques are used to minimize an objective function, corresponding to the difference between the model output and the experimental results. In the context of parameter estimation, one can also note that, given the experimental data and the model, it is not always possible to find unique values for all parameters. The relationship between the experimental data and the model can be such that the system is nonidentifiable.[35,37]

A question related to parameter estimation is that of robustness. It has been argued that biological signaling cascades with similar functions can work in very different contexts, such as different cells, animals, or environments, and therefore have to be robust against changes in, for example, rate constants and other parameters.[38,39] This implies that changes in these parameters do not have a large effect on the system output. Within the modeling field described here, tests for parameter robustness have mostly been performed by changing parameters one-at-a-time, and then recording the output. If the output after perturbations in a certain parameter is similar to what it was before, the model is considered to be robust against perturbations to this parameter alone. Here, one can note that the system cannot be robust against everything, as a signaling cascade has to be sensitive to some signals.

Considerable system insights can also be retrieved by analyzing how the model output depends on the model parameters by the means of local or global sensitivity analysis.[40] The local sensitivity of the system output o_i, with respect to the parameter input p_j, is measured by the first-order partial derivative, $\partial o_i / \partial p_j$, or derivations thereof,[40] and describes how the model output depends on different parameters in the neighborhood of a specific point in parameter space. For biological systems, with a large uncertainty in the parameters, it might be more interesting to perform a global sensitivity analysis, where the aim is to investigate a larger part of the parameter space. Global methods are most often based on analyses of random parameter samples, and performed by statistical methods, for example, by decomposing the variance of the output into different parts that can be attributed to a single input parameter or combinations thereof.[41,42] Within the systems biology field, sensitivity analysis has started to become an important part of the modeling process,[43] but also within the field of neuroscience, some studies have been performed.[44]

3. A MODELING EXAMPLE

Here, we illustrate a modeling procedure by a simple model example describing calcium-dependent activation of calmodulin (CaM), calcineurin, and calcium/calmodulin-dependent protein kinase II (CaMKII). Calmodulin is a calcium-binding protein which is involved in various signaling processes and is strongly implicated in synaptic plasticity.[45] Calmodulin contains four calcium-binding domains and each of them binds to one calcium ion.[46] The binding of calcium with calmodulin is a cooperative process.[47] Calcium-bound calmodulin activates protein phosphatase 2B (PP2B), also known as calcineurin (CaN).[48] This protein is also highly implicated in the molecular processes related to learning. As we see in Section 4, this protein has a role in striatal signaling. CaMKII is a kinase which is activated by the binding of calcium/calmodulin. CaMKII molecules exist as a dodecamer,[49] with two juxtaposed hexamers. If two of its neighboring subunits in a hexamer are active by calcium/calmodulin, then one unit can phosphorylate the other one at Thr-286.[50] The phosphorylated unit can remain active even in the absence of calcium/calmodulin.[51] A reaction scheme for the model of activation of calmodulin and subsequent activation of PP2B and CaMKII is shown in Fig. 12.2.

The model has been implemented using a deterministic approach with Simbiology toolbox in Matlab. The activation of calmodulin by calcium is modeled as a four-step reaction. In order to obtain the dissociation constants for each of these calcium–calmodulin binding steps, we have used the Adair–Klotz equation (Ref. 52 equation 8). According to this, there should be one dissociation constant for each step. So, there should be four constants for this particular binding reaction. The steady-state data are taken from Ref. 47. The estimated dissociation constant, K_d, and the dissociation rate constants, k(off), obtained from additional experimental measurement[53] are then used to calculate the forward rate constants of the reactions, k(on), using the relation, $K_d = k(\text{off})/k(\text{on})$. At each calcium-binding step, a calcium–calmodulin complex can bind to calmodulin-target-proteins, namely, PP2B and CaMKII (Fig. 12.2). The parameters for these binding steps are estimated by fitting the model against experimental data for CaMKII and PP2B activation.[50,53–56] The fitting of the model to these measurements is shown in Fig. 12.3. It could be noted that there are a number of cyclic reactions in the model, for example, a complex containing calcium, calmodulin, and PP2B can be formed via two routes. Either it can be formed by the association of a calcium-bound calmodulin with a PP2B or calcium binding to a calmodulin–PP2B complex. Thermodynamically, the net free energy change for both these paths should be the same. This thermodynamic constraint is taken into consideration while estimating the reaction parameters. The input to the final model is a calcium transient which is modeled as a double-exponential by fitting the intracellular calcium measurement using fluorescence,[57] Fig. 12.3F, with a maximum amplitude of 700 nM. The phosphorylation of CaMKII is calculated with an approach similar to the one used by Li et al.,[58] where the possible concentration of active CaMKII to be phosphorylated is calculated by taking the average number of times two active CaMKII appear as a neighbor in a hexamer for 1000 random samples.

Next, we investigate how this simple model behaves for different frequencies of calcium inputs, two of them are shown in Fig. 12.4B. The output, which we are interested in, is the difference in the activity of CaMKII and PP2B. This variable is of physiological relevance as described in Section 4. The model is being simulated for 10 s with 10 calcium transients for frequencies ranging from 1 to 25 Hz. The initial concentrations of CaM, PP2B, and CaMKII are 5, 1, and 5 μM, respectively. The activation of PP2B

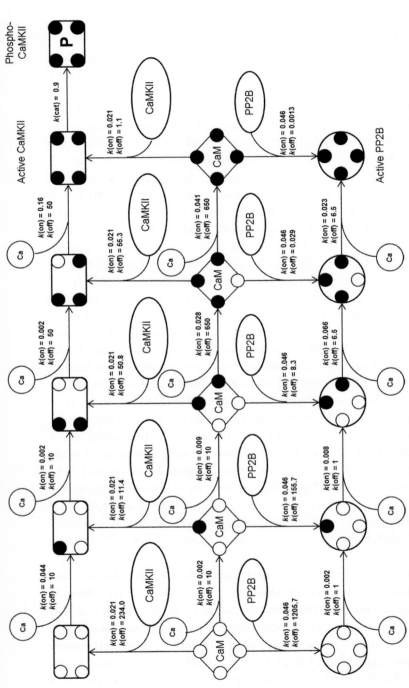

Figure 12.2 Reaction scheme for the model example along with the reaction rate constants where $k(on)$ is the forward rate constant, $k(off)$ is the reverse rate constant, and $k(cat)$ is the catalytic rate constant. All the reactions having a $k(on)$ and a $k(off)$ are reversible reactions. This represents the calcium binding to the calmodulin (CaM) as a four-step process. The species at each step can bind to its target, either CaMKII or PP2B. The first, second, and third row in the diagram represents the calcium binding to CaM–CaMKII complex, CaM, and CaM-PP2B complex, respectively. The CaM–CaMKII complex (rectangle with four smaller circles, first row) or CaM–PP2B complex (circle with four smaller circles, third row) can bind to calcium until all the four calcium-binding sites of calmodulin are occupied. Active CaMKII can be phosphorylated to phospho-CaMKII. The unit for $k(on)$ is nM^{-1} s^{-1}, $k(off)$ is s^{-1}, and $k(cat)$ is s^{-1}.

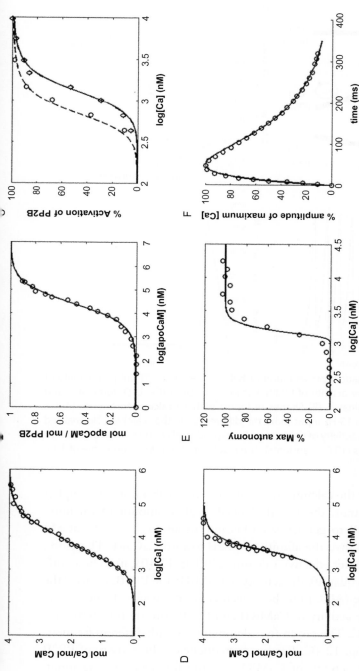

Figure 12.3 Model fitting to the experimental measurements. (A) Number of moles of calcium (Ca) bound to each mole calmodulin (CaM) versus calcium concentration; the line (—) represents the model and the empty circle (○) represents the data from Ref. 47. (B) Number of moles of apo calmodulin (apoCaM, calmodulin without calcium bound to it) bound to each mole PP2B versus the concentration of apoCaM; the line (—) represents the model and the empty circle (○) represents the data from Ref. 54. (C) Percentage activation of PP2B versus calcium concentration at various concentrations of CaM: 30 nM (experiment ◇, model —) and 300 nM (experiment ○, model – –).[47] (D) Number of moles calcium bound to per mole CaM against calcium concentration in the presence of CaMKII; the line (—) represents the model and the empty circle (○) represents the data from Ref. 56. (E) Percentage of phosphorylated CaMKII (autonomy in CaMKII activity) versus calcium concentration; the line (—) represents the model and the empty circle (○) represents the data from Ref. 50. (F) Intracellular calcium concentration in terms of percentage change of calcium fluorescence for a single synaptic activation as measured by Sabatini et al.[57] This calcium transient is used as the model input as shown in Fig. 12.4.

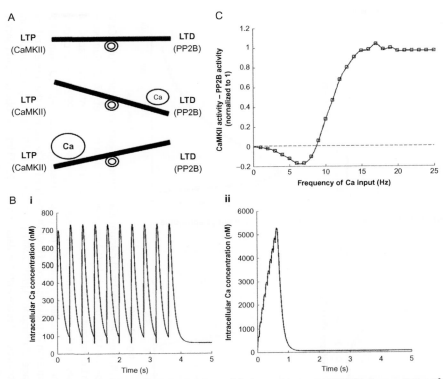

Figure 12.4 (A) Schematic representation of balance between LTP (higher activity of CaMKII) and LTD (higher activity of PP2B). For lower calcium (Ca) LTD dominates and for higher Ca LTP dominates. (B) Shapes of inputs with 10 calcium transients: (i) is a 2.5-Hz input and (ii) is a 15-Hz input (note the differences in the scale). (C) Normalized difference between the activity of CaMKII and PP2B produced by the model for different frequencies of Ca inputs (10 Ca transients). The results are for a simulation time of 10 s.

follows qualitatively the calcium transient and eventually saturates at higher frequencies. In contrast, the CaMKII needs higher amount of calcium–calmodulin (CaCaM) to get activated because its affinity toward CaCaM is lower (higher K_d) than the affinity of PP2B toward CaCaM. Therefore, CaMKII requires higher effective calcium as in Fig. 12.4(B, ii) to be significantly activated. To quantify the activity of CaMKII and PP2B, we used the area under the curve (AUC) for the trajectory of the species. It means that the difference in the activity of CaMKII and PP2B corresponds to the difference in their AUC. The results for calcium inputs with different frequencies are shown in Fig. 12.4C. For lower frequencies, the activity of PP2B dominates over the activity of CaMKII. As the frequency of the input, and thereby the effective calcium concentration, increases, this leads to the activation of CaMKII and in turn its autophosphorylation. Due to the increased autophosphorylation, the decay in the CaMKII activity, after the calcium spike ends, is slow. Since the initial concentration of CaMKII is higher than that of PP2B and there is an increased phosphorylation, its activity dominates for higher frequencies in our model. This frequency-dependent differential activity level has been studied previously as well.[58] This shows how even a small model can enhance our mechanistic understanding about the dynamic behavior of a signaling module.

4. SUBCELLULAR MODELS REPRESENTING STRIATAL SIGNALING

The glutamatergic projections from prefrontal cortex and dopaminergic projections from the midbrain innervate many of the neuronal types in the striatum, in particular, the two GABAergic projection neurons (MSN), which together comprise about 95% of the striatal neurons. They are distinguished by the differential expression of dopamine D1 and D2 receptors (D1R and D2R), and there is a small population coexpressing both receptors.[59,60] The convergence of corticostriatal glutamate and dopamine from the midbrain on these striatal neurons triggers synaptic plasticity processes that underlie reinforcement learning and pathological conditions such as psychostimulant addiction. The phasic changes in striatal dopamine signal encompass different timescales in these two situations. A transient (subsecond) but high amplitude dopamine signal results from a reward in reinforcement learning trials,[61] while an hour long elevation in striatal dopamine level occurs upon administration of psychostimulants.[62] This prolonged dopamine increase is likely to produce abnormal synaptic plasticity, whereas the lack of dopamine in Parkinson's disease leads to a different abnormality in synaptic plasticity. Therefore, understanding the intracellular signaling pathways activated by convergent glutamate and dopamine is crucial to understanding aberrant synaptic plasticity associated with these disorders.

4.1. Postsynaptic signaling models

DARPP32 is a phosphoprotein, abundantly expressed in MSNs, which integrates the dopaminergic and glutamatergic inputs. It acts as one of the major points of interaction between the two inputs (Fig. 12.5). The physiological effects of the inputs depend on the level of phosphorylation at different

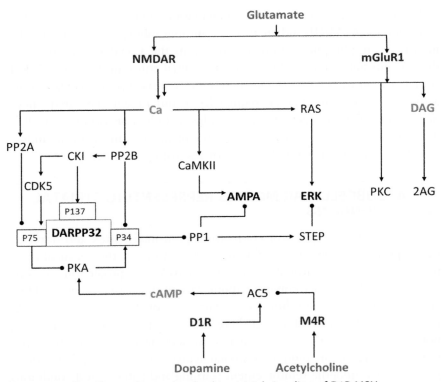

Figure 12.5 Signaling pathways involved in striatal signaling of D1R MSN.

residues of this molecule.[63] This has been the focus of many modeling studies because of its involvement in pathological conditions.[6,64] One of the earliest efforts to model the DARPP32 dynamics, by Kötter, captured some interesting aspects of the glutamatergic and dopaminergic signal integration but it was challenged by the scarcity of detailed data about the other proteins involved.[65]

The first two sufficiently detailed intracellular signaling models for striatum were single compartment mass action models, which independently addressed the integration by DARPP32 of dopamine and glutamate triggered signaling cascades in MSN carrying D1R dopamine receptors.[66,67] DARPP32 when phosphorylated by PKA at Thr34 is turned into a nanomolar inhibitor of the phosphatase PP1. As this phosphatase acts on many of the proteins phosphorylated by protein kinase A (PKA), DARPP32 is expected to act as a booster of D1R-mediated activation of PKA by dopamine. Besides the phosphorylation at threonine (Thr) 34, there are at least three other residues phosphorylated by different kinases. The modification of Thr75 by Cdk5 turns DARPP32 into a micromolar inhibitor of PKA. On the other hand, the Ca^{2+} entering the cell by the glutamatergic activation of NMDAR receptors activates the phosphatases PP2B and PP2A, which are known to dephosphorylate DARPP32. These opposing changes on DARPP32 Thr34 phosphorylation levels were thought to be the regulator of the kinase–phosphatase ratio. Prolonged treatment with Ca^{2+} and dopamine, separately, has opposite effect on Thr34. D1R stimulation increases the phosphorylation of Thr34, whereas increased Ca^{2+}, induced by glutamate receptor agonists, decreases the phosphorylation of Thr34.

One of the early models on this system[67] was built to understand the response evoked by transient dopamine and Ca^{2+} inputs. In this model, it is assumed that Thr34 and Thr75 are the two important phosphorylation sites on DARPP32 majorly affected by the inputs. A conclusion drawn by this model is that the transient inputs produce different downstream effects compared with the prolonged inputs. The model investigated how a combined input of dopamine and Ca^{2+}, which represents a combined activation of nigrostriatal and corticostriatal afferents, could produce a larger activation of PKA and inhibition of PP1 as compared to separate activation of these afferents.

The understanding of DARPP32 regulation was further enhanced by a model with an increased number of DARPP32-phosphorylation sites.[66] In this model, an additional, well-characterized, phosphorylation site, at Serine (Ser) 137, was also considered along with the other two previously mentioned sites (Fig. 12.5). Ser137 is phosphorylated by casein kinase 1 (CK1) and dephosphorylated by protein phosphatase-2C (PP2C). Phosphorylation at this site decreases the rate of dephosphorylation at Thr34 by PP2B. The activity of CK1 itself is regulated by autophosphorylation and dephosphorylation by PP2B. An autophosphorylated state of the kinase is inactive. Activation of PP2B removes this inhibitory phosphorylation and renders kinase activity to it. This incoherent type feed-forward loop has a significant effect on the sharpness and the duration of the response evoked by dopamine at Thr34. They went further and studied the effect of Ser137 mutation on the function of DARPP32 as a signal integrator and concluded that the function was impaired.

A pathway which affects DARPP32 phosphorylation state at Thr75 is the Cdk5 pathway (Fig. 12.5). This was stressed in one of the later models.[68] Cdk5 is activated by CK1 which in turn is activated by Ca^{2+}. Cdk5 then phosphorylates Thr75 of DARPP32 which is dephosphorylated by PP2A.

PP2A can also be activated by calcium. A weak calcium input leads to a dominant CK1–Cdk5 arm of the signaling cascade which is surpassed by the PP2A arm in response to strong calcium.

A more recent model for the dopamine D1R and ionotropic glutamate receptor-induced cascade looks into considerable details of downstream interactions[69] (Fig. 12.5). It presents a hypothesis of physical segregation between different pathways, namely, the Ca^{2+}-dependent ERK phosphorylation and the dopamine-dependent GluR1 phosphorylation. The model tries to explain how the two different arms of the signaling network interact with each other via the striatal-enriched tyrosine phosphatase. This model is an example of how integrating relatively large amounts of data could bring new insights about the plausible properties of a reaction compartment, physical segregation in this case. It also takes into account the phosphorylation state of different proteins, like ERK and GluR1, in the psychostimulant paradigm and elaborates on the effects of knocking down key signaling molecules, like DARPP32. Cross talk between the two signaling modules is one important aspect and how different cross talk schemes can affect the measured outputs is illustrated. Even though the complexity of this model is high, still there is room for improvement to explain in more detail the multitude of physiological behaviors manifested by these complex signaling networks.

While modeling a small volume, like a dendritic spine, it should be noted that the actual number of molecules present may be too low to safely assume the well mixed assumption of chemical kinetics.[70] In this case, the stochastic nature of chemical reactions and diffusion of different signaling molecules should be taken into account.[22] By taking into account the stochastic and diffusive effects for the signaling molecules, it has been possible to model the effects of subcellular localization on the efficiency of information transmission between signaling proteins connected through a second messenger, for example, the colocalization of PKA with adenylyl cyclase in dendritic spines leads to a higher Thr34 phosphorylation.[71]

The signaling pathways aforementioned are involved in long-term synaptic plasticity in the striatum, and over the years, it has become clear that synaptic plasticity such as LTP (long-term potentiation) or LTD (long-term depression) underlies not only learning and memory in the healthy brain but also the aberrations in the diseased brain.[72] Disturbances in synaptic plasticity can explain the onset of drug addiction, progression of neurodegenerative disease symptoms, as well as side effects following their prolonged treatment.[73,74] DARPP32 is involved in these plasticity events but the induction of LTP and LTD also depends on the differential activation of a variety of signaling molecules other than DARPP32. This induction requires certain criteria to be satisfied, for example, NMDAR calcium current seems to be important for LTP induction, whereas release of endocannabinoids may be important for LTD induction[75] (Fig. 12.6). Endocannabinoids are small lipid soluble molecules released into the synaptic cleft by the postsynaptic site. They activate the presynaptic cannabinoid receptors which in turn reduces the neurotransmission.[80] Several other neurotransmitters, like acetylcholine and GABA, also affect synaptic plasticity.[81,82] Similar to that observed in the hippocampus, the selection between LTD and LTP sometimes depends on the frequency and time duration of the input signal[83,83a] (Ronesi & Lovinger, 2005).

One of the recent striatal models tries to explain the selection between LTP and LTD based on the temporal patterns of the synaptic input.[84] The LTP and LTD markers used in this model are protein kinase C (PKC) and encannabinoids, respectively. Both of these molecules require activation of metabotropic glutamate receptor type I (mGluR) and calcium. The inputs to the model are mGluR agonist and calcium. The simulation results highlight that an extended 20-Hz input signal induces LTD in corticostriatal synapses, while a theta burst protocol induces LTP due to differential level of

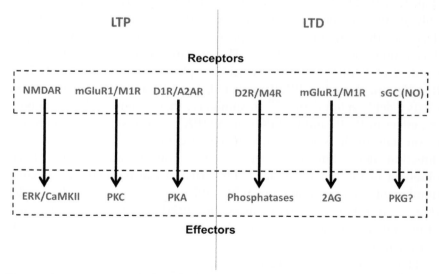

Figure 12.6 Receptors and downstream effectors involved in synaptic plasticity. The effectors represented here are markers for LTD and LTP.[76–79]

effector activation. Specifically, theta burst protocol resulted in an increase in the ratio of PKC to endocannabinoid production while the 20-Hz stimulation had the reverse effect. These themes of competing pathways and balance between effectors, for example, kinases and phosphatases, have been looked at with interest in a number of studies where one pathway leads to LTP and the other leads to LTD. A balance between CaMKII and PP2B has been used as a factor involved in the selection between LTP and LTD in various studies of Li et al.[58] and Stefan et al.[85] The balance between CaMKII and PP2B controls the phosphorylation state of ionotropic glutamate receptors.[68,86] At a low calcium elevation, such as produced by a weak stimulation, the PP2B activity dominates, leading to a reduction in the number of synaptic ionotropic glutamate receptors, mainly the GluR1 subunit of the AMPA type receptor. At a higher calcium level, such as produced by strong stimulation, CaMKII activity takes over,[58] producing an increase in these receptors. This phenomenon has been highlighted as important for the striatal system as well,[68] where the AMPA type receptor concentration in the postsynaptic membrane was considered as a marker for synaptic plasticity. Our example model, Fig. 12.4A, also schematically describes this balance.

For our example model, the balance between phosphatase and kinase is expressed as the difference between the activity of PP2B and CaMKII (Fig. 12.4B). The balance between PP2B and CaMKII is of relevance not only in the hippocampus but also to striatal signaling, because PP2B decreases the inhibition of phosphatase PP1 by dephosphorylating the T34 residue on DARPP32.[16,87] This in turn increases the dephosphorylation of GluR1 subunits, which are phosphorylated by CaMKII.[88] In addition, there is a PP2B-mediated, but PP1-independent effect, on GluR1 dephosphorylation.[89] In summary, there seems to be an opposing effect of PP2B and CaMKII on AMPA receptor subunit phosphorylation state in the striatum, which would control the synaptic plasticity direction depending upon the calcium concentration as has been observed in the hippocampus.[90] This balance can be seen by the same number of calcium inputs but with different frequencies to the model (Fig. 12.4C). For lower frequencies, the PP2B activity is higher because of its high affinity with calcium/calmodulin and vice versa for CaMKII, as explained above.

Here, we have reviewed different aspects of existing postsynaptic signaling network models in MSNs thought to be involved in synaptic plasticity. These models provide us with a better mechanistic understanding of the functioning of this complex system and can, hopefully, serve as tools to understand the aberrations under pathological conditions.

4.2. Modeling dopamine metabolism and energy demand in midbrain dopaminergic neurons

It is not only the postsynaptic signaling events in MSNs which have been formalized using modeling approaches. Signaling in the presynaptic dopamine neurons has also been modeled because of its implications in pathological conditions such as Parkinson disease which results from the death of midbrain dopaminergic neurons. The innervation of the striatum by the axonal arbor of these neurons is so dense that most of the cell volume is located there.[91] There has been an intense debate on the cause of demise of these neurons and several lines of evidence point to oxidative stress as one potential factor. To explore possible factors contributing to dopaminergic cell death, two different groups have modeled physiological aspects of midbrain dopaminergic neurons.

The synthesis, release, uptake, and degradation of dopamine have been modeled using the biochemical system theory.[44] In this framework, the rate of change of the system species is represented as a power law of its concentrations. This is an approximation to mechanistic steady-state formalisms like Michaelis–Menten or Hill equations.[92,93] Interestingly, while the model parameters were set to match metabolite amounts according to "experts," the model reproduced with relatively high accuracy experimental results obtained with mutants and knockouts of several enzymes in the system. Assisted by sensitivity analysis, the authors suggested the possible therapeutic intervention of several enzymes for increasing dopamine level in Parkinsonian patients. For example, while the inhibition of monoamine oxidase does increase the dopamine level by reducing its degradation, it also increases the level of reactive metabolites that could contribute to the demise of dopaminergic neurons characteristic of Parkinson disease. This analysis suggested that the enhancement of the vesicular transporter of dopamine (VMAT2) and addressing multiple targets were better strategies. The analysis also illustrates a low sensitivity of the output to most of parameters in the model, which was interpreted as the model being robust to moderate perturbations.[44,94]

Another study addressed the energy demand of electrical conduction along the uniquely extensive and nonmyelinated axon of these neurons.[95] The authors developed a multicompartment single neuron model based on previous models for these neurons[96] but with a more detailed representation of the axonal arbor and including intracellular Ca^{2+} buffers. With this model, they estimated the amount of ions that moved down the electrochemical gradient upon an action potential. These ions are pumped back to reestablish resting conditions with an energy cost that was estimated from the stoichiometry of ions per ATP by the Na^+–K^+ and Ca^{2+} pumps. The authors predicted that these neurons are subjected to extraordinary energy demands which grow exponentially with the complexity of the axonal arbor. These demands would stress the cell with the higher amount of reactive oxygen species generated by mitochondrial respiration. The dependency on axonal complexity would explain why the disease appears in larger animals like humans but not in smaller animals. While the conclusions of this work are connected to intracellular energy generation pathways, they are not modeled explicitly and the core of the argument rests on the cellular model.

5. THE WAY FORWARD: MULTISCALE MODELING AND DATA INTEGRATION

The basal ganglia are both experimentally and computationally studied on several levels, from subcellular, cellular, and synaptic properties to involvement in behavior in health and disease. The macroscopic properties of the neurons and synapse, for example, membrane excitability and synaptic strength, can be controlled by molecular events, like dopamine receptor activation.[97,98] In order to integrate data from these different levels and sources in a manner that supports a quantitative understanding, modeling spanning the subcellular—microcircuit—systems level is necessary.

To reach a thorough quantitative understanding of how phenomena on one level, for example, the subcellular level, affects another level, for example, the cellular or network levels, one needs eventually to combine "classic" computational neuroscience with modeling of subcellular processes (which rather can be classified as systems biology). Such multiscale approaches are necessary to trace the casual chain of events in neural systems.[99] There have been attempts to integrate the different levels of abstraction.[29,100] However, such integration between the scales, if done at runtime, is computationally demanding.

To integrate neural modeling levels in practice implies that models built by different research groups at different scales or representing different brain structures need to be specified in a manner that makes it possible to combine and extend to eventually construct simulations on how different scales interact. Historically, however, models of individual neurons/networks or subcellular systems have been developed in single labs, possibly using a standard simulator or a custom code in a high-level language. What methods are available for integrating models of several brain areas or description levels of a certain brain systems? One approach of integrating scales would be to rework all the models into a single simulation environment. However, this is enormously wasteful of previous coding effort and debugging and is fraught with potential for misrepresenting the model and obtaining the wrong behavior. In addition, in a multiscale approach, it is unlikely that a single simulation environment will work for all scales of analysis. One solution is to reuse the code bases for all the individual contributions and tie them together with a "middleware" which deals transparently with interlanguage problems, simulation timing issues, and other technical issues which should not have to be solved at every step. Fortunately, such middleware exists. MUSIC, the Multi-Simulation Coordinator,[101] is software which allows multiple neuronal simulators operating in parallel to communicate online in a super-computer. This allows large-scale simulations to be built in a modular way and allows for the composition of models written for different simulation software such as Neuron, MOOSE, and Nest. It could provide one of the technologies when moving toward multiscale and multiphysics simulations.[102]

In addition to runtime model integration discussed earlier, computational predictions obtained at different scales can be used as constraints during the model building process, for example, the molecular structures of the proteins involved can put constraints on reaction schemes for a biochemical signaling network. This is illustrated with the following example. If it is known that protein A and protein B interact and protein C disrupts this interaction, two types of inhibitory interactions may occur: (a) a competitive inhibition, which consists of two separate binary complexes, namely, AB and AC or (b) a noncompetitive inhibition model, which also contains a ternary complex ABC. With protein structural information in hand, a computational protein–protein docking exercise[103] can help to resolve which type of inhibition occurs. As mentioned earlier, one of the most important steps in building signaling models is to identify the interactions between different molecules. Despite all efforts, like high-throughput pull-down techniques, it is difficult to capture all the protein–protein interactions due to their transient nature.[104] Here, the protein structural details come in handy.[105] Apart from providing the details of how to model a reaction qualitatively, computational methods can also help with the estimation of quantitative parameters of an interaction, such as association rate constants.[106] If the affinity constant is known for an interacting homologous protein pair, then a range of affinity constants can be estimated using structural information and programs like FoldX.[104,107–109] Thus, the rate constants and affinities determined using these structural models may be incorporated into systems biology models.

It is not just the protein interaction or structural data that could be used for model building at the subcellular level, but high-throughput data could also assist this process. In the past decade, proteomics has contributed considerably to the field of neuroscience in delineating the machineries involved in synaptic dynamics and nervous system disorders.[110–112] Efforts have also been made to develop new techniques and resources to generate gene expression data[113,114] and organize them into databases like GENSAT and Allen Brain Atlas.[115,116] Not surprisingly, the transcriptome for the striatal MSN also exists.[114,117] This information can help us greatly to extend the already existing models in the field. For example, the transcriptome data show that a number of signal transduction proteins currently not included in any models and molecules like CalDAGGEF and EPAC are guanosine exchange factors known to be involved in the MAPK/ERK activation, etc. These components could be included into the existing models to get a better understanding of the system. We therefore feel that multiscale modeling together with high-throughput data integration should be an important way forward.

ACKNOWLEDGMENTS

We would like to thank NIAAA (Grant 2R01AA016022), the Swedish Research council (Grants 2010-3149 and 2010-4429), Stockholm Brain Institute, and the Swedish e-Science Research Center (SeRC) for financial support.

REFERENCES

1. Liljeholm M, O'Doherty JP. Contributions of the striatum to learning, motivation, and performance: an associative account. *Trends Cogn Sci.* 2012;16(9):467–475.
2. Redgrave P, Rodriguez M, Smith Y, et al. Goal-directed and habitual control in the basal ganglia: implications for Parkinson's disease. *Nat Rev Neurosci.* 2010;11:760–772.
3. Macdonald ME, Ambrose CM, Duyao MP, et al. A novel gene containing a trinucleotide that is expanded and unstable on Huntington's disease chromosomes. *Cell.* 1993;72:971–983.
4. Simpson EH, Kellendonk C, Kandel E. A possible role for the striatum in the pathogenesis of the cognitive symptoms of schizophrenia. *Neuron.* 2010;65(5):585–596.
5. Philibin SD, Hernandez A, Self DW, Bibb JA. Striatal signal transduction and drug addiction. *Front Neuroanat.* 2011;5:60.
6. Santini E, Valjent E, Fisone G. Parkinson's disease: levodopa-induced dyskinesia and signal transduction. *FEBS J.* 2008;275:1392–1399.

7. Geerts H. Mechanistic disease modeling as a useful tool for improving CNS drug research and development. *Drug Dev Res*. 2011;72(1):66–73.
8. Nutt D, Goodwin G. ECNP Summit on the future of CNS drug research in Europe 2011: report prepared for ECNP by David Nutt and Guy Goodwin. *Eur Neuropsychopharmacol*. 2011;21(7):495–499.
9. Alexander G, Crutcher M, DeLong M. Basal ganglia-thalamocortical circuits: parallel substrates for motor, oculomotor, "prefrontal" and "limbic" functions. *Prog Brain Res*. 1990;85:119–146.
10. Grillner S, Hellgren J, Ménard A, Saitoh K, Wikström MA. Mechanisms for selection of basic motor programs-roles for the striatum and pallidum. *Trends Neurosci*. 2005;28(7):364–370.
11. Rual J-F, Venkatesan K, Hao T, et al. Towards a proteome-scale map of the human protein-protein interaction network. *Nature*. 2005;437(7062):1173–1178.
12. Stelzl U, Worm U, Lalowski M, et al. A human protein-protein interaction network: a resource for annotating the proteome. *Cell*. 2005;122(6):957–968.
13. Ge H, Walhout AJM, Vidal M. Integrating "omic" information: a bridge between genomics and systems biology. *Trends Genet*. 2003;19(10):551–560.
14. Hernández Patiño CE, Jaime-Muñoz G, Resendis-Antonio O. Systems biology of cancer: moving toward the integrative study of the metabolic alterations in cancer cells. *Front Physiol*. 2012;3:481.
15. Bibb JA, Snyder GL, Nishi A, et al. Phosphorylation of DARPP-32 by Cdk5 modulates dopamine signalling in neurons. *Nature*. 1999;402:669–671.
16. Hemmings HC, Greengard P, Tung HYL, Cohen P. DARPP-32 is a potent inhibitor of protein phosphatase-1. *Nature*. 1984;310(9):503–505.
17. Croft D. Building models using reactome pathways as templates. *Methods Mol Biol*. 2013;1021:273–283.
18. Kanehisa M, Goto S, Kawashima S, Okuno Y, Hattori M. The KEGG resource for deciphering the genome. *Nucleic Acids Res*. 2004;32(Database issue):D277–D280.
19. Li C, Donizelli M, Rodriguez N, et al. BioModels database: an enhanced, curated and annotated resource for published quantitative kinetic models. *BMC Syst Biol*. 2010;4:92.
20. Hines ML, Morse T, Migliore M, Carnevale NT, Shepherd GM. ModelDB: a database to support computational neuroscience. *J Comput Neurosci*. 2004;17(1):7–11.
21. Sivakumaran S, Hariharaputran S, Mishra J, Bhalla US. The Database of Quantitative Cellular Signaling: management and analysis of chemical kinetic models of signaling networks. *Bioinformatics*. 2003;19(3):408–415.
22. Bhalla US. Signaling in small subcellular volumes. II. Stochastic and diffusion effects on synaptic network properties. *Biophys J*. 2004;87(2):745–753.
23. Vilar JMG, Kueh HY, Barkai N, Leibler S. Mechanisms of noise-resistance in genetic oscillators. *Proc Natl Acad Sci USA*. 2002;99(9):5988–5992.
24. Gillespie DT, Hellander A, Petzold LR. Perspective: stochastic algorithms for chemical kinetics. *J Chem Phys*. 2013;138(17):170901-1–170901-14.
25. Hepburn I, Chen W, Wils S, De Schutter E. STEPS: efficient simulation of stochastic reaction-diffusion models in realistic morphologies. *BMC Syst Biol*. 2012;6:36.
26. Oliveira RF, Terrin A, Di Benedetto G, et al. The role of type 4 phosphodiesterases in generating microdomains of cAMP: large scale stochastic simulations. *PLoS One*. 2010;5(7):e11725.
27. Drawert B, Engblom S, Hellander A. URDME: a modular framework for stochastic simulation of reaction-transport processes in complex geometries. *BMC Syst Biol*. 2012;6(76):1–17.
28. Santamaria F, Antunes G, Schutter ED. Breakdown of mass-action laws in biochemical computation. [chapter 4]. In: Le Novère N, ed. *Computational Systems Neurobiology*. Dordrecht: Springer Netherlands; 2012:119–132.

29. Bhalla US. Multiscale interactions between chemical and electric signaling in LTP induction, LTP reversal and dendritic excitability. *Neural Netw.* 2011;24(9):943–949.

30. Leier A, Marquez-Lago TT. Correction factors for boundary diffusion in reaction-diffusion master equations. *J Chem Phys.* 2011;135(13):134109-3–134109-11.

31. Ajay SM, Bhalla US. A role for ERKII in synaptic pattern selectivity on the time-scale of minutes. *Eur J Neurosci.* 2004;20(10):2671–2680.

32. Brown K, Sethna JP. Statistical mechanical approaches to models with many poorly known parameters. *Phys Rev E.* 2003;68(2):1–9.

33. Gutenkunst RN, Waterfall JJ, Casey FP, Brown KS, Myers CR, Sethna JP. Universally sloppy parameter sensitivities in systems biology models. *PLoS Comput Biol.* 2007;3(10): e189.

34. Chen WW, Schoeberl B, Jasper PJ, et al. Input-output behavior of ErbB signaling pathways as revealed by a mass action model trained against dynamic data. *Mol Syst Biol.* 2009;5(239):239.

35. Ashyraliyev M, Fomekong-Nanfack Y, Kaandorp JA, Blom JG. Systems biology: parameter estimation for biochemical models. *FEBS J.* 2009;276(4):886–902.

36. Lillacci G, Khammash M. Parameter estimation and model selection in computational biology. *PLoS Comput Biol.* 2010;6(3):e1000696.

37. Raue A, Kreutz C, Maiwald T, et al. Structural and practical identifiability analysis of partially observed dynamical models by exploiting the profile likelihood. *Bioinformatics.* 2009;25(15):1923–1929.

38. Morohashi M, Winn AE, Borisuk MT, Bolouri H, Doyle J, Kitano H. Robustness as a measure of plausibility in models of biochemical networks. *J Theor Biol.* 2002;216(1):19–30.

39. Von Dassow G, Meir E, Munro EM, Odell GM. The segment polarity network is a robust developmental module. *Nature.* 2000;406(6792):188–192.

40. Saltelli A, Tarantola S, Campolongo F, Ratto M. *Sensitivity Analysis in Practice: A Guide to Assessing Scientific Models.* Hoboken, NJ: John Wiley & Sons; 2004.

41. Homma T, Saltelli A. Importance measures in global sensitivity analysis of nonlinear models. *Reliab Eng Syst Saf.* 1996;52:1–17.

42. Sobol' I. Global sensitivity indices for nonlinear mathematical models and their Monte Carlo estimates. *Math Comput Simul.* 2001;55(1–3):271–280.

43. Zi Z. Sensitivity analysis approaches applied to systems biology models. *IET Syst Biol.* 2011;5(6):336–346.

44. Qi Z, Miller GW, Voit EO. Computational systems analysis of dopamine metabolism. *PLoS One.* 2008;3(6):e2444.

45. Xia Z, Storm DR. The role of calmodulin as a signal integrator for synaptic plasticity. *Nat Rev Neurosci.* 2005;6(4):267–276.

46. Weinstein H, Mehler EL. Ca(2+)-binding and structural dynamics in the functions of calmodulin. *Annu Rev Physiol.* 1994;56:213–236.

47. Stemmer PM, Klee CB. Dual calcium ion regulation of calcineurin by calmodulin and calcineurin B. *Biochemistry.* 1994;33(22):6859–6866.

48. King MM, Huangs CY. The calmodulin-dependent activation and deactivation of the phosphoprotein phosphatase, calcineurin, and the effect of nucleotides, pyrophosphate, and divalent metal ions. Identification of calcineurin as a Zn and Fe metalloenzyme. *J Biol Chem.* 1984;259(14):8847–8856.

49. Rosenberg OS, Deindl S, Sung R-J, Nairn AC, Kuriyan J. Structure of the auto-inhibited kinase domain of CaMKII and SAXS analysis of the holoenzyme. *Cell.* 2005;123(5):849–860.

50. Bradshaw JM, Kubota Y, Meyer T, Schulman H. An ultrasensitive Ca^{2+}/calmodulin-dependent protein kinase II-protein phosphatase 1 switch facilitates specificity in post-synaptic calcium signaling. *Proc Natl Acad Sci USA.* 2003;100(18):10512–10517.

51. Yang E. Structural examination of autoregulation of multifunctional calcium/calmodulin-dependent protein kinase II. *J Biol Chem*. 1999;274(37):26199–26208.
52. Klotz IM. Ligand-receptor complexes: origin and development of the concept. *J Biol Chem*. 2004;279(1):1–12.
53. Martin SR, Andersson Teleman A, Bayley PM, Drakenberg T, Forsen S. Kinetics of calcium dissociation from calmodulin and its tryptic fragments. A stopped-flow fluorescence study using Quin 2 reveals a two-domain structure. *Eur J Biochem*. 1985;151(3):543–550.
54. O'Donnell SE, Yu L, Fowler CA, Shea MA. Recognition of β-calcineurin by the domains of calmodulin: thermodynamic and structural evidence for distinct roles. *Proteins*. 2011;79(3):765–786.
55. Quintana AR, Wang D, Forbes JE, Waxham MN. Kinetics of calmodulin binding to calcineurin. *Biochem Biophys Res Commun*. 2005;334(2):674–680.
56. Shifman JM, Choi MH, Mihalas S, Mayo SL, Kennedy MB. Ca(2+)/calmodulin-dependent protein kinase II (CaMKII) is activated by calmodulin with two bound calciums. *Proc Natl Acad Sci USA*. 2006;103(38):13968–13973.
57. Sabatini BL, Oertner TG, Svoboda K. The life cycle of Ca(2+) ions in dendritic spines. *Neuron*. 2002;33(3):439–452.
58. Li L, Stefan MI, Le Novère N. Calcium input frequency, duration and amplitude differentially modulate the relative activation of calcineurin and CaMKII. *PLoS One*. 2012;7(9):e43810.
59. Bertran-Gonzalez J, Bosch C, Maroteaux M, et al. Opposing patterns of signaling activation in dopamine D1 and D2 receptor-expressing striatal neurons in response to cocaine and haloperidol. *J Neurosci*. 2008;28(22):5671–5685.
60. Bertran-Gonzalez J, Hervé D, Girault J-A, Valjent E. What is the degree of segregation between striatonigral and striatopallidal projections? *Front. Neuroanat*. 2010;4(136):1–9.
61. Schultz W. Multiple dopamine functions at different time courses. *Annu Rev Neurosci*. 2007;30:259–288.
62. Volkow ND, Wang G-J, Fowler JS, Tomasi D, Telang F. Addiction: beyond dopamine reward circuitry. *Proc Natl Acad Sci USA*. 2011;108(37):15037–15042.
63. Svenningsson P, Nishi A, Fisone G, Girault J-A, Nairn AC, Greengard P. DARPP-32: an integrator of neurotransmission. *Annu Rev Pharmacol Toxicol*. 2004;44:269–296.
64. Håkansson K, Lindskog M, Pozzi L, Usiello A, Fisone G. DARPP-32 and modulation of cAMP signaling: involvement in motor control and levodopa-induced dyskinesia. *Parkinsonism Relat Disord*. 2004;10(5):281–286.
65. Kötter R. Postsynaptic integration of glutamatergic and dopaminergic signals in the striatum. *Prog Neurobiol*. 1994;44(2):163–196.
66. Fernandez E, Schiappa R, Girault J-A, Le Novere N. DARPP-32 is a robust integrator of dopamine and glutamate signals. *PLoS Comput Biol*. 2006;2(12):e176.
67. Lindskog M, Kim M, Wikström MA, Blackwell KT, Kotaleski JH. Transient calcium and dopamine increase PKA activity and DARPP-32 phosphorylation. *PLoS Comput Biol*. 2006;2(9):e119.
68. Nakano T, Doi T, Yoshimoto J, Doya K. A kinetic model of dopamine- and calcium-dependent striatal synaptic plasticity. *PLoS Comput Biol*. 2010;6(2):e1000670.
69. Gutierrez-Arenas O, Eriksson O, Hellgren Kotaleski J. Segregation and crosstalk of D1 receptor-mediated activation of ERK in striatal medium spiny neurons upon acute administration of psychostimulants. PLoS Comput Biol (In press).
70. Bhalla US. Signaling in small subcellular volumes. I. Stochastic and diffusion effects on individual pathways. *Biophys J*. 2004;87(2):733–744.
71. Oliveira RF, Kim M, Blackwell KT. Subcellular location of PKA controls striatal plasticity: stochastic simulations in spiny dendrites. *PLoS Comput Biol*. 2012;8(2):e1002383.

72. Picconi B, Pisani A, Barone I, et al. Pathological synaptic plasticity in the striatum: implications for Parkinson's disease. *Neurotoxicology*. 2005;26(5):779–783.
73. Kasanetz F, Deroche-Gamonet V, Berson N, et al. Transition to addiction is associated with a persistent impairment in synaptic plasticity. *Science*. 2010;328(5986):1709–1712.
74. Picconi B, Centonze D, Håkansson K, et al. Loss of bidirectional striatal synaptic plasticity in L-DOPA-induced dyskinesia. *Nat Neurosci*. 2003;6(5):501–506.
75. Malenka RC, Bear MF. LTP and LTD: an embarrassment of riches. *Neuron*. 2004;44(1):5–21.
76. Calabresi P, Picconi B, Tozzi A, Di Filippo M. Dopamine-mediated regulation of corticostriatal synaptic plasticity. *Trends Neurosci*. 2007;30(5):211–219.
77. Gerfen CR, Surmeier DJ. Modulation of striatal projection systems by dopamine. *Annu Rev Neurosci*. 2011;34:441–466.
78. Lovinger DM. Neurotransmitter roles in synaptic modulation, plasticity and learning in the dorsal striatum. *Neuropharmacology*. 2010;58(7):951–961.
79. Tritsch NX, Sabatini BL. Dopaminergic modulation of synaptic transmission in cortex and striatum. *Neuron*. 2012;76(1):33–50.
80. Lovinger DM, Mathur BN. Endocannabinoids in striatal plasticity. *Parkinsonism Relat Disord*. 2012;18(Suppl. 1):S132–S134.
81. Bonsi P, Martella G, Cuomo D, et al. Loss of muscarinic autoreceptor function impairs long-term depression but not long-term potentiation in the striatum. *J Neurosci*. 2008;28(24):6258–6263.
82. Paille V, Fino E, Du K, et al. GABAergic circuits control spike-timing-dependent plasticity. *J Neurosci*. 2013;33(22):9353–9363.
83. Charpier S, Mahon S, Deniau JM. In vivo induction of striatal long-term potentiation by low-frequency stimulation of the cerebral cortex. *Neuroscience*. 1999;91(4):1209–1222.
83a. Ronesi J, Lovinger DM. Induction of striatal long-term synaptic depression by moderate frequency activation of cortical afferents in rat. *J. Physiol*. 2005;562:245–256.
84. Kim B, Hawes SL, Gillani F, Wallace LJ, Blackwell KT. Signaling pathways involved in striatal synaptic plasticity are sensitive to temporal pattern and exhibit spatial specificity. *PLoS Comput Biol*. 2013;9(3):e1002953.
85. Stefan MI, Edelstein SJ, Le Nove N, et al. *Proc Natl Acad Sci USA*. 2008;105(48):10768–10773.
86. Hayashi Y. Driving AMPA receptors into synapses by LTP and CaMKII: requirement for GluR1 and PDZ domain interaction. *Science*. 2000;287(5461):2262–2267.
87. Svenningsson P, Lindskog M, Ledent C, et al. Regulation of the phosphorylation of the dopamine- and cAMP-regulated phosphoprotein of 32 kDa in vivo by dopamine D1, dopamine D2, and adenosine A2A receptors. *Proc Natl Acad Sci USA*. 2000;97(4):1856–1860.
88. Snyder GL, Allen PB, Fienberg AA, et al. Regulation of phosphorylation of the GluR1 AMPA receptor in the neostriatum by dopamine and psychostimulants in vivo. *J Neurosci*. 2000;20(12):4480–4488.
89. Snyder GL, Galdi S, Fienberg AA, Allen P, Nairn AC, Greengard P. Regulation of AMPA receptor dephosphorylation by glutamate receptor agonists. *Neuropharmacology* 2003;45(6):703–713.
90. Castellani GC, Quinlan EM, Bersani F, Cooper LN, Shouval HZ. A model of bidirectional synaptic plasticity: from signaling network to channel conductance. *Learn Mem*. 2005;12(4):423–432.
91. Matsuda W, Furuta T, Nakamura KC, et al. Single nigrostriatal dopaminergic neurons form widely spread and highly dense axonal arborizations in the neostriatum. *J Neurosci*. 2009;29(2):444–453.
92. Voit EO. Biochemical systems theory: a review. *ISRN Biomath*. 2013;2013:1–53.

93. Voit EO, Savageau MA. Accuracy of alternative representations for integrated biochemical systems. *Biochemistry*. 1987;1987(26):6869–6880.

94. Qi Z, Miller GW, Voit EO. Mathematical models of dopamine metabolism in Parkinson's disease. In: Wellstead P, Cloutier M, eds. *Systems Biology of Parkinson's Disease*. New York, NY: Springer New York; 2012:151–171.

95. Pissadaki EK, Bolam JP. The energy cost of action potential propagation in dopamine neurons: clues to susceptibility in Parkinson's disease. *Front Comput Neurosci*. 2013;7:13.

96. Canavier CC, Landry RS. An increase in AMPA and a decrease in SK conductance increase burst firing by different mechanisms in a model of a dopamine neuron in vivo. *J Neurophysiol*. 2006;96(5):2549–2563.

97. Nicola SM, Surmeier DJ, Malenka RC. Dopaminergic modulation of neuronal excitability in the striatum and nucleus accumbens. *Annu Rev Neurosci*. 2000;23:185–215.

98. Szalisznyó K, Müller L. Dopamine induced switch in the subthreshold dynamics of the striatal cholinergic interneurons: a numerical study. *J Theor Biol*. 2009;256(4):547–560.

99. Kotaleski JH, Blackwell KT. Modelling the molecular mechanisms of synaptic plasticity using systems biology approaches. *Nat Rev Neurosci*. 2010;11(4):239–251.

100. Mattioni M, Le Novère N. Integration of biochemical and electrical signaling-multiscale model of the medium spiny neuron of the striatum. *PLoS One*. 2013;8(7): e66811.

101. Djurfeldt M, Hjorth J, Eppler JM, et al. Run-time interoperability between neuronal network simulators based on the MUSIC framework. *Neuroinformatics*. 2010;8(1):43–60.

102. Brandi M, Brocke E, Talukdar H, et al. Connecting MOOSE and NeuroRD through MUSIC: towards a communication framework for multi-scale modeling. *BMC Neurosci*. 2011;12(Suppl. 1):P77.

103. Dominguez C, Boelens R, Bonvin AMJJ. HADDOCK: a protein-protein docking approach based on biochemical or biophysical information. *J Am Chem Soc*. 2003;125(7):1731–1737.

104. Beltrao P, Kiel C, Serrano L. Structures in systems biology. *Curr Opin Struct Biol*. 2007;17(3):378–384.

105. Aloy P, Russell RB. Structural systems biology: modelling protein interactions. *Nat Rev Mol Cell Biol*. 2006;7(3):188–197.

106. Kiel C, Aydin D, Serrano L. Association rate constants of ras-effector interactions are evolutionarily conserved. *PLoS Comput Biol*. 2008;4(12):e1000245.

107. Guerois R, Nielsen JE, Serrano L. Predicting changes in the stability of proteins and protein complexes: a study of more than 1000 mutations. *J Mol Biol*. 2002;320(2):369–387.

108. Schymkowitz J, Borg J, Stricher F, Nys R, Rousseau F, Serrano L. The FoldX web server: an online force field. *Nucleic Acids Res*. 2005;33(Web Server issue): W382–W388.

109. Schymkowitz JWH, Rousseau F, Martins IC, Ferkinghoff-Borg J, Stricher F, Serrano L. Prediction of water and metal binding sites and their affinities by using the Fold-X force field. *Proc Natl Acad Sci USA*. 2005;102(29):10147–10152.

110. Grant SGN, Marshall MC, Page K-L, Cumiskey MA, Armstrong JD. Synapse proteomics of multiprotein complexes: en route from genes to nervous system diseases. *Hum Mol Genet*. 2005;14 Spec No(2):R225–R234.

111. Husi H, Grant SG. Proteomics of the nervous system. *Trends Neurosci*. 2001;24(5):259–266.

112. Micheva KD, Busse B, Weiler NC, O'Rourke N, Smith SJ. Single-synapse analysis of a diverse synapse population: proteomic imaging methods and markers. *Neuron*. 2010;68(4):639–653.

113. Doyle JP, Dougherty JD, Heiman M, et al. Application of a translational profiling approach for the comparative analysis of CNS cell types. *Cell.* 2008;135(4):749–762.
114. Heiman M, Schaefer A, Gong S, et al. A translational profiling approach for the molecular characterization of CNS cell types. *Cell.* 2008;135(4):738–748.
115. Gong S, Zheng C, Doughty ML, et al. A gene expression atlas of the central nervous system based on bacterial artificial chromosomes. *Nature.* 2003;425(6961):917–925.
116. Lein ES, Hawrylycz MJ, Ao N, et al. Genome-wide atlas of gene expression in the adult mouse brain. *Nature.* 2007;445(7124):168–176.
117. Lobo MK, Karsten SL, Gray M, Geschwind DH, Yang XW. FACS-array profiling of striatal projection neuron subtypes in juvenile and adult mouse brains. *Nat Neurosci.* 2006;9(3):443–452.

CHAPTER THIRTEEN

Data-Driven Modeling of Synaptic Transmission and Integration

Jason S. Rothman, R. Angus Silver

Department of Neuroscience, Physiology & Pharmacology, University College London, London, UK

Contents

Abstract

In this chapter, we describe how to create mathematical models of synaptic transmission and integration. We start with a brief synopsis of the experimental evidence underlying our current understanding of synaptic transmission. We then describe synaptic transmission at a particular glutamatergic synapse in the mammalian cerebellum, the mossy fiber to granule cell synapse, since data from this well-characterized synapse can provide a benchmark comparison for how well synaptic properties are captured by different mathematical models. This chapter is structured by first presenting the simplest mathematical description of an average synaptic conductance waveform and then introducing methods for incorporating more complex synaptic properties such as nonlinear voltage dependence of ionotropic receptors, short-term plasticity, and stochastic fluctuations. We restrict our focus to excitatory synaptic transmission, but most of the modeling approaches discussed here can be equally applied to inhibitory

Progress in Molecular Biology and Translational Science, Volume 123
ISSN 1877-1173
http://dx.doi.org/10.1016/B978-0-12-397897-4.00004-8

synapses. Our data-driven approach will be of interest to those wishing to model synaptic transmission and network behavior in health and disease.

1. INTRODUCTION
1.1. A brief history of synaptic transmission

Some of the first intracellular voltage recordings from the neuromuscular junction (NMJ) revealed the presence of spontaneous miniature end plate potentials with fast rise and slower decay kinetics.[1] The similarity of these "mini" events to the smallest events evoked by nerve stimulation, together with the discrete nature of the fluctuations in the amplitude of the end plate potentials,[2] lead to the hypothesis that transmitter was released probabilistically in discrete all-or-none units called "quanta,"[3] units that were subsequently shown to be vesicles containing neurotransmitter. The quantum hypothesis is an elegantly simple yet extremely powerful statistical model of transmitter release: the average number of quanta released at a synapse per stimulus (quantal content, m) is simply the product of the total number of quanta available for release (N_T) and their release probability (P):

$$m = N_T P \qquad (13.1)$$

Quantitative comparison of the predictions of the quantum hypothesis against experimental measurements confirmed the hypothesis,[3] albeit under nonphysiological conditions of low release probabilities. Subsequent electron micrograph studies revealed presynaptic vesicles clustered at active zones,[4–7] providing compelling morphological equivalents for the quanta and their specialized release sites. Other work around the same time revealed the dynamic nature of synaptic transmission at the NMJ, providing the first concepts for activity-dependent short-term changes in synaptic strength.[8,9] Further work by Katz and colleagues lead to the concept of Ca^{2+}-dependent vesicular release and the refinement of ideas regarding the activation of postsynaptic receptors.[3] Together, this early body of work on the NMJ provided the basis for our current understanding of the intricate signaling cascade underlying synaptic transmission. The basic mechanisms underlying synaptic transmission are summarized in Fig. 13.1: an action potential, propagating down the axon of the presynaptic neuron, invades synaptic terminals. The brief depolarization of the terminals causes voltage-gated Ca^{2+} channels (VGCCs) to open, leading to Ca^{2+} influx and a transient increase in the intracellular Ca^{2+} concentration ($[Ca]_i$) in the vicinity of the VGCCs.

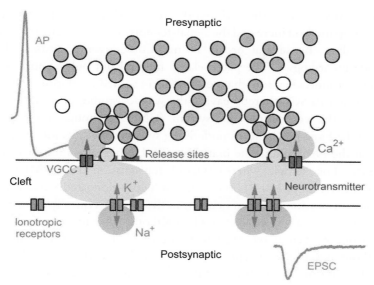

Figure 13.1 Cartoon illustrating the basic sequence of events underlying synaptic transmission. The sequence starts with an action potential (AP) invading a presynaptic terminal, leading to the opening of voltage-gated Ca^{2+} channels (VGCCs), some of which are located near vesicle release sites within one or more active zones. For those release sites containing a readily releasable vesicle, the local rise in $[Ca^{2+}]_i$ causes the fusion of the vesicle with the terminal's membrane, resulting in the release of neurotransmitter packed inside the vesicle. The neurotransmitter diffuses across the synaptic cleft to reach the postsynaptic membrane where it binds to ionotropic receptors, causing the channels to open and pass Na^+ and K^+. The permeation of these ions through the ionotropic receptors leads to a local injection of current, known as the EPSC. The EPSC often contains fast and slow components due to the fast activation of receptors immediately opposite to the vesicle release site and the slower activation of receptors further away (i.e., extrasynaptic). Kinetics of the EPSC will also depend on the receptor's affinity for the neurotransmitter and the receptor's gating properties, which may include blocked and desensitization states. (See the color plate.)

For those vesicles docked at a release site near one or more VGCCs, the local increase in $[Ca]_i$ triggers the vesicles to fuse with the terminal membrane and release their content of neurotransmitter into the synaptic cleft. The released neurotransmitter diffuses across the narrow synaptic cleft and binds to post-synaptic ionotropic receptors, transiently increasing their open probability. The resulting flow of Na^+ and K^+ through the receptors' ion channels results in an excitatory postsynaptic potential (EPSP) or excitatory postsynaptic current (EPSC) depending on whether the intracellular recording is made under a current– or voltage–clamp configuration.

Some 20 years after the early work on the NMJ, development of the patch–clamp method increased the signal-to-noise ratio of electrophysiolog-ical recordings by several orders of magnitude over traditional sharp-electrode recordings.[10] The patch–clamp method not only confirmed the existence of individual ion channels but also enabled resolution of signifi-cantly smaller EPSCs, thereby paving the way for studies of synaptic trans-mission in the central nervous system (CNS). Although these studies revealed the basic mechanisms underlying synaptic transmission are largely similar at the NMJ and in the brain (Fig. 13.1), there are a number of key differences. For example, whereas synaptic transmission in the NMJ is medi-ated by the release of 100–1000 vesicles[2] at highly elongated active zones,[11] synaptic transmission between neurons in the brain is typically mediated by the release of just a few vesicles at a handful of small active zones.[12,13] The number of postsynaptic receptors is also quite different: vesicle release acti-vates thousands of postsynaptic receptors in the NMJ[14] but only a few (~ 10–100) at central excitatory synapses.[15,16] These differences in scale link directly to synaptic function: the large potentials generated at the NMJ ensure a reliable relay of motor command signals from presynaptic neuron to postsynaptic muscle. In contrast, the much smaller potentials generated by central synapses require spatiotemporal summation in order to trigger action potentials.

Another important distinction between the NMJ and central synapses is the difference in neurotransmitter (acetylcholine at the NMJ vs. glutamate, GABA, glycine, etc., in the CNS) and the diversity in postsynaptic receptors and their function. Here we focus on excitatory central synapses, where two major classes of ionotropic glutamate receptors, AMPA and NMDA recep-tors (AMPARs and NMDARs), are colocalized.[17,18] These two receptor types have different gating kinetics and current–voltage relations and there-fore play distinct roles in synaptic transmission. The majority of AMPARs, for example, have relatively fast kinetics and a linear (ohmic) current–voltage relation, often expressed as:

$$I_{AMPAR} = G_{AMPAR}(V - E_{AMPAR}) \qquad (13.2)$$

where V is the membrane potential and E_{AMPAR} is the reversal potential of the AMPAR conductance (G_{AMPAR}), which is typically 0 mV. Both of these properties, that is, fast kinetics and a linear current–voltage relation, make AMPARs well suited for mediating temporally precise signaling and setting synaptic weight. NMDARs, in contrast, have slower kinetics and a nonlinear current–voltage relation, the latter caused by Mg^{2+} block at hyperpolarized potentials.[19] These properties make NMDARs well suited for coincidence detection and plasticity, since presynaptic glutamate release *and* postsynaptic depolarization are required for NMDAR activation.[20] Certain subtypes of NMDARs, however, show a weaker Mg^{2+} block (i.e., those containing the GluN2C and GluN2D subunits) and therefore create substantial synaptic current at hyperpolarized potentials.[21,22] These types of NMDARs are thought to enhance synaptic transmission by enabling temporal integration of low-frequency inputs.[22] Of course, numerous other differences exist between the NMJ and central synapses, including those pertaining to stochasticity- and time-dependent plasticity. These are discussed further in the next section where we introduce the MF–to–GC synapse, our synapse of choice for providing accurate data for the synaptic models presented in this chapter.

1.2. The cerebellar MF–GC synapse as an experimental model system

The input layer of the cerebellum receives sensory and motor signals via MFs[23] which form large *en passant* synapses, each of which contacts several GCs (Fig. 13.2A). Although GCs are the smallest neuron in the vertebrate brain, they account for more than half of all neurons. Each GC receives excitatory synaptic input from 2 to 7 MFs, and each synaptic connection consists of a handful of active zones.[27,28] The small number of synaptic inputs, along with a small soma and electrically compact morphology, makes GCs particularly suitable for studying synaptic transmission.[15,18] In Fig. 13.2B, we show representative examples of EPSCs recorded at a single MF–GC synaptic connection under resting basal conditions (gray traces). Here, fluctuations in the peak amplitude of the EPSCs highlight the stochastic behavior of synaptic transmission introduced above. Analysis of such fluctuations using multiple-probability fluctuation analysis (MPFA), a technique based on a multinomial statistical model, has provided estimates for N_T, P and the postsynaptic response to a quantum of transmitter (Q), for single MF–GC connections. MPFA indicates that at low frequencies synaptic transmission is meditated by 5–10 readily releasable vesicles (or, equivalently the number of functional release sites N_T), with each vesicle or site having a vesicular release probability (P) of ~ 0.5.[26,29] Experiments with rapidly equilibrating AMPAR antagonists suggest that release is predominantly univesicular at this synapse (one vesicle released per synaptic contact), an interpretation that is supported by the finding that at some weak MF–GC

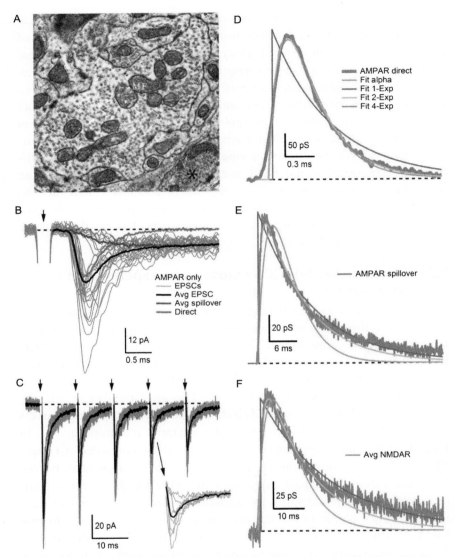

Figure 13.2 Synaptic transmission at the cerebellar MF–GC synapse. (A) Electron micrograph of a cerebellar MF terminal filled with thousands of synaptic vesicles and a few large mitochondria. Synaptic contacts with GC dendrites appear along the contours of the MF membrane at several locations, evident by the wider and darker appearance of the membrane due to clustering of proteins within the presynaptic active zone and postsynaptic density. (B) Superimposed AMPAR-mediated EPSCs (gray) recorded from a single MF–GC connection, showing considerable variability in amplitude and time course from trial to trial. On some trials, failure of direct release revealed a spillover current with slow rise time. Such trials were separated using the rise time criteria of Ref. 24.

connections a maximum of only one vesicle is released even when P is increased to high levels.[15,26]

Synaptic responses to low-frequency presynaptic stimuli (e.g., those in Fig. 13.2B) provide useful information about N_T, P, and Q under resting conditions. To explore how these quantal synaptic parameters change in an activity–dependent manner, however, paired–pulse stimulation protocols or high-frequency trains of stimuli are required. Figure 13.2C shows an example of the latter, where responses of a single MF–GC connection to the same 100 Hz train of stimuli are superimposed (gray traces). Here, fluctuations in the peak amplitude of the EPSCs can still be seen (see inset), but successive peaks between stimuli also show clear signs of depression. The average of all responses (black trace) reveals the depression more clearly. Although by eye, signs of facilitation are not apparent in Fig. 13.2C, facilitation at this synapse most likely exists. We know this since lowering P at this synapse, by lowering the extracellular Ca^{2+} concentration, has revealed the presence of both depression and facilitation; however, because depression predominates under normal conditions, facilitation is not always apparent.[29] As described in detail later in this chapter, mathematical models have been developed to simulate synaptic depression and facilitation. If used appropriately, these models can provide useful insights into the underlying mechanisms of synaptic transmission. Such models have revealed, for

---

The average direct-release component (green) was computed by subtracting the average spillover current (blue) from the average total EPSC (black). Arrow denotes time of extracellular MF stimulation, which occurred at a slow frequency of 2 Hz; most of the stimulus artifact has been blanked for display purposes. (C) Superimposed AMPAR-mediated EPSCs (gray) recorded from a single MF–GC connection and their average (black). The MF was stimulated at 100 Hz with an external electrode (arrows at top). Successive EPSCs show clear signs of depression. Inset shows EPSC responses to fourth stimulus on expanded timescale, showing the variation in peak amplitude. Stimulus artifacts have been blanked. (D) Average direct-release AMPAR conductance waveform (gray) fit with $G_{syn}(t)$ defined by the following functions: alpha (Eq. 13.5), one-exponential (Eq. 13.4), two-exponential (Eq. 13.6), multiexponential (4-Exp, Eq. 13.7). Most functions gave a good fit except the one-exponential function (blue). The conductance waveform was computed from the average current waveform in (B) via Eq. (13.3). (E) Same as (D) but for the average spillover component in (B). Most functions gave a good fit except the alpha function (green). (F) Same as (D) but for an average NMDAR-mediated conductance waveform computed from four different MF–GC connections. Again, most functions gave a good fit except the alpha function (green). Dashed lines denote 0. *(A) Image from Palay and Chan-Palay[25] with permission. (B) Data from Sargent et al.[26] with permission. (See the color plate.)*

example, a rapid rate of vesicle reloading at the MF–GC synapse ($k_1 = 60$–80 ms^{-1}) as well as a large pool of vesicles that can be recruited rapidly at each release site (~ 300[29–31]). These findings offer an explanation as to how the MF–GC synapse can sustain high-frequency signaling for prolonged periods of time.

The MF–GC synapse forms part of a glomerular-type synapse, which also occur in the thalamus and dorsal spinocerebellar tract. While the purpose of the glomerulus has not been determined definitively, experimental evidence from the MF–GC synapse indicates this glial-ensheathed structure promotes transmitter spillover between excitatory synaptic connections[24,32] and between excitatory and inhibitory synaptic connections.[33] AMPAR-mediated EPSCs recorded from a MF–GC connection, therefore, exhibit both a fast "direct" component arising from quantal release at the MF–GC connection under investigation (Fig. 13.2B, green trace) and a slower component mediated by glutamate spillover from neighboring MF–GC connections (blue trace). While direct quantal release is estimated to activate about 50% of postsynaptic AMPARs at the peak of the EPSC,[34] spillover is estimated to activate a significantly smaller fraction. However, because spillover produces a prolonged presence of glutamate in the synaptic cleft, activation of AMPARs by spillover can contribute as much as 50% of the AMPAR-mediated charge delivered to GCs.[24]

Glutamate spillover also activates NMDARs, but mostly at mature MF–GC synapses when the NMDARs occupy a perisynaptic location.[35] At a more mature time of development, MF–GC synapses also exhibit a weak Mg^{2+} block due to the expression of GluN2C and/or GluN2D subunits.[22,36,37] The weak Mg^{2+} block allows NMDARs to pass a significant amount of charge at subthreshold potentials, thereby creating a spillover current comparable in size to the AMPAR-mediated spillover current. Using several of the modeling techniques discussed in this chapter, we were able to show the summed contribution from both AMPAR and NMDAR spillover currents enables GCs to integrate over comparatively long periods of time, thereby enabling transmission of low-frequency MF signals through the input layer of the cerebellum.[22]

In the following sections, we describe how to capture the various properties of synaptic transmission recorded at the MF–GC synapse in mathematical forms that can be used in computer simulations. We start with the most basic features of the synapse, the postsynaptic conductance waveform, and the resulting postsynaptic current, and add biological detail from there. However, several aspects of synaptic transmission are beyond the scope of this chapter. These include long-term plasticity (i.e., Hebbian learning) and presynaptic Ca^{2+} dynamics. Mathematical models of these synaptic processes can be found elsewhere.[38–42]

2. CONSTRUCTING SYNAPTIC CONDUCTANCE WAVEFORMS FROM VOLTAGE-CLAMP RECORDINGS

The time course of a synaptic conductance, denoted $G_{syn}(t)$, can be computed from the synaptic current, $I_{syn}(t)$, measured at a particular holding potential (V_{hold}) using the whole-cell voltage-clamp technique. If the synapse under investigation is electrotonically close to the somatic patch pipette, as is the case with the MF–GC synapse, then adequate voltage clamp can be achieved and the measured $I_{syn}(t)$ will have relatively small distortions due to poor space clamp. On the other hand, if the synapse under investigation is electrotonically distant to the somatic patch pipette, for example, at the tip of a spine several hundred micrometers from the soma, then significant errors due to poor space clamp will distort nearly all aspects of $I_{syn}(t)$, including its amplitude, kinetics, and reversal potential.[43] To overcome this problem, a technique using voltage jumps can be used to extract the decay time course under conditions of poor space clamp, or dendritic patching can be used to reduce the electrotonic distance between the synapse and recording site.[44]

When measuring $I_{syn}(t)$ under voltage clamp, individual current components (e.g., the AMPAR and NMDAR current components, I_{AMPAR} and I_{NMDAR}) can be cleanly separated using selective antagonists (e.g., APV or NBQX), and the reversal potential of the currents (e.g., E_{AMPAR} and E_{NMDAR}) can be established by measuring the current–voltage relation and correcting for the liquid junction potential of the recording pipette. The synaptic current component can then be converted to conductance using the following variant of Eq. (13.2):

$$G_{syn}(t) = I_{syn}(t) / (V_{hold} - E_{syn})$$ (13.3)

where E_{syn} denote the reversal potential of the synaptic conductance under investigation. The next step is to find a reasonable mathematical expression for $G_{syn}(t)$. The simplest way to do this is to first remove stochastic fluctuations in the amplitude and timing of $G_{syn}(t)$ by averaging many EPSCs recorded under low-frequency conditions (e.g., see Fig. 13.2B) and then fit one of the waveforms described below (Eqs. 13.4–13.7) to the averaged EPSC. Later in the chapter, we discuss methods for incorporating stochastic fluctuations into the mathematical representation of $G_{syn}(t)$.

Exponential functions are typically used to represent $G_{syn}(t)$. If computational overhead is a major consideration, for example, in large-scale network modeling, single-exponential functions can be used since they are described by only two parameters, the peak conductance g_{peak} and a single decay time constant τ_d:

$$G_{syn}(t) = g_{peak}\, e^{-t'/\tau_d} \tag{13.4}$$

where $t' = t - t_j$. Here, the arrival of the presynaptic action potential at $t = t_j$ leads to an instantaneous jump in $G_{syn}(t)$ from 0 to g_{peak}, after which $G_{syn}(t)$ decays back to zero (note, here and below $G_{syn}(t) = 0$ for $t < t_j$; for consistency, a notation similar to that of Ref. 41 has been used). This mathematical description of $G_{syn}(t)$ may be sufficient if the decay time is much larger than the rise time. However, if the precise timing of individual synaptic inputs is important, as in the case of an auditory neuron performing synaptic coincidence detection, then a realistic description of the rise time should be included in $G_{syn}(t)$. In this case, the simplest description is to use the alpha function, which has an exponential-like rise time course:

$$G_{syn}(t) = g_{peak}\frac{t'}{\tau}\, e^{1-t'/\tau} \tag{13.5}$$

where t' is defined as in Eq. (13.4). The convenience of the alpha function is that it only contains two parameters, g_{peak} and τ, which directly set the peak value and the time of the peak. However, the alpha function only fits waveforms with a rise time constant (τ_r) and τ_d of similar magnitude, which is not usually the case for synaptic conductances. When τ_r and τ_d are of different magnitude, then a double-exponential function is more appropriate for capturing the conductance waveform:

$$G_{syn}(t) = g_{peak}\left[-e^{-t'/\tau_r} + e^{-t'/\tau_d}\right]/a_{norm} \tag{13.6}$$

Here, the constant a_{norm} is a scale factor that normalizes the expression in square brackets so that the peak of $G_{syn}(t)$ equals g_{peak} (see Ref. 41 for an analytical expression of a_{norm}). Still, Eq. (13.6) may not be suitable for some conductance waveforms. Synaptic AMPAR conductance waveforms, for example, typically exhibit a sigmoidal rise time course, which can usually be neglected, but there are certain instances when it is important to accurately capture this component.[26,32] In this case, a multiexponential function with an $m^x h$ formalism can be used to fit the conductance waveform[45]:

$$G_{syn}(t) = g_{peak} \left[1 - e^{-t/\tau_r} \right]^x \left[d_1 e^{-t/\tau_{d1}} + d_2 e^{-t/\tau_{d2}} + d_3 e^{-t/\tau_{d3}} \right] / a_{norm}$$

$$(13.7)$$

Here, the first expression in square brackets describes the rise time course, which, when raised to a power $x > 1$, exhibits sigmoidal activation. The second expression in square brackets describes the decay time course and includes three exponentials for flexibility, one or two of which can be removed if unnecessary. This function is flexible in fitting synaptic current or conductance waveforms and has produced good fits to the time course of miniature EPSCs recorded in cultured hippocampal neurons[45] and AMPAR and NMDAR currents recorded from cerebellar GCs.[24,32,46] With nine free parameters, however, Eq. (13.7) is not only computationally expensive but also has the potential to cause problems when used in curve-fitting algorithms. We have found the best technique for fitting Eq. (13.7) to EPSCs is to begin with x fixed at 1 (no sigmoidal activation) and one or two decay components fixed to zero ($d_2 = 0$ and/or $d_3 = 0$). If the initial fits under these simplified assumptions are inadequate, then one by one the fixed parameters can be allowed to vary to improve the fit. The scale factor a_{norm} can be calculated by computing the product of the expressions in square brackets at high temporal resolution and setting a_{norm} equal to the peak of the resulting waveform.

To illustrate how well the different mathematical functions capture synaptic conductance waveforms in practice, we fit Eqs. (13.4)–(13.7) to the average direct-release AMPAR conductance component of the MF–GC synapse (computed from currents in Fig. 13.2B) and plotted the fits together in Fig. 13.2D. The single-exponential function (Eq. 13.4) fit neither the rise nor decay time course. The two-exponential function (Eq. 13.6) fit well, except for the initial onset period, which lacked a sigmoidal rise time course. The alpha function (Eq. 13.5) fit both the rise and decay time course well since τ_r and τ_d of the direct-release component are of similar magnitude. The multiexponential function (Eq. 13.7) showed the best overall fit. The same comparison was computed for the average spillover AMPAR conductance component (Fig. 13.2E). This time only the multiexponential function provided a good fit to both the rise and decay time course. The two-exponential function also fit well except for a small underestimate of the decay time course; an additional exponential decay component would improve this fit. The one-exponential function provided a suitable fit to the decay time course but not the rise time course. The alpha function fit neither the rise or decay time course. Finally, the same comparison was made for an average NMDAR conductance waveform computed from four MF–GC connections (Fig. 13.2F). These results were similar to those of the spillover AMPAR conductance. Hence, as the results in Fig. 13.2D–F highlight, Eqs. (13.4)–(13.7) can reproduce a $G_{syn}(t)$ with different rise and decay time courses. These differences may or may not be consequential depending on the computer simulation at hand. In most instances, it is always preferable to choose the simplest level of description, but it is also important to verify the simplification does not significantly alter the outcome or conclusions of the study.

As a general guide, the direct AMPAR current typically has a rise time course of 0.2 ms and a decay time course between 0.3 and 2.0 ms at physiological temperatures, depending on the AMPAR subunit composition at the synapse type under investigation.[18,47,48] The spillover AMPAR current typically has a rise time course of 0.6 ms and decay time course of 6.0 ms, measured at the MF–GC synapse.[24] The NMDAR current has the slowest kinetics, with a rise time course of ∼10 ms and a decay time course anywhere between 30 and 70 ms, but can even be longer than 500 ms depending on the NMDAR subunit composition.[49,50]

3. EMPIRICAL MODELS OF VOLTAGE-DEPENDENT Mg^{2+} BLOCK OF THE NMDA RECEPTOR

The voltage dependence of the synaptic AMPAR component can usually be modeled with the simple linear current–voltage relation described in Eq. (13.2). In contrast, the synaptic NMDAR component exhibits strong voltage dependence due to Mg^{2+} binding inside the receptor's ion channel.[19] The block is strongest near the neuronal resting potential and becomes weaker as the membrane potential becomes more depolarized. This unique characteristic of NMDARs allows them to behave like logical AND gates: the receptors conduct current only when they are in the glutamate-bound state AND when the postsynaptic neuron is depolarized. It is this AND-gate property combined with their high Ca^{2+} permeability that enables NMDARs to play such a pivotal role in long-term plasticity, learning and memory.[20,51–53] Here, we consider how to model the electrophysiological AND-gate properties of synaptic NMDARs.

As mentioned in the previous section, the time course of the NMDAR component can be captured with a multiexponential function. The key additional step required for modeling the NMDAR component is the nonlinear voltage-dependent scaling of the conductance waveform, here referred to as $\varphi(V)$, which is the fraction of the NMDAR conductance that is *unblocked*. This scaling can be easily incorporated into a current–voltage relation as follows:

$$I_{NMDAR} = g_{NMDAR}\,\varphi(V)(V - E_{NMDAR}) \tag{13.8}$$

Typically, a Boltzmann function is used to describe $\varphi(V)$, which takes on values from 0 at the most hyperpolarized potentials (all blocked) to 1 at the most depolarized potentials (all unblocked), and is commonly written as:

$$\varphi(V) = \frac{1}{1 + e^{-(V - V_{0.5})/k}} \tag{13.9}$$

where $V_{0.5}$ is the potential at which half the NMDAR channels are blocked and k is the slope factor that determines the steepness of the voltage dependence around $V_{0.5}$. While the Boltzmann function is simple and easy to use, its free parameters $V_{0.5}$ and k do not directly relate to any physical aspect of the Mg^{2+} blocking mechanism. The two-state Woodhull formalism,[54] in contrast, is derived from a kinetic model of extracellular Mg^{2+} block, in which case its free parameters have more of a physical meaning. In this two-state kinetic model, an ion channel is blocked when an ion species, in this case extracellular Mg^{2+}, is bound to a binding site somewhere inside the channel, or open when the ion species is unbound (Fig. 13.3A). If the rate of binding and unbinding of the ion species is denoted by k_1 and k_{-1}, respectively, then $\varphi(V)$ will equal:

$$\varphi(V) = \frac{k_{-1}}{k_{-1} + k_1} = \frac{1}{1 + k_1/k_{-1}} \tag{13.10}$$

where

$$k_1 = \left[Mg^{2+}\right]_o K_1 e^{-\delta\phi V/2}$$
$$k_{-1} = K_{-1} e^{\delta\phi V/2}$$
$$\phi = zF/RT$$

Here, K_1 and K_{-1} are constants, δ is the fraction of the membrane voltage that Mg^{2+} experiences at the blocking site, z is the valence of the blocking ion (here, $+2$), F is the Faraday constant, R is the gas constant, and T is the absolute temperature. Dividing through terms, Eq. (13.10) can be expressed in a more familiar notation that includes a dissociation constant (K_d):

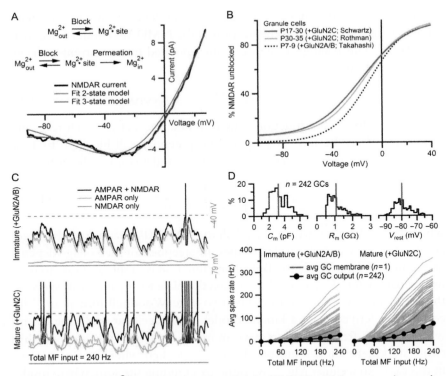

Figure 13.3 Weak Mg^{2+} block in GluN2C-containing NMDARs. (A) Current–voltage relation of an NMDAR current from a mature GC (black) fit to Eq. (13.8) ($E_{NMDAR} = 0$ mV) where $\varphi(V)$ was defined by either a two-state kinetic model (blue; Eq. 13.11) or a three-state kinetic model that includes Mg^{2+} permeation (red; Eq. 13.12). The latter kinetic model produced the better fit. Kinetic models are shown at top. (B) Percent of unblocked NMDARs, $\varphi(V)$, from the three-state kinetic model fit in (A) (red), compared to $\varphi(V)$ derived from fits to the same model for another data set of mature GCs (purple; data from Ref. 46) and immature GCs (black; data from Ref. 50). At nearly all potentials, NMDARs from mature GCs show weaker Mg^{2+} block than those from immature GCs. This difference is presumably due to the developmental maturation switch in GCs from GluN2A/B-containing receptors to GluN2C-containing receptors, discussed in text. (C) IAF simulations (Eq. 13.20) of a GC with immature (top, +GluN2A/B) and mature (bottom, +GluN2C) NMDARs, using $\varphi(V)$ functions in (B) (black and red, respectively), demonstrating the enhanced depolarization and spiking under mature NMDAR conditions. Identical simulations were repeated with $G_{NMDAR} = 0$ (yellow) and $G_{AMPAR} = 0$ (green) to compare the contribution of AMPARs and NMDARs to depolarizing the membrane. G_{AMPAR} consisted of a simulated direct and spillover component, both with depression, as described in Fig. 13.5F. G_{NMDAR} was simulated with both depression and facilitation, as described in Fig. 13.5G. The peak value of the G_{NMDAR} waveform equaled that of the G_{AMPAR} waveform, giving an amplitude ratio of unity, which is in the physiological range for GCs. The total synaptic current consisted of the sum of four

$$\varphi(V) = \frac{1}{1 + [\mathrm{Mg}^{2+}]_o / K_d} \tag{13.11}$$

$$K_d = K_{d0} e^{\delta \phi V}$$

where K_{d0} is the dissociation constant at 0 mV and equals K_{-1}/K_1. This equation, like the Boltzmann function (Eq. 13.9), has two free parameters, K_{d0} and δ. However, unlike the Boltzmann function, both parameters now directly relate to the Mg^{2+} blocking mechanism: K_{d0} quantifies the strength or affinity of Mg^{2+} binding and δ quantifies the location of the Mg^{2+} binding site within the channel. On the other hand, Eqs. (13.9) and (13.11) are formally equivalent since their free parameters are directly convertible via the following relations: $k = (\delta\phi)^{-1}$ and $V_{0.5} = \delta\phi \cdot \ln([\mathrm{Mg}^{2+}]_o / K_{d0})$. Under physiological $[\mathrm{Mg}^{2+}]_o$, Eq. (13.11) is also equivalent to a more complicated three-state channel model with an open, closed, and blocked state.[55]

While the simple Boltzmann function and the equivalent two-state Woodhull formalism are often used to describe $\varphi(V)$, the two functions have not always proved adequate in describing experimental data. Single-channel recordings of NMDAR currents, for example, have indicated there are actually two binding sites for Mg^{2+}: one that binds external Mg^{2+} and one that binds internal Mg^{2+}.[56–60] Moreover, there are indications Mg^{2+} permeates through the NMDAR channel.[19,57] Hence, more complicated expressions of $\varphi(V)$ have been adopted. The three-state Woodhull formalism depicted

independent I_{syn}, each representing a different MF input. Spike times for each MF input were generated for a constant mean rate of 60 Hz (Eq. 13.16), producing a total MF input of 240 Hz. Total I_{syn} also contained the following tonic GABA-receptor current not discussed in this chapter: $I_{GABAR} = 0.438(V + 75)$. IAF membrane parameters matched the average values computed from a population of 242 GCs: $C_m = 3.0$ pF, $R_m = 0.92$ GΩ, $V_{rest} = -80$ mV. Action potential parameters were: $V_{thresh} = -40$ mV (gray dashed line), $V_{peak} = 32$ mV, $V_{reset} = -63$ mV, $\tau_{AR} = 2$ ms. Action potentials were truncated to -15 mV for display purposes. (D) Average output spike rate of the IAF GC model as shown in (C) as a function of total MF input rate for immature (bottom left) and mature (bottom right) NMDARs, again demonstrating the enhanced spiking caused by GluN2C subunits. A total of 242 simulations were computed using C_m, R_m, V_{rest} values derived from a data base of 242 real GCs (top distributions, red lines denote average population values), with the average output spike rate plotted as black circles. Red line denotes one GC simulation whose C_m, R_m, V_{rest} matched the average population values shown at top, which are the same parameters used in (C). Note, the output spike rate of this "average GC" simulation is twice as large as the average of all 242 GC simulations due to the nonlinear behavior of the IAF model. *Data in this figure is from Schwartz et al.[22] with permission.* (See the color plate.)

in Fig. 13.3A, for example, has been used to describe Mg^{2+} block.[57,59] This model includes Mg^{2+} permeation through the NMDAR channel, described by k_2, which is assumed to be nonreversible (i.e., $k_{-2}=0$), in which case $\varphi(V)$ equals:

$$\varphi(V) = \frac{k_{-1}+k_2}{k_{-1}+k_2+k_1} = \frac{1}{1+k_1/(k_{-1}+k_2)} \tag{13.12}$$

$$k_2 = K_2 e^{-\delta_2 \phi V/2}$$

This equation reduces to Eq. (13.11) but with K_d as follows:

$$K_d = K_{d0}e^{(\delta_1+\delta_{-1})\phi V/2} + K_{p0}e^{(\delta_1-\delta_2)\phi V/2} \tag{13.13}$$

$$K_{p0} = K_2/K_1$$

Here, separate δ have been used for each k (δ_1, δ_{-1}, δ_2) to conform to the more general notation of Kupper and colleagues. If the original Woodhull assumptions are used ($\delta_1=\delta_{-1}=\delta$ and $\delta_2=1-\delta$), then Eq. (13.13) reduces to:

$$K_d = K_{d0}e^{\delta\phi V} + K_{p0}e^{(2\delta-1)\phi V/2} \tag{13.14}$$

which has three free parameters: δ, K_{d0}, and K_{p0}. In previous work, we found this latter expression of $\varphi(V)$ (Eqs. 13.12 and 13.13) gives a better empirical fit to the Mg^{2+} block of NMDARs at the MF–GC synapse than the two-state Woodhull formalism (Fig. 13.3A; Ref. 22). At this synapse, the Mg^{2+} block of NMDARs is incomplete at potentials near the resting potential of mature GCs (Fig. 13.3B), presumably due to the presence of GluN2C subunits.[21,36,37] Using simple models as described in this chapter, we were able to show the incomplete Mg^{2+} block at subthreshold potentials boosts the efficacy of low-frequency MF inputs by increasing the total charge delivered by NMDARs, consequently increasing the output spike rate (Fig. 13.3C and D). Hence, these modeling results suggested the incomplete Mg^{2+} block of NMDARs plays an important role in enhancing low-frequency rate-coded signaling at the MF–GC synapse.

Characterization of the Mg^{2+} block of NMDARs is still ongoing. Besides the potential existence of two binding sites, and Mg^{2+} permeation, it has been shown that Mg^{2+} block is greatly affected by permeant monovalent cations.[60,61] This latter finding has the potential to resolve a longstanding paradox referred to as the "crossing of δ's," where the two internal and external Mg^{2+} binding site locations (i.e., their δ's), estimated using the

Woodhull formalisms described above, puzzlingly cross each other within the NMDAR.[61] Other details of Mg^{2+} block have been added by studies investigating the response of NMDARs to long steps of glutamate application.[62,63] These studies have revealed multiple blocked and desensitization states, and slow Mg^{2+} unblock due to inherent voltage-dependent gating, all of which are best described by more complicated kinetic-scheme models. Hence, given the added complexities from these more recent studies, it is all the more apparent that the often-used equations for $\varphi(V)$ described above are really only useful for providing empirical representations of the blocking action of Mg^{2+} (i.e., for setting the correct current–voltage relation described in Eq. 13.8), rather than characterizing the biophysical mechanisms of the Mg^{2+} block. In this case, parameters for $\varphi(V)$ are best chosen to give a realistic overall current–voltage relation of the particular NMDAR under investigation. Because the voltage dependence of NMDARs is known to vary with age, temperature, subunit composition and expression (i.e., native vs. recombinant receptors), care must be taken when selecting these parameters. Ideally, one should select parameters from studies of NMDARs in the neuron of interest, at the appropriate age and temperature.

4. CONSTRUCTION OF PRESYNAPTIC SPIKE TRAINS WITH REFRACTORINESS AND PSEUDO-RANDOM TIMING

To simulate the temporal patterns of activation that a synapse is likely to experience *in vivo*, it is necessary to construct trains of discrete events that can be used to activate model synaptic conductance events, $G_{syn}(t)$, as described in Eqs. (13.4)–(13.7), at specific times (i.e., t_j). These trains can then be used to mimic the timing of presynaptic action potentials as they reach the synaptic terminals. Real presynaptic spike trains can exhibit a wide range of statistics. The statistical properties of the spike trains reflect the manner in which information is encoded. Often, sensory information conveyed by axons entering the CNS is encoded as firing rate, and the interval between spikes has a Poisson-like distribution.[64–66] Other types of sensory input may signal discrete sensory events as bursts of action potentials.[67] In sensory cortex, information is typically represented as a sparse code and the firing rate of individual neurons is low on average (<1 Hz [68]). The inter-spike interval of cortical neurons can exhibit a higher variance than expected for a Poisson process where the variance equals the mean. Here, we describe how to generate spike trains with specific statistics; however, another approach would be to use spike times measured directly from single-cell *in vivo* recordings.

To compute an arbitrary train of random spike times t_j ($j = 1, 2, 3, \ldots$) with instantaneous rate $\lambda(t)$, a series of interspike intervals (Δt_j) can be generated from a series of random numbers (u_j) uniformly distributed over the interval (0, 1] by solving for Δt_j in the following equation[69,70]:

$$- \ln \left(u_j \right) = \int_{t_{j-1}}^{t_{j-1}+\Delta t_j} \lambda(\sigma) \, d\sigma \qquad (13.15)$$

where σ is the integration variable. The right-hand side of this equation represents the cumulative distribution function of finding a spike after t_{j-1}, in which case $\lambda(t)$ is the probability density function. Since $\lambda(t)$ can be any arbitrary function of time, Eq. (13.15) is extremely flexible in generating any number of random spike trains. Here, we outline a few examples.

First, we consider the simplest case of generating a random spike train with constant instantaneous rate: $\lambda(t) = \lambda_0$. In this case, Eq. (13.15) reduces to:

$$\Delta t_j = - \ln \left(u_j \right) / \lambda_0 \qquad (13.16)$$

Plugging a series of random numbers u_j into Eq. (13.16) results in a series of Δt_j with exponential distribution (i.e., Poisson) and mean $1/\lambda_0$. Since the solution contains no memory of the previous spike time (i.e., there are no terms containing t_{j-1}), Δt_j can be computed independently and then converted to a final t_j series.

Next, we consider the case of generating a random spike train with an exponential instantaneous rate of decay: $\lambda(t) = \lambda_0 \exp(-t/\tau)$. In this case, Eq. (13.15) reduces to:

$$\Delta t_j = -\tau \ln \left[1 + \frac{\ln \left(u_j \right)}{\tau \lambda \left(t = t_{j-1} \right)} \right] \qquad (13.17)$$

Unlike Eq. (13.16), this solution contains memory of the previous spike in the term $\lambda(t = t_{j-1})$, in which case values for Δt_j and t_j must be computed in consecutive order.

One problem with Eqs. (13.16) and (13.17) is that they do not take into account the spike refractoriness of a neuron, which can be on the order of 1–2 ms at physiological temperatures. A solution to this problem is to reject any Δt_j that are less than the absolute refractory period (τ_{AR}) or evaluate the integral in Eq. (13.15) from $t_{j-1} + \tau_{AR}$ to $t_{j-1} + \Delta t_j$. However, both procedures will increase the average of Δt_j in which case the final instantaneous rate of the t_j series will not match $\lambda(t)$. To produce a t_j series with instantaneous rate matching $\lambda(t)$, one can correct $\lambda(t)$ for refractoriness via the following equation[71]:

$$\Lambda(t) = \frac{1}{\lambda(t)^{-1} - \tau_{AR}} \tag{13.18}$$

where $\lambda(t)^{-1} > \tau_{AR}$, which should be the case if both $\lambda(t)$ and τ_{AR} are derived from experimental data. As a simple example, if we consider the case of a constant instantaneous rate, where $\lambda(t) = \lambda_0 = 0.25$ kHz and $\tau_{AR} = 1$ ms, then $\Lambda(t) = 0.333$ kHz. Another simple example is shown in Fig. 13.4A1, where 200 spike trains (four shown at top) were computed for $\lambda(t)$ that exhibits an exponentially decaying time course (bottom, solid red line) and $\tau_{AR} = 1$ ms. $\Lambda(t)$, the corrected rate function used to compute the spike trains, is plotted as the dashed red line, which only shows significant deviation from $\lambda(t)$ at rates above 100 Hz. Computing the peri-stimulus time histogram (PSTH, noisy black line) from the 200 spike trains confirmed the instantaneous rate of the trains matched that of $\lambda(t)$, and computing the interspike interval histogram (ISIH; Fig. 13.4A2) confirmed the spike intervals had an exponential distribution with $\tau_{AR} = 1$ ms.

A more complicated scenario arises when τ_{AR} is followed by a relative refractory period (τ_{RR}). In this case, one will need to multiply the instantaneous rate by a probability density function for refractoriness, $H(t)$, similar to a hazard function, which takes on values between 0 and 1. A simple $H(t)$ would be one that starts at 0 and rises exponentially to 1, in which case a t_j series could be computed via the following:

$$-\ln\left(u_j\right) = \int_{t_{j-1} + \tau_{AR}}^{t_{j-1} + \Delta t_j} \Lambda(\sigma) H(\sigma) d\sigma \tag{13.19}$$

$$\Lambda(t) = \frac{1}{\lambda(t)^{-1} - \tau_{AR} - \tau_{RR}}$$

$$H(t) = 1 - e^{-t'/\tau_{RR}}$$

where $t' = t - t_{j-1} - \tau_{AR}$. Examples of 200 spike trains computed via Eq. (13.19) are shown in Fig. 13.4B1 (top), where $\lambda(t)$ was a half-wave

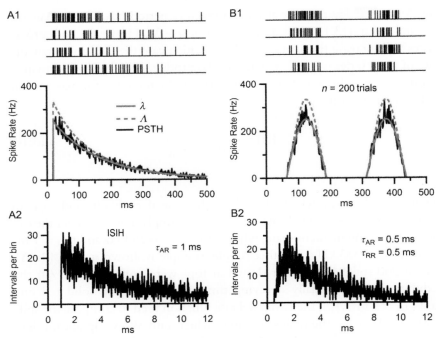

Figure 13.4 Simulated spike trains with refractoriness and pseudorandom timing. (A1) Trains of spike event times (top) computed for an instantaneous rate function $\lambda(t)$ with exponential decay time constant of 150 ms (bottom, solid red line) and absolute refractory period (τ_{AR}) of 1 ms. To compute the trains, a refractory-corrected rate function $\Lambda(t)$ (dashed red line) was first derived from Eq. (13.18) and then used in the integral of Eq. (13.15) to compute the spike intervals in sequence. The PSTH (black, 2-ms bins) computed from 200 such trains closely matches $\lambda(t)$. (A2) Interspike interval histogram (ISIH) computed from the same 200 trains in (A1), showing the 1 ms absolute refractory period. The overall exponential decay of the ISIH is a hallmark sign of a random Poisson process. (B1) and (B2) Same as (A1) and (A2) except $\lambda(t)$ was a half-wave rectified sinusoid with 250 ms period, and refractoriness was both absolute and relative: $\tau_{AR} = 0.5$ ms and $\tau_{RR} = 0.5$ ms. Intervals were computed via Eq. (13.19). (See the color plate.)

rectified sinusoid (solid red line), $\tau_{AR} = 0.5$ ms and $\tau_{RR} = 0.5$ ms. Also shown is $\Lambda(t)$ (dashed red line) which again only shows significant deviation from $\lambda(t)$ at rates above 100 Hz. Computing the PSTH of the 200 spike trains again confirmed the instantaneous rate matched that of $\lambda(t)$, and computing the ISIH confirmed the spike intervals had an exponential distribution with $\tau_{AR} = 0.5$ ms and $\tau_{RR} = 0.5$ ms (Fig. 13.4B2).

The above solutions for a simple $\lambda(t)$ described in Eqs. (13.16) and (13.17) were relatively easy to compute since Eq. (13.15) could be solved analytically. If an analytical solution is not possible, however, then Eq. (13.15) (or Eq. 13.19) must be obtained numerically with suitably small time step $d\sigma$. Ideally, this can be achieved using an integration routine with built-in mechanism to halt integration based on evaluation of an arbitrary equality. If the integration routine does not have such a built-in halt mechanism, then integration will have to proceed past $t = t_{j-1} + \Delta t_j$, perhaps to a predefined simulation end time, and Δt_j computed via a search routine that evaluates the equality defined in Eq. (13.15). To improve computational efficiency, an iterative routine can be written which computes the integration over small chunks of time, and the search routine implemented after each integration step. The length of the consecutive integration windows could be related to $\Lambda(t = t_{j-1})$, such as $3/\Lambda$.

5. SYNAPTIC INTEGRATION IN A SIMPLE CONDUCTANCE-BASED INTEGRATE-AND-FIRE NEURON

Once we have built a train of presynaptic spike times (t_j) and synapses with realistic conductance waveforms (G_{AMPAR} and G_{NMDAR}) and current–voltage relations (I_{AMPAR} and I_{NMDAR}), we are well on our way to simulating synaptic integration in a simple point neuron like the GC, which is essentially a single RC circuit with a battery. The simplest neuronal integrator is the integrate-and-fire (IAF) model.[72] Most modern versions of the IAF model act as a leaky integrator with a voltage threshold and reset mechanism to simulate an action potential.[73,74] The equation describing the subthreshold voltage of such a model is as follows:

$$C_{m} \frac{dV}{dt} = \frac{V - V_{rest}}{R_{m}} + I_{syn}(V, t) \qquad (13.20)$$

where C_m, R_m, and V_{rest} are the membrane capacitance, resistance, and resting potential, and $I_{syn}(V,t)$ is the sum of all synaptic current components, such as I_{AMPAR} and I_{NMDAR} (e.g., Eqs. 13.2 and 13.8) which are usually both voltage and time dependent. Spikes are generated the moment integration of Eq. (13.20) results in a V greater than a predefined threshold value (V_{thresh}). At this time, integration is halted and V is set to the peak value of an action potential (V_{peak}) for one integration time step. V is then set to a reset potential (V_{reset}) for a period of time defined by an absolute refractory period

(τ_{AR}) after which integration of Eq. (13.20) is resumed. To produce realistic spiking behavior, the parameters can be tuned to match the particular neuron under investigation. V_{thresh}, V_{peak}, and V_{reset}, for example, can be set to the average onset inflection point, peak value, and minimum after-hyperpolarization of experimentally recorded action potentials. τ_{AR} can be set to the minimum interspike interval observed during periods of high spike activity, and further tuned using input–output curves (e.g., matching plots of spike rate vs. current injection for experimental and simulated data). Due to the complexity of $I_{syn}(V,t)$, the solution of Eq. (13.20) will most likely require numerical integration. The integration can be implemented in a similar manner as that described for $\lambda(t)$ above, using a built-in integration routine to solve Eq. (13.20) over small chunks of time, and searching for V the moment it exceeds V_{thresh}, or using an integration routine with built-in mechanism to halt integration once V exceeds V_{thresh}. Usually, all of the above procedures can be implemented using few lines of code.

Due to their electrically compact morphology and simple subthreshold integration properties, GCs are particularly well suited for modeling with an IAF modeling approach.[22,46] Example simulations of an IAF model tuned to match the firing properties of an average GC can be found in Fig. 13.3C. Here, the model was driven by $I_{syn}(V,t)$ that contained either a mixture of I_{AMPAR} and I_{NMDAR} or the two currents in isolation. To simulate the convergence of four MF inputs, four different trains of $I_{syn}(V,t)$ with independent spike timing were computed and summed together before integration of Eq. (13.20). Because repetitive stimulation of the MF–GC synapse at short time intervals often results in depression and/or facilitation of I_{AMPAR} and I_{NMDAR}, plasticity models of the two currents were included in the simulations. These plasticity models are described in detail in the next section.

One consideration often overlooked in simulations of synaptic integration is the variability in C_m, R_m, and V_{rest}. We have found, for example, that the natural variability of these parameters in GCs can produce dramatically different output spike rates for a given synaptic input rate, as shown in Fig. 13.3D (gray curves). Moreover, due to the nonlinear nature of spike generation, using average values of C_m, R_m, and V_{rest} in a simulation (red) does not replicate the average output behavior of the total population (black): the spike rate of the "average GC" simulation in Fig. 13.3D is twice the average population spike rate. Hence, control simulations that include variation in C_m, R_m, and V_{rest} should be considered when simulating synaptic integration.

If the neuron under investigation has extended dendrites that are not electrically compact, then a multicompartmental model may be required. In this case, Eq. (13.20) can be used to describe the change in voltage within the various compartments where synapses are located, with an additional term on the right-hand side of the equation denoting the flow of current between individual compartments. Spike generation is then computed as described above but occurs only in the compartment designated as the soma. Also, an additional current due to spike generation can be added to the soma. Further details about multicompartment IAF modeling can be found in Gerstner and Kistler.[38] More often than not, however, multicompartmental models are simulated with Hodgkin–Huxley-style Na^+ and K^+ conductances to generate realistic action potentials.[75] Popular simulation packages developed to solve these more complex multicompartmental models with Hodgkin–Huxley-style conductances include NEURON and GENESIS, which are discussed further below.

6. SHORT-TERM SYNAPTIC DEPRESSION AND FACILITATION

So far, we have only considered the simulation of fixed amplitude synaptic conductances recorded under basal conditions. At synapses with a relatively high release probability, repetitive stimulation at short time intervals often results in depression of the postsynaptic response (see, e.g., Fig. 13.2C). This kind of synaptic depression was first described by Eccles et al.[8] for endplate potentials at the NMJ and has since been described for synapses in the CNS. Because recovery from synaptic depression takes a relatively short time, on the order of tens of milliseconds to seconds, it is referred to as short-term depression, distinguishing it from the longer-lasting forms of depression, including long-term depression that is believed to play a central role in learning and memory. Here, we refer to short-term depression as simply depression or synaptic depression.

Since the discovery of synaptic depression, numerous studies have sought to determine its underlying mechanisms and potential roles in neural signaling (for review, see Refs. 76,77). Often these studies have employed mathematical models to test and verify their hypotheses. The first model of presynaptic depression was described by Liley and North in 1953, before the discovery of synaptic vesicles. At the time, depression was thought to reflect a depletion of a finite pool of freely diffusing neurotransmitter available for release, and recovery from depression was thought to reflect a replenishment of the depleted pool, via synthesis from a freely diffusing chemical precursor. This explanation fit well with the observation that increasing the initial release of neurotransmitter produced a larger degree of depression, and the recovery from depression followed an exponential time course. Liley and North formalized this hypothesis by a simple mathematical treatment of synaptic transmission at the NMJ, known as a "depletion model," whereby a size-limited pool of readily releasable neurotransmitter (N) is in equilibrium with a large store of precursor (N_s) with forward and backward rate constants k_1 and k_{-1} (Fig. 13.5A). In response to stimulation of the nerve, say at time t_j, a fraction (P) of N is released into the synaptic cleft, disturbing the equilibrium with N_s. The change of N with respect to time after the stimulus can then be described by the following differential equation:

$$\frac{dN}{dt} = k_1 N_s - k_{-1} N \tag{13.21}$$

If N_s is relatively large, one can assume N_s is constant and Eq. (13.21) has the following solution:

$$N = N_\infty + \left(N_{j+\varepsilon} - N_\infty\right)e^{-(t-t_j)/\tau_r} \tag{13.22}$$

where $N_\infty = N_s k_1/k_{-1}$ and $\tau_r = 1/k_{-1}$. Here, N_∞ is the steady-state value of N, τ_r is the recovery time constant, and $N_{j+\varepsilon}$ is the value of N immediately after transmitter release at time t_j. This solution means that, after a sudden depletion of N due to a stimulus, N will exponentially increase from $N_{j+\varepsilon}$ to N_∞ with time constant τ_r. By comparing their experimental data to predictions of their mathematical model, Liley and North were able to estimate the steady-state value of P was 0.45, as well as reveal subtle signs of potentiation during a short train of stimuli, a conditioning tetanus, which they speculated was due to a brief period of temporarily raised P. At the time, such facilitation had long been reported[8] and was thought to be due to an increase in size of the nerve action potential, or an increase in the extracellular K^+ concentration. Today, facilitation is thought to be largely due to a rise in the intracellular Ca^{2+} concentration ($[Ca^{2+}]_i$) as described further below.

Subsequent to Liley and North's study, Betz[78] modified the depletion model to account for vesicular release. More recently, Heinemann et al.[79] added to the depletion model a pulsatile increase in $[Ca^{2+}]_i$ leading to a steep increase in P from a near zero value. While the latter addition made the

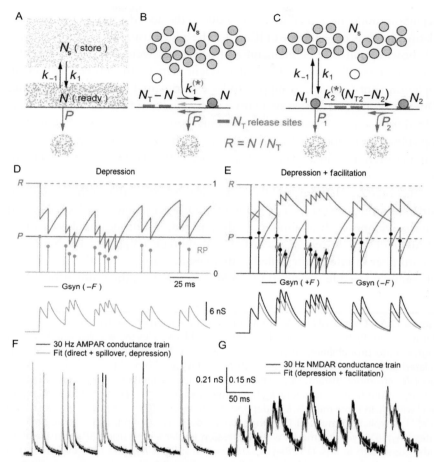

Figure 13.5 Modeling short-term depression and facilitation. (A) Original depletion model of Liley and North[9] describing release of freely diffusing transmitter (N). N is in equilibrium with a large store of precursor molecules (N_s), governed by forward and backward rate constants k_1 and k_{-1}. The arrival of an action potential causes a rise in $[Ca^{2+}]_i$, triggering a fraction (P) of N to be released (NP) into the synaptic cleft (red), disrupting the balance between N and N_s. N recovers back to its steady-state value (N_∞) with an exponential time course (τ_r), where N_∞ and τ_r are set by k_1 and/or k_{-1}. (B) A modern version of the depletion model with a large store of synaptic vesicles (N_s, gray circles) and a fixed number of vesicle release sites (N_T, blue), where N now represents the number of vesicles docked at a release site and are therefore readily releasable (orange circle). The arrival of an action potential now triggers a certain fraction of the readily releasable vesicles to be released (NP), freeing release sites. The number of free release sites at any given time is equal to $N_T - N$. Variations of this model include a k_1 that is dependent on residual $[Ca^{2+}]_i$ (red star), in which case $[Ca^{2+}]_i$ is explicitly simulated, and the inclusion of a backwards rate constant k_{-1} (gray arrow) representing the undocking of a vesicle, that is, the return

(Continued)

depletion model more realistic, it introduced the added complication of simulating the time dependence of $[Ca^{2+}]_i$. The added complication proved useful, however, in that Heinemann and colleagues were able to explore the consequences of adding a Ca^{2+}-dependent step to the process of vesicle replenishment (i.e., k_1), as supported by experimental evidence at the time. One such consequence was an increase in the number of readily releasable vesicles during a spike–plateau Ca^{2+} transient, thereby enhancing secretion during subsequent stimuli. More recent experimental evidence supports such a link between increased levels of $[Ca^{2+}]_i$ and enhanced vesicle replenishment (k_1).[80–84]

As noted by Heinemann et al.[79], k_{-1} was introduced into their model in order to limit the steady-state value of readily releasable vesicles (N_∞). An alternative approach to limit N_∞, they noted, would be to allow a finite number of vesicle release sites at the membrane (N_T), and let N denote the number of release sites filled with a vesicle, or equivalently the number of readily releasable vesicles (Fig. 13.5B). In this case, the number of empty

---

Figure 13.5—Cont'd of N to N_s. (C) A more recent version of the depletion model, similar to that in (B), has two pools of readily releasable vesicles (N_1 and N_2) with low- and high-release probabilities, respectively (P_1 and P_2). The difference in probabilities is related to the distance vesicles in pools N_1 and N_2 are from VGCCs, where vesicles in pool N_2 are more distant. Here, the model includes a maturation process where N_2 emerges from N_1 at a rate set by k_2, but some models have N_2 emerging from N_s in parallel with N_1. In some models, k_2 is dependent on residual $[Ca^{2+}]_i$ (red star), in which case $[Ca^{2+}]_i$ is explicitly simulated. Only the second pool has a fixed number of vesicle release sites (N_{T2}). (D) Synaptic model with depression implemented using RP recursive algorithm described in Eqs. (13.28)–(13.32) ($\tau_r = 20$ ms; $R = N/N_T$). The time evolution of R and P are shown at top (blue and red), where $P_\infty = 0.4$. Since there is no facilitation ($\Delta_P = 0$), P is constant. At the arrival of an action potential at t_j, the fraction of vesicles released (R_e) is computed: $R_e = RP$ (gray circles). R_e is then used to scale a synaptic conductance waveform $G_{syn}(t=t_j)$ (Eq. 13.30) and also subtracted from R (Eq. 13.31). The time evolution of the sum of all $G_{syn}(t=t_j)$ is shown at the bottom (gray). (E) The same simulation in (D), except the synaptic model includes facilitation ($\Delta_P = 0.5$, $\tau_f = 30$ ms). For comparison, R_e and the sum of all $G_{syn}(t=t_j)$ are plotted in black (+F) along with their values in (D) (−F, gray). (F) Fit of a synaptic model with depression (yellow) to a 30 Hz MF–GC AMPAR conductance train (black). The fit consisted of the sum of two separate components, the direct and spillover components, where each component had its own depression parameters. Parameters for the fit can be found in Schwartz et al.[22] (G) Same as (F) but for a corresponding 30 Hz MF–GC NMDAR conductance train. This time the fit (green) consisted of a single component that had depression and facilitation. Scale bars are for (F) and (G), with two different y-scale values denoted on the left and right, respectively. Data in (F) and (G) is from Schwartz et al.[22] with permission. (See the color plate.)

release sites will equal $N_T - N$, and the rate at which the empty release sites are filled will equal $k_1(N_T - N)$. Hence, Eq. (13.21) can be rewritten as:

$$\frac{dN}{dt} = k_1(N_T - N) \tag{13.23}$$

This equation has the same solution defined in Eq. (13.22) except $N_\infty = N_T$ and $\tau_r = 1/k_1$. Because N_T now directly defines N_∞, k_{-1} is no longer necessary. Although the backward reaction defined by k_{-1} may very well exist, its rate is usually presumed small and neglected. In most depletion models, however, it is customary to express Eq. (13.23) with respect to the fraction of release sites filled with a vesicle (i.e., N/N_T), also known as site occupancy, which is assumed to be unity under resting conditions (i.e., low stimulus frequencies). To be consistent with these other models, therefore, we define a fractional "resource" variable $R = N/N_T$. Substituting terms, Eq. (13.23) becomes:

$$\frac{dR}{dt} = k_1(1 - R) \tag{13.24}$$

which has the following solution based on Eq. (13.22):

$$R = 1 + (R_{j+\varepsilon} - 1)e^{-(t-t_j)/\tau_r} \tag{13.25}$$

where $\tau_r = 1/k_1$. This is the expression one often sees in depletion models (e.g., Ref. 85); however, sometimes R is denoted as D,[86] x,[87] n,[88] or as the ratio N/N_T.[80]

To simulate vesicle release, many depletion models treat the process of spike generation, Ca^{2+} channel gating and vesicle release as instantaneous events (Fig. 13.5B, red P). To do this, one first computes the fraction of the resource of vesicles released (R_e) at the time of a stimulus: $R_e = RP$. Next, R_e is used to compute the amplitude of the postsynaptic response, for example: $g_{peak} = QN_TR_e$ (see Eqs. 13.4–13.7), where Q is the peak quantal conductance. Finally, R_e is subtracted from R ($R \rightarrow R - R_e$) increasing the number of empty release sites. A few variations in this release algorithm are worth noting. First, in some models, the latter decrement in R is expressed with respect to a depression scale factor (D). However, the result is the same since D can be expressed as $D = 1 - P$, in which case, $R \rightarrow RD = R(1 - P) = R - R_e$. Second, in some models, a synaptic delay is added to the postsynaptic response. However, if the same delay is added to each response, then the result is the same with only an added time shift.

Third, in some models, the release sites become inactive after a vesicle is released.[87] This requires the addition of an inactive state, after release and before the empty state. Transition from the inactive state to the empty state (i.e., recovery from inactivation) is then determined by an extra time constant. Hence, in this three-state model, the recovery of N will have a double-exponential time course. Because of the added complexity, differential equations of the three-state model will most likely have to be solved using numerical methods. Finally, in the more detailed models that simulate $[Ca^{2+}]_i$, such as the Heinemann model discussed above, the stimulus (i.e., action potential) will often cause an instantaneous increase in $[Ca^{2+}]_i$ followed by a slower decay. Since P is a nonlinear function of $[Ca^{2+}]_i$, it may remain elevated above zero for some time following an action potential, causing a delayed component of vesicular release. This scenario therefore requires the added complication of calculating release continuously as a function of $[Ca^{2+}]_i$, which may have to be solved via numerical methods.

More recent studies investigating vesicle release in the calyx of Held[42] and cerebellar MF[30] have reported success in replicating experimental data using a depletion model with two pools of releasable vesicles (N_1 and N_2), one with a low release probability (P_1, reluctantly releasable), the other with a high release probability (P_2, readily releasable; Fig. 13.5C). In this two-pool model, the size of N_1 is not limited by a fixed number of release sites, but rather is limited by the forward and backward rate constants k_1 and k_{-1}. The size of N_2, on the other hand, is limited by a finite number of release sites (N_{T2}). As depicted in Fig. 13.5C, N_2 emerges from N_1 via a maturation process that is Ca^{2+} independent (i.e., k_2). However, Trommershäuser and colleagues modeled N_2 emerging from N_s in parallel with N_1, where k_2 was Ca^{2+} dependent. The success of both models in replicating experimental data may indicate true differences in the synapse types under investigation, or may indicate a need for more experimental data to constrain this type of model. To simulate two different release probabilities, P_1 and P_2 are defined according to a biophysical model that places the vesicles of pools N_1 and N_2 at different distances from VGCCs (Ref. 42; see also Ref. 89). The result is individual $[Ca^{2+}]_i$ expressions for each pool of vesicles, which adds to the complexity of this type of depletion model.

As noted above, Liley and North[9] observed signs of facilitation at the NMJ which they attributed to a brief period of temporarily raised P after stimulation of the presynaptic terminal. Today, there is considerable evidence the raised P is due to an accumulation of residual Ca^{2+} in the synaptic terminal following an action potential (for review, see Ref. 77). Although facilitation may well be a universal characteristic of all chemical synapses, it has not always been readily apparent at some synapse types, for example, the climbing fiber synapse.[80] The lack of observable facilitation at some synapse types is thought to be due to a presence of strong depression that dominates over facilitation (due to a higher release probability), or the presence of intracellular Ca^{2+} buffers that significantly speed the decay of residual $[Ca^{2+}]_i$, or some molecular difference in the vesicle release machinery. The lack of observable facilitation at some synapse types has meant facilitation has not always been included in depletion models. However, for those depletion models that have included facilitation, the typical implementation of facilitation has been to instantaneously increase P after the arrival of an action potential and let P decay back to its steady-state value. In this case, the change of P with respect to time after an action potential can be described by the following differential equation:

$$\frac{dP}{dt} = \frac{P_\infty - P}{\tau_f} \tag{13.26}$$

which has the following solution:

$$P = P_\infty + \left(P_{j+\varepsilon} - P_\infty\right)e^{-\left(t-t_j\right)/\tau_f} \tag{13.27}$$

where P_∞ is the steady-state value of P (the probability of release during resting conditions, sometimes referred to as P_0), τ_f is the time constant for recovery from facilitation, and $P_{j+\varepsilon}$ is the value of P immediately after an action potential at time t_j. More complicated models that simulate Ca^{2+} dynamics will equate P as a function of $[Ca^{2+}]_i$.

If the differential equations that describe synaptic plasticity have exact analytical solutions, then a simple recursive algorithm can be used to compute a solution for a given spike t_j series. As a simple example, if the change in R and P after t_j are described in Eqs. (13.25) and (13.27), then a solution can be obtained by executing the following three steps at each spike time t_j ($j = 1, 2, 3, \ldots$). In step 1, values for R and P at the arrival of a spike at t_j are computed via the following equations derived from Eqs. (13.25) and (13.27):

$$R_{j-\varepsilon} = 1 + \left(R_{j-1+\varepsilon} - R_\infty\right)e^{-\Delta t_j/\tau_r} \tag{13.28}$$

$$P_{j-\varepsilon} = P_\infty + \left(P_{j-1+\varepsilon} - P_\infty\right)e^{-\Delta t_j/\tau_f} \tag{13.29}$$

where j denotes the current spike, $j-1$ is the previous spike, and Δt_j is the inter-spike time ($\Delta t_j = t_j - t_{j-1}$). Since both R and P change instantaneously at t_j, it is necessary to distinguish their values immediately before and after a spike. Here, this is accomplished with the notation $-\varepsilon$ and $+\varepsilon$, respectively. Note that for the first spike ($j=1$) $P_{j-\varepsilon} = P_\infty$. In step 2, values for R and P derived from step 1 are used to compute the amplitude of the postsynaptic response at t_j:

$$g_{\text{peak}} = Q N_T R_{j-\varepsilon} P_{j-\varepsilon} \tag{13.30}$$

g_{peak} can then be used in Eqs. (13.4)–(13.7). In step 3, values for R and P immediately after the arrival of a spike are computed:

$$R_{j+\varepsilon} = R_{j-\varepsilon}\left(1 - P_{j-\varepsilon}\right) \tag{13.31}$$

$$P_{j+\varepsilon} = P_{j-\varepsilon} + \Delta_P\left(1 - P_{j-\varepsilon}\right) \tag{13.32}$$

where Δ_P is a facilitation factor with values between 0 and 1. Varela et al.[86] use a slightly different approach to step 3 that disconnects the usage dependency of N from P:

$$R_{j+\varepsilon} = R_{j-\varepsilon} \Delta_R \tag{13.33}$$

$$P_{j+\varepsilon} = P_{j-\varepsilon} + \Delta_P \tag{13.34}$$

where Δ_R and Δ_P are their depression (D) and facilitation (F) factors. Equation (13.34) is similar to Eq. (13.32) in spirit; however, $P_{j+\varepsilon}$ in Eq. (13.34) has the potential to grow without bound at high spike rates, in which case g_{peak} in Eq. (13.30) can become larger than $Q \cdot N_T$, the maximum amplitude possible for N_T release sites. $P_{j+\varepsilon}$ in Eq. (13.32), on the other hand, is limited to going no higher than 1, and therefore g_{peak} no higher than $Q \cdot N_T$.

An example of a $G_{\text{syn}}(t)$ train computed with the above RP recursive algorithm is shown in Fig. 13.5D (bottom, gray), along with the time evolution of R (top, blue), P (red), and RP (gray circles). In this example, there is no facilitation ($\Delta_P = 0$) so P is constant. To show the effects of facilitation, the same $G_{\text{syn}}(t)$ train is shown in Fig. 13.5E now with facilitation (bottom, black; $\Delta_P = 0.5$, $\tau_f = 30$ ms). Comparison of $G_{\text{syn}}(t)$ with and without facilitation (black vs. gray) shows the enhancement of g_{peak} due to facilitation. However, the comparison also shows the signs of facilitation in this example are subtle and might not be readily apparent by visual inspection of the $G_{\text{syn}}(t)$ train.

A more realistic example of a $G_{\text{syn}}(t)$ train computed with the above RP recursive algorithm is shown in Fig. 13.5F. Here, parameters for R and P were optimized to fit a 30 Hz G_{AMPAR} train computed from recordings from four GCs (black). Because G_{AMPAR} of GCs contains a direct and spill-over component (Fig. 13.2B), the fit (yellow) consisted of two separate components simultaneously summed together. Furthermore, because a good fit could be achieved without facilitation, only depression of the two components was considered. A similar fit to a 30 Hz G_{NMDAR} train computed from recordings from the same four GCs is shown in Fig. 13.5G. This time a good fit (green) was achieved using one component that had both depression and facilitation.

There is one caveat, however, about the fits in Fig. 13.5F: studies have shown most of the depression of the AMPAR conductance at the MF–GC synapse at 100 Hz is not due to the depletion of presynaptic readily releasable vesicles, but to the desensitization of postsynaptic AMPARs.[29,34] Hence, while the depletion model has accurately captured the overall mean behavior of the MF–GC synapse, it has done so by lumping presynaptic and postsynaptic sources of depression. This could be the case for the fit to G_{NMDAR} as well. The technique of fitting a depletion model to the data is still valid, however, since the intended goal of the fits was to create realistic conductance waveforms that could be used in a simple IAF model, as reported elsewhere.[22] An alternative approach would be to simulate each source of depression and facilitation as independent scale factors, which are then used to compute g_{peak} in Eq. (13.30). Whether to lump the various sources of plasticity into single components or to split them apart into individual components ultimately depends on the purpose of the plasticity model. If the aim of the plasticity model is to generate mean synaptic conductance trains for driving a neural network, or for injecting into the cell body a real neuron via dynamic clamp, then the simple lumping approach can be taken. On the other hand, if the aim of the plasticity model is to reproduce the mean and variance of the synaptic trains (see below), or gain insight into and construct hypotheses about one or more of the various components of synaptic transmission, then a "splitting" approach is perhaps better. The splitting approach will, of course, require extra experimental data to constrain the various parameters of the independent components. A more detailed description on how to model the various components of synaptic depression and facilitation independently can be found in a recent review by Hennig.[88] This review also describes other sources of synaptic plasticity not discussed here, including sources of slow modulation of P, temporal enhancement of vesicle replenishment and the longer-lasting forms of synaptic plasticity, augmentation and post-tetanic potentiation. LTP at the MF–GC synapse has also been modeled in detail elsewhere.[90]

7. SIMULATING TRIAL-TO-TRIAL STOCHASTICITY

Up until now, the synaptic models we have presented are deterministic. However, as mentioned in Section 1, synapses exhibit considerable variability in their trial-to-trial response (see, e.g., Fig. 13.2B) due to the probabilistic nature of the mechanisms underlying synaptic transmission, from the release of quanta to the binding and opening of postsynaptic ionotropic receptors (Fig. 13.1). Here, we discuss the simulation of three sources of stochastic variation that account for the bulk of the variance exhibited by EPSCs recorded at central synapses: variation in the number of vesicles released, variation in the amplitude of the postsynaptic quantal response, and variation in the timing of vesicular release.

The main source of synaptic variation arises from the stochastic nature of vesicular release at an active zone, a process that lead Katz[3] to the quantum hypothesis. Since the nomenclature of quantal release can be confusing, it is useful to define terms. Here, we use the term "release sites" to mean *functional* release sites (i.e., N_T). This is equivalent to the maximum number of vesicles that can be released by a single action potential under resting conditions when all release sites are occupied. Synapses may have one or more than one release site per anatomical synaptic contact or active zone. Multivesicular release refers to the situation where multiple vesicles are released per active zone.[91] A Poisson model is typically used to describe stochastic quantal release at the NMJ under conditions of low-release probability.[2] This model works well since the number of release sites is large at this synapse. In contrast, a simple binomial model is typically used to describe stochastic quantal release at central synapses,[92] which have relatively few release sites with intermediate release probabilities. The simple binomial model assumes the vesicular release probability P and the amplitude of the postsynaptic response to a single quantum (quantal peak amplitude, Q) are uniform across release sites. Under these assumptions, and the *proviso* that release is perfectly synchronous, the mean (μ), variance (σ^2), and frequency (f) of the postsynaptic response can be expressed as:

$$\mu = N_T P Q \tag{13.35}$$

$$\sigma^2 = Q^2 N_T P(1-P) = Q\mu - \mu^2/N_T \tag{13.36}$$

$$f(k; N_T, P) = \frac{N_T!}{k!(N_T - k)!} P^k (1-P)^{N_T - k} \tag{13.37}$$

where k denotes the number of quanta released from a maximum of N_T release sites. The parabolic $\sigma^2-\mu$ relation described in Eq. (13.36) has proved particularly useful as it defines how the variance of the EPSC changes with P. MPFA, or variance mean analysis, uses a related multinomial model that includes nonuniform release probability and quantal variability to estimate Q, P, and N_T from synaptic responses recorded at different P. This approach is discussed in detail elsewhere.[29,93,94]

This statistical framework makes simulation of a simple binomial synapse with N_T independent release sites, each with release probability P and quantal size Q, relatively straightforward. For the simulations presented in this section, Q pertains to the peak amplitude of the quantal excitatory postsynaptic conductance but could also pertain to the peak amplitude of the EPSC or EPSP. On arrival of an action potential at time t_j, a random number is drawn from the interval $[0, 1]$ for each release site. If the random number is greater than P, then release at the site is considered a failure and the site is ignored; otherwise release is considered a success and a synaptic conductance waveform with amplitude Q is generated for that site (e.g., Eqs. 13.4–13.7, $g_{peak} = Q$). After computing the release at each site, the synaptic conductances at each site are summed together giving $G_{total}(t)$, which is used as the conductance waveform at t_j. On the arrival of the next action potential, the above steps are repeated. Figure 13.6A1 shows results of such simulation (blue traces, superimposed at each t_j) where values for N_T, P, and Q matched those of an average GC ($N_T = 5$, $P = 0.5$, $Q = 0.2$ nS) and stimulation was at low enough frequency that there was no residual conductance from previous events. The synaptic conductance waveform was a direct G_{AMPAR}, similar to that in Fig. 13.2D (red trace), and the number of trials (i.e., action potentials) was 1000, 20 of which are displayed. As expected for a binomial process with $N_T = 5$, peak values of $G_{total}(t)$ consisted of six different combinations of Q, including 0 for the case of failures at all sites. Furthermore, the mean, variance, and frequency of the peak amplitudes (Fig. 13.6A1 and A2) matched the expected values of a random variable with binomial distribution computed via Eqs. (13.35)–(13.37). When the same synapse was simulated with a low–release probability ($P = 0.1$, red), most action potentials resulted in failure of release (Fig. 13.6A2). Hence, μ and σ^2 of the peak values of $G_{total}(t)$ were both low (Fig. 13.6A1). In contrast, when the release probability was high ($P = 0.9$, green) most action potentials resulted in release at 4 or 5 sites (Fig. 13.6A2), resulting in high μ but low σ^2 (Fig. 13.6A1). A final comparison of μ and σ^2 across P values showed μ and

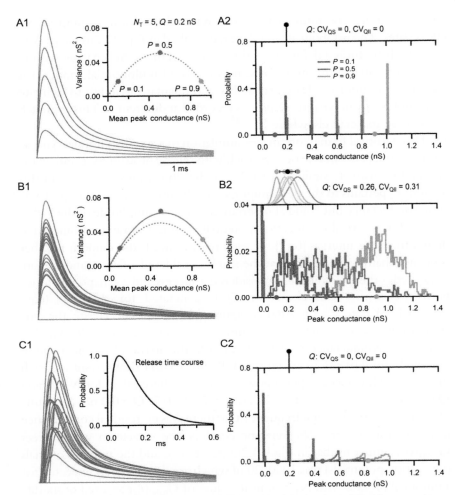

Figure 13.6 Simulating trial-to-trial variability using a binomial model with quantal variability and asynchronous release. (A1) Simulations from a binomial model of a typical MF–GC connection with five release sites (N_T) each with 0.5 release probability (P) and 0.2 nS peak conductance response (Q). Q was used to scale a G_{AMPAR} waveform with only a direct component. A total of 1000 trials were computed, 20 of which are displayed (blue). Inset shows $\sigma^2 - \mu$ relation computed from the peak amplitudes of all 1000 trials (blue circle), matching the theoretical expected value computed from Eq. (13.36) (dashed line). Repeating the simulations using a low P (0.1, red) and high P (0.9, green) confirmed the parabolic $\sigma^2 - \mu$ relation of the binomial model. (A2) Frequency distribution (bottom, blue) of the 1000 peak amplitudes computed in (A1), which closely matched the expected distribution computed via Eq. (13.37) (not shown). Circles on x-axis denote μ. Distributions for low and high P are also shown (red and green). Top graph shows Q which lacked variation. (B1) Same as (A1) except Q included intrasynaptic

σ^2 matched the parabolic relation predicted in Eq. (13.36) (Fig. 13.6A1), the hallmark sign of a simple binomial model.

The second main source of synaptic variability arises from variation in the quantal size Q. This can arise from trial-to-trial variation in Q at a single release site (intrasite or Type I variance) or from differences in the mean Q between release sites (intersite or Type II variance). Sources of intrasite quantal variability include variation in the amount of transmitter released per vesicle and the stochastic gating of postsynaptic receptor channels. Sources of intersite variability, on the other hand, include variation in the cleft geometry, number of postsynaptic receptors and electrotonic distance to the soma (for somatically recorded responses). Whereas the variance arising from intrasite quantal variation increases linearly with release probability, the variance arising from intersite quantal variance shows a parabolic relation as a function of release probability, since its origins are combinatorial. Both intrasite and intersite quantal variability are often expressed as coefficients of variation, CV_{QS} and CV_{QII}, respectively, where $CV = \sigma/\mu$, using the notation of Silver[94] for consistency. For GCs, CV_{QS} was estimated to be 0.26 and CV_{QII} to be 0.31.[26] Incorporating these two sources of quantal variance into a synaptic model is again relatively straightforward. First, an average Q for each release site i, denoted Q_i, is computed by drawing a random value from a Gaussian distribution with μ and σ defined by CV_{QII}, where μ equals the desired final average peak synaptic conductance and $\sigma = \mu \cdot CV_{QII}$. For simulations with a small N_T, however, the small number of samples from the Gaussian distribution may produce a CV_{QII} and/or μ that are relatively distant from their intended values. To avoid this problem, one can repeat the sampling of Q_i until both CV_{QII} and μ fall within a predefined tolerance range, such as 1% of their intended values. The binomial simulation for

---

variation ($CV_{QS} = 0.26$) and intersynaptic variation ($CV_{QII} = 0.31$), creating a larger combination of peak amplitudes and therefore larger variance. Inset shows theoretical $\sigma^2 - \mu$ relation with (solid line) and without (dashed line) variation in Q, the former computed using Eq. 11 of Silver.[94] (B2) Same as (A2) but for the simulations in (B1). Top graph shows the distribution of the average Q at each site i (i.e., Q_i, colored circles) with $\mu \pm \sigma = 0.20 \pm 0.62$ nS (black circle), as defined by CV_{QII}. Gaussian curves show distribution of Q at each site, defined by Q_i and CV_{QS}. (C1) Same as (A1) except a delay, or release time ($t_{release}$), was added to each quantal release event. Values for $t_{release}$ were randomly sampled from the release time course shown in the inset, which is typical for a single release site at a MF–GC connection. (C2) Same as (A2) but for the simulations in (C1). Peaks were measured over the entire simulation window. (See the color plate.)

N_T and P can then be executed as described in the previous paragraph, except for the additional step of adding intrasite variability to Q at the arrival of each action potential. To do this, a value for Q is randomly drawn from a Gaussian distribution with μ and σ defined by CV_{QS}, where μ now equals Q_i and $\sigma = Q_i \cdot CV_{QS}$. Figure 13.6B1 shows the effects of adding intrasite and intersite quantal variance to the binomial model simulations in Fig. 13.6A1 ($CV_{QS} = 0.26$ and $CV_{QII} = 0.31$). In this case, the peak amplitude of $G_{total}(t)$ showed a significant increase in variation (Fig. 13.6B1), resulting in a smearing of the peak-amplitude frequency distribution (Fig. 13.6B2). Hence, as these results demonstrate, a nonuniform Q tends to obscure the underlying binomial process, especially under circumstances when a single value of P is investigated. Only by varying P does the underlying binomial process become apparent in the $\sigma^2 - \mu$ relation and frequency distribution of peak values (i.e., red vs. blue vs. green data points).

The third source of synaptic variation arises from the asynchronous release of synaptic vesicles, which can be considered another source of intrasite variation, or Type I variance, unless release sites are far apart and axon conduction introduces significant delays between sites. Asynchronous release is usually quantified by a function known as the release time course (RTC), computed by measuring the latency of individual quantal events in postsynaptic recordings at low-release probability,[95] or computed by deconvolution methods.[26] A detailed discussion on how to estimate the RTC for synapses either with a few release sites or with many release sites is given by Minneci and colleagues.[95] In GCs, the RTC for a single site has a gamma-like distribution that rises abruptly from 0 and peaks near 0.1 ms,[26] similar to that shown in Fig. 13.6C1. With such a RTC distribution at hand, simulating asynchronous vesicular release only requires the following two steps. First, a release time ($t_{release}$) is randomly drawn from the RTC distribution for each release site at the arrival of an action potential at t_j. Second, each site's $t_{release}$ is added to t_j when computing the quantal waveform (e.g., $t' = t - t_j - t_{release}$ in Eqs. 13.4–13.7) thereby creating asynchronous release. Figure 13.6C1 shows such a simulation using the RTC of a GC, where the delay in quantal release is evident in the smearing of peaks. Here, in order to show the sole effects of asynchronous release, there was no variation in Q. Computing the frequency of peak amplitudes, where peaks are computed over the entire simulation window, shows asynchronous release has no effect on release events composed of less than two quanta, as expected. However, for release events composed of two or more quanta, asynchronous release causes a reduction in peak amplitudes, that is, a smearing of peak amplitudes toward smaller values, where the smearing effect is more pronounced for those events composed of the largest number of quanta (Fig. 13.6C2). On the other hand, if the peak amplitude is computed by averaging over a fixed time window centered on the mean peak EPSC, as for MPFA, a slightly different effect is observed. In this case, variability is also observed at those release events composed of a single quantum, since the peak of these events may fall outside the measurement window.[94]

Deviation from the variance predicted from binomial models of vesicular release can also arise from nonuniformities in release probability P. Dispersion in the release probability across release sites, which can also be quantified by the coefficient of variation (CV_P), tends to reduce the variance of the synaptic response. Since this source of variation has the largest impact at high-release probabilities, its incorporation into a binomial model may not be necessary when simulating synapses with low P. If one wishes to include variability in the release probability, however, methods for doing so can be found elsewhere.[94,95]

Finally, we consider the simulation of a binomial synapse with short-term depression and facilitation, similar to that described previously.[96] The most flexible way to simulate such a synapse is to treat each release site independently, as this will allow one to add variability to the quantal size Q and vesicle release time. For simplicity, we here assume the differential equations that describe depression and facilitation have exact analytical solutions, in which case the RP recursive algorithm described in Eqs. (13.28)–(13.32) can be used to simulate release at each release site. However, since each site is now simulated independently, R represents the probability a release site contains a vesicle, rather than the fraction of filled release sites. Addition of binomial release then requires the following two modifications of the RP recursive algorithm. First, step 2 described in Eq. (13.30) is modified as follows: a random number is drawn from the interval [0, 1]; if the random number is greater than $R_{j-\varepsilon} \cdot P_{j-\varepsilon}$ (i.e., the value R times P at the arrival time of the action potential), then release at the site is considered a failure and the site is ignored; otherwise release is considered a success and a synaptic conductance waveform with amplitude Q is added to $G_{total}(t)$. Second, step 3 described in Eq. (13.31) is modified so that the decrement in R only occurs in the event of vesicle release (the increment in P described in Eq. (13.32) is always implemented since facilitation is assumed to be linked to the arrival of the action potential, here assumed to always occur at the time of the stimulus); furthermore, in the event of vesicle release, R is now decremented to zero (i.e., $R_{j+\varepsilon} = 0$). Figure 13.7A1 shows results of such a simulation for a synapse with five release sites, each with $P=0.5$ and $Q=0.2$ nS. To demonstrate the effects of depression and facilitation at moderate levels, the synapse was driven with a 100 Hz train of action potentials. Here, the binomial nature of vesicle release can be seen in the time course of R at each release site (blue traces), and $G_{total}(t)$ computed from the sum of all quanta released

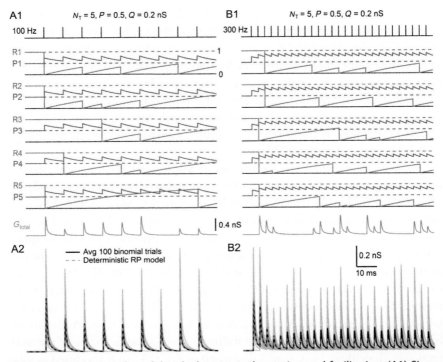

Figure 13.7 A binomial model with short-term depression and facilitation. (A1) Simulation of a binomial synapse with $N_T=5$, $P=0.5$, and $Q=0.2$ nS using the *RP* recursive algorithm described in Eqs. (13.28)–(13.32) ($\tau_r=50$ ms, $\Delta_P=0.5$, $\tau_f=12$ ms). The stimulus was a 100 Hz train of action potentials (top). Bottom graphs show time evolution of R (blue) and P (red) for each release sight during the train. Note, action potentials always caused facilitation, but only caused depression when there was success of vesicle release. At each release site, vesicle release was a success if a random number drawn from [0, 1] was less than the product *RP*. Gray trace (bottom) shows the sum of resulting quantal waveforms from all five sites. The quantal waveform was a G_{AMPAR} waveform with only a direct component, the same used in Fig. 13.6. (A2) Conductance trains of the same simulation in (A1) for 100 trials (gray). Black trace shows average of the 100 trials, which closely matches the time course of the same simulation computed for a deterministic *RP* model (green dashed line). (B1 and B2) Same as (A1) and (A2) but for a 300 Hz train of action potentials. (See the color plate.)

at each site (gray trace). The time course of P, on the other hand, is not binomial but shows an incremental increase at the arrival of each action potential. Figure 13.7A2 shows the average of 100 such simulations (black trace), which closely matches the same simulation computed for a deterministic synapse (green trace). Figure 13.7B1 and B2 shows the same analysis for a 300 Hz train of action potentials to demonstrate the effect of a larger level of synaptic depression and facilitation. Note that the simulations in Fig. 13.7 include no variability in Q or the vesicle release times. To add variability to Q, one only needs to compute a nonuniform Q for each release site as described above, using CV_{QS} and CV_{QII}. One can also add variability to the vesicle release times by simulating asynchronous release at each release site using a RTC function as described above.

8. GOING MICROSCOPIC

The models discussed in this chapter are intended to capture the basic macroscopic features of synaptic transmission, mainly the time and voltage dependence of the transfer of charge into the postsynaptic neuron. These types of models are useful for investigating signal processing at the cellular and network level but are generally not as useful for investigating the mechanism underlying signal transmission at a microscopic level. Hence, other modeling approaches are usually adopted when studying microscopic aspects of synaptic transmission. These approaches typically use partial differential equations to describe one or more aspect of the system under investigation, which might include the diffusion of ions, buffers, and vesicles, reactions between these entities, and conformational changes of protein structures (i.e., state transitions). The equations that describe the biological system under investigation are then solved numerically. Early examples of simulating presynaptic Ca^{2+} dynamics, including Ca^{2+} influx, diffusion, buffering, and extrusion in three dimensions, include the studies of Fogelson and Zucker[97] and Roberts.[98] More recent studies attempt to simulate the mechanisms underlying Ca^{2+}-secretion coupling.[39,99] Finite-difference methods have also been used to simulate glutamate dynamics within the synaptic cleft, including glutamate release and diffusion in three dimensions.[32,34] A Monte Carlo approach can also be used[100]; in this case, a simulator such as MCell should be considered (http://www.mcell.cnl.salk.edu/). On the postsynaptic side, kinetics of ionotropic receptors such as AMPARs and NMDARs can be described by Markov models with multiple states, including open, closed, blocked, and desensitized states.[32,34,63,91] Solutions to Markov models can be computed via transition matrixes, or stochastically (see Ref. 41).

9. SIMULATORS AND STANDARDIZED MODEL DESCRIPTIONS

A number of options exist for creating computer simulations of the synaptic models presented in this chapter. Generic simulation and analysis packages like MATLAB (http://www.mathworks.co.uk/products/matlab/) and Igor Pro (http://www.wavemetrics.com) are commonly used for simulating relatively simple models. These packages have the advantage that the user is completely in control of the model structure and can perform analysis in the same scripting language as that used in the model description. However, these packages are less useful when the models become so complex that the user's scripts start to reproduce the functionality of freely available neuronal simulation packages. The more complex scripts can also be difficult for others to understand and adapt for their own needs.

Packages like NEURON[101] and GENESIS[102] have been used for many years to simulate neural systems from single cells to complex neuronal networks, with inbuilt features to assist in the development of multicompartmental neurons, membrane conductances, synapse models, and spiking networks. These packages, along with a recent reimplementation of GENESIS called MOOSE (http://moose.ncbs.res.in/), are freely available, well documented, and supported by user communities. They are particularly useful for physiological data-driven simulations, for example, investigating the effects of synaptic integration in complex neuronal morphologies.[46] They also have a number of inbuilt synaptic model types which users can incorporate into their own custom models, particularly in NEURON using its NMODL language. A number of published models using these simulation packages are available online at ModelDB (http://senselab.med.yale.edu/modeldb/). NEST (http://www.nest-initiative.org) and Brian (http://www.briansimulator.org) are two other popular simulation packages which focus more on spiking neural networks. While the synapse models of these packages are more phenomenological than biophysical, they have been used in several investigations into the effects of synaptic properties on network function.

While it is useful having multiple simulators to build synaptic and network models, a disadvantage is that the simulators often have different languages for specifying the models, making exchange of models between investigators difficult. NeuroML (http://www.NeuroML.org; Ref. 103) is an initiative to define a simulator-independent language to define cell, ion channel, synapse, and network models and to facilitate exchange of these models. A range of synapse model types are supported, including single- and double-exponential synapses, voltage-dependent synaptic conductances, short-term plasticity, and spike-timing-dependent plasticity. NeuroML v2.0 adds greater support for users to extend the markup language and define their own synapse models in a simulator-independent manner. The Open Source Brain repository (http://www.opensourcebrain.org) is a recent initiative to encourage collaborative development of a range of models in computational neuroscience. Moreover, making the models available in NeuroML ensures transparency, accessibility and cross-simulator portability.

10. SUMMARY

In this chapter, we discussed how mathematical models can capture various aspects of synaptic transmission. At their most basic level, the models are simple empirical descriptions of the average conductance waveform and current–voltage relation of postsynaptic receptors. Above this basic level, the models can be extended to capture more and more of the behavior of real synapses, including stochasticity and short-term plasticity. Throughout the chapter, we examined how well the different models replicate the experimental behavior of the cerebellar MF–GC synapse, an extensively characterized excitatory central synapse. The techniques we used can equally be applied to other excitatory and inhibitory synapses in the CNS. In our data-driven approach, we hoped to have highlighted a few key principles about modeling synaptic transmission, and modeling neurons in general. These include considering the balance between accurately replicating the biological processes under investigation and simplifying model descriptions to reduce variables and computational overhead. We urge modelers to obtain and directly compare their models to raw data (e.g., EPSCs and current–voltage relations) from the system under investigation wherever possible. This is important for ensuring the system being modeled is operating within physiologically relevant regimes. A data-driven approach also enables higher dimensional synaptic, neuronal, and network models to be effectively constrained and the effects of natural variability of presynaptic and postsynaptic elements to be explored.

ACKNOWLEDGMENTS

We thank Bóris Marin, Eugenio Piasini, Arnd Roth, and Stefan Hallermann for their comments on the manuscript, and Padraig Gleeson for his contribution to the section on simulator packages. This work was funded by the Wellcome Trust (086699) and ERC. R. A. S. holds a Wellcome Trust Principal Research Fellowship (095667) and an ERC Advanced Grant (294667).

REFERENCES

1. Fatt P, Katz B. Spontaneous subthreshold activity at motor nerve endings. *J Physiol.* 1952;117(1):109–128.
2. del Castillo J, Katz B. Quantal components of the end-plate potential. *J Physiol.* 1954;124(3):560–573.
3. Katz B. *The Release of Neural Transmitter Substances.* Liverpool: Liverpool University Press; 1969.
4. Couteaux R, Pécot-Dechavassine M. Synaptic vesicles and pouches at the level of "active zones" of the neuromuscular junction. *C R Acad Sci Hebd Seances Acad Sci D.* 1970;271(25):2346–2349.
5. De Robertis ED, Bennett HS. Some features of the submicroscopic morphology of synapses in frog and earthworm. *J Biophys Biochem Cytol.* 1955;1(1):47–58.
6. Palade GE, Palay SL. Electron microscope observations of interneuronal and neuro-muscular synapses. *Anat Rec.* 1954;118:335–336.
7. Palay SL. Synapses in the central nervous system. *J Biophys Biochem Cytol.* 1956;2:193–202.
8. Eccles JC, Katz B, Kuffler SW. Nature of the 'end-plate potential' in curarized muscle. *J Neurophysiol.* 1941;5:362–387.
9. Liley AW, North KA. An electrical investigation of effects of repetitive stimulation on mammalian neuromuscular junction. *J Neurophysiol.* 1953;16(5):509–527.
10. Neher E, Sakmann B, Steinbach JH. The extracellular patch clamp: a method for resolving currents through individual open channels in biological membranes. *Pflugers Arch.* 1978;375(2):219–228.
11. Harlow ML, Ress D, Stoschek A, Marshall RM, McMahan UJ. The architecture of active zone material at the frog's neuromuscular junction. *Nature.* 2001;409(6819):479–484.
12. Biró AA, Holderith NB, Nusser Z. Quantal size is independent of the release proba-bility at hippocampal excitatory synapses. *J Neurosci.* 2005;25(1):223–232.
13. Silver RA, Lübke J, Sakmann B, Feldmeyer D. High-probability uniquantal transmis-sion at excitatory synapses in barrel cortex. *Science.* 2003;302(5652):1981–1984.
14. Salpeter MM, Loring RH. Nicotinic acetylcholine receptors in vertebrate muscle: properties, distribution and neural control. *Prog Neurobiol.* 1985;25(4):297–325.
15. Silver RA, Cull-Candy SG, Takahashi T. Non-NMDA glutamate receptor occupancy and open probability at a rat cerebellar synapse with single and multiple release sites. *J Physiol.* 1996;494(Pt. 1):231–250.
16. Traynelis SF, Silver RA, Cull-Candy SG. Estimated conductance of glutamate receptor channels activated during EPSCs at the cerebellar mossy fiber-granule cell synapse. *Neuron.* 1993;11(2):279–289.
17. Bekkers JM, Stevens CF. NMDA and non-NMDA receptors are co-localized at indi-vidual excitatory synapses in cultured rat hippocampus. *Nature.* 1989;341(6239):230–233.
18. Silver RA, Traynelis SF, Cull-Candy SG. Rapid-time-course miniature and evoked excitatory currents at cerebellar synapses *in situ. Nature.* 1992;355(6356):163–166.

19. Ascher P, Nowak L. The role of divalent cations in the N-methyl-D-aspartate responses of mouse central neurones in culture. *J Physiol.* 1988;399:247–266.

20. Bliss TV, Collingridge GL. A synaptic model of memory: long-term potentiation in the hippocampus. *Nature.* 1993;361(6407):31–39.

21. Monyer H, Burnashev N, Laurie DJ, Sakmann B, Seeburg PH. Developmental and regional expression in the rat brain and functional properties of four NMDA receptors. *Neuron.* 1994;12(3):529–540.

22. Schwartz EJ, Rothman JS, Dugué GP, et al. NMDA receptors with incomplete Mg^{2+} block enable low-frequency transmission through the cerebellar cortex. *J Neurosci.* 2012;32(20):6878–6893.

23. Eccles JC, Ito M, Szentágothai J. *The Cerebellum as a Neuronal Machine.* New York: Springer; 1967.

24. DiGregorio DA, Nusser Z, Silver RA. Spillover of glutamate onto synaptic AMPA receptors enhances fast transmission at a cerebellar synapse. *Neuron.* 2002;35(3):521–533.

25. Palay SL, Chan-Palay V. *Cerebellar Cortex: Cortex and Organization.* Berlin: Springer-Verlag; 1974.

26. Sargent PB, Saviane C, Nielsen TA, DiGregorio DA, Silver RA. Rapid vesicular release, quantal variability, and spillover contribute to the precision and reliability of transmission at a glomerular synapse. *J Neurosci.* 2005;25(36):8173–8187.

27. Cathala L, Holderith NB, Nusser Z, DiGregorio DA, Cull-Candy SG. Changes in synaptic structure underlie the developmental speeding of AMPA receptor-mediated EPSCs. *Nat Neurosci.* 2005;8(10):1310–1318.

28. Jakab RL, Hámori J. Quantitative morphology and synaptology of cerebellar glomeruli in the rat. *Anat Embryol (Berl).* 1988;179(1):81–88.

29. Saviane C, Silver RA. Fast vesicle reloading and a large pool sustain high bandwidth transmission at a central synapse. *Nature.* 2006;439(7079):983–987.

30. Hallermann S, Fejtova A, Schmidt H, et al. Bassoon speeds vesicle reloading at a central excitatory synapse. *Neuron.* 2010;68(4):710–723.

31. Hallermann S, Silver RA. Sustaining rapid vesicular release at active zones: potential roles for vesicle tethering. *Trends Neurosci.* 2013;36(3):185–194.

32. Nielsen TA, DiGregorio DA, Silver RA. Modulation of glutamate mobility reveals the mechanism underlying slow-rising AMPAR EPSCs and the diffusion coefficient in the synaptic cleft. *Neuron.* 2004;42(5):757–771.

33. Mitchell SJ, Silver RA. GABA spillover from single inhibitory axons suppresses low-frequency excitatory transmission at the cerebellar glomerulus. *J Neurosci.* 2000;20(23):8651–8658.

34. DiGregorio DA, Rothman JS, Nielsen TA, Silver RA. Desensitization properties of AMPA receptors at the cerebellar mossy fiber granule cell synapse. *J Neurosci.* 2007;27(31):8344–8357.

35. Cathala L, Brickley S, Cull-Candy SG, Farrant M. Maturation of EPSCs and intrinsic membrane properties enhances precision at a cerebellar synapse. *J Neurosci.* 2003;23(14):6074–6085.

36. Cathala L, Misra C, Cull-Candy SG. Developmental profile of the changing properties of NMDA receptors at cerebellar mossy fiber-granule cell synapses. *J Neurosci.* 2000;20(16):5899–5905.

37. Farrant M, Feldmeyer D, Takahashi T, Cull-Candy SG. NMDA-receptor channel diversity in the developing cerebellum. *Nature.* 1994;368(6469):335–339.

38. Gerstner W, Kistler WM. *Spiking Neuron Models. Single Neurons, Populations, Plasticity.* Cambridge, UK: Cambridge University Press; 2002.

39. Meinrenken CJ, Borst JG, Sakmann B. Calcium secretion coupling at calyx of held governed by nonuniform channel-vesicle topography. *J Neurosci.* 2002;22(5):1648–1667.

40. Pan B, Zucker RS. A general model of synaptic transmission and short-term plasticity. *Neuron.* 2009;62(4):539–554.
41. Roth A, van Rossum MCW. Modeling synapses. In: De Schutter E, ed. *Computational Modeling Methods for Neuroscientists.* Cambridge, Massachusetts: The MIT Press; 2009.
42. Trommershäuser J, Schneggenburger R, Zippelius A, Neher E. Heterogeneous presynaptic release probabilities: functional relevance for short-term plasticity. *Biophys J.* 2003;84(3):1563–1579.
43. Williams SR, Mitchell SJ. Direct measurement of somatic voltage clamp errors in central neurons. *Nat Neurosci.* 2008;11(7):790–798.
44. Häusser M, Roth A. Estimating the time course of the excitatory synaptic conductance in neocortical pyramidal cells using a novel voltage jump method. *J Neurosci.* 1997;17(20):7606–7625.
45. Bekkers JM, Stevens CF. Cable properties of cultured hippocampal neurons determined from sucrose-evoked miniature EPSCs. *J Neurophysiol.* 1996;75(3):1250–1255.
46. Rothman JS, Cathala L, Steuber V, Silver RA. Synaptic depression enables neuronal gain control. *Nature.* 2009;457(7232):1015–1018.
47. Jonas P, Spruston N. Mechanisms shaping glutamate-mediated excitatory postsynaptic currents in the CNS. *Curr Opin Neurobiol.* 1994;4(3):366–372.
48. Silver RA, Colquhoun D, Cull-Candy SG, Edmonds B. Deactivation and desensitization of non-NMDA receptors in patches and the time course of EPSCs in rat cerebellar granule cells. *J Physiol.* 1996;493(Pt. 1):167–173.
49. Feldmeyer D, Cull-Candy S. Functional consequences of changes in NMDA receptor subunit expression during development. *J Neurocytol.* 1996;25(12):857–867.
50. Takahashi T, Feldmeyer D, Suzuki N, et al. Functional correlation of NMDA receptor epsilon subunits expression with the properties of single-channel and synaptic currents in the developing cerebellum. *J Neurosci.* 1996;16(14):4376–4382.
51. Nevian T, Sakmann B. Spine Ca^{2+} signaling in spike-timing-dependent plasticity. *J Neurosci.* 2006;26:11001–11013.
52. Shouval HZ, Bear MF, Cooper LN. A unified model of NMDA receptor-dependent bidirectional synaptic plasticity. *Proc Natl Acad Sci USA.* 2002;99(16):10831–10836.
53. Wang XJ. Synaptic basis of cortical persistent activity: the importance of NMDA receptors to working memory. *J Neurosci.* 1999;19(21):9587–9603.
54. Woodhull AM. Ionic blockage of sodium channels in nerve. *J Gen Physiol.* 1973;61(6):687–708.
55. Jahr CE, Stevens CF. Voltage dependence of NMDA-activated macroscopic conductances predicted by single-channel kinetics. *J Neurosci.* 1990;10(9):3178–3182.
56. Johnson JW, Ascher P. Voltage-dependent block by intracellular Mg^{2+} of N-methyl-D-aspartate-activated channels. *Biophys J.* 1990;57(5):1085–1090.
57. Kupper J, Ascher P, Neyton J. Internal Mg^{2+} block of recombinant NMDA channels mutated within the selectivity filter and expressed in Xenopus oocytes. *J Physiol.* 1998;507(Pt. 1):1–12.
58. Li-Smerin Y, Johnson JW. Kinetics of the block by intracellular Mg^{2+} of the NMDA-activated channel in cultured rat neurons. *J Physiol.* 1996;491(Pt. 1):121–135.
59. Wollmuth LP, Kuner T, Sakmann B. Adjacent asparagines in the NR2-subunit of the NMDA receptor channel control the voltage-dependent block by extracellular Mg^{2+}. *J Physiol.* 1998;506(1):13–32.
60. Yang YC, Lee CH, Kuo CC. Ionic flow enhances low-affinity binding: a revised mechanistic view into Mg^{2+} block of NMDA receptors. *J Physiol.* 2010;588(Pt. 4):633–650.
61. Antonov SM, Johnson JW. Permeant ion regulation of N-methyl-D-aspartate receptor channel block by Mg^{2+}. *Proc Natl Acad Sci USA.* 1999;96(25):14571–14576.

62. Clarke RJ, Glasgow NG, Johnson JW. Mechanistic and structural determinants of NMDA receptor voltage-dependent gating and slow Mg^{2+} unblock. *J Neurosci.* 2013;33(9):4140–4150.

63. Kampa BM, Clements J, Jonas P, Stuart GJ. Kinetics of Mg^{2+} unblock of NMDA receptors: implications for spike-timing dependent synaptic plasticity. *J Physiol.* 2004;556(Pt. 2):337–345.

64. Adrian ED, Zotterman Y. The impulses produced by sensory nerve-endings. Part II. The response of a single end-organ. *J Physiol.* 1926;61(2):151–171.

65. Goldberg JM, Adrian HO, Smith FD. Response of neurons of the superior olivary complex of the cat to acoustic stimuli of long duration. *J Neurophysiol.* 1964;27:706–749.

66. Rieke F, Warland D, de Ruyter van Steveninck R, Bialek W. *Spikes: Exploring the Neural Code.* Cambridge, Massachusetts: The MIT Press; 1999.

67. Rancz EA, Ishikawa T, Duguid I, Chadderton P, Mahon S, Häusser M. High-fidelity transmission of sensory information by single cerebellar mossy fibre boutons. *Nature.* 2007;450(7173):1245–1248.

68. Margrie TW, Brecht M, Sakmann B. *In vivo*, low-resistance, whole-cell recordings from neurons in the anaesthetized and awake mammalian brain. *Pflugers Arch.* 2002;444(4):491–498.

69. Johnson DH, Tsuchitani C, Linebarger DA, Johnson MJ. Application of a point process model to responses of cat lateral superior olive units to ipsilateral tones. *Hear Res.* 1986;21(2):135–159.

70. Rothman JS, Young ED, Manis PB. Convergence of auditory nerve fibers onto bushy cells in the ventral cochlear nucleus: implications of a computational model. *J Neurophysiol.* 1993;70(6):2562–2583.

71. Young ED, Barta PE. Rate responses of auditory nerve fibers to tones in noise near masked threshold. *J Acoust Soc Am.* 1986;79(2):426–442.

72. Lapicque L. Recherches quantitatives sur l'excitation électrique des nerfs traitée comme une polarisation. *J Physiol Pathol Gen.* 1907;9:620–635.

73. Knight BW. Dynamics of encoding in a population of neurons. *J Gen Physiol.* 1972;59(6):734–766.

74. Stein RB. A theoretical analysis of neuronal variability. *Biophys J.* 1965;5(2):173–194.

75. Hodgkin AL, Huxley AF. A quantitative description of membrane current and its application to conduction and excitation in nerve. *J Physiol.* 1952;117(4):500–544.

76. Abbott LF, Regehr WG. Synaptic computation. *Nature.* 2004;431:796–803.

77. Zucker RS, Regehr WG. Short-term synaptic plasticity. *Annu Rev Physiol.* 2002;64:355–405.

78. Betz WJ. Depression of transmitter release at the neuromuscular junction of the frog. *J Physiol.* 1970;206(3):629–644.

79. Heinemann C, von Rüden L, Chow RH, Neher E. A two-step model of secretion control in neuroendocrine cells. *Pflugers Arch.* 1993;424(2):105–112.

80. Dittman JS, Regehr WG. Calcium dependence and recovery kinetics of presynaptic depression at the climbing fiber to Purkinje cell synapse. *J Neurosci.* 1998;18(16):6147–6162.

81. Lipstein N, Sakaba T, Cooper BH, et al. Dynamic control of synaptic vesicle replenishment and short-term plasticity by Ca^{2+}-calmodulin-Munc13-1 signaling. *Neuron.* 2013;79(1):82–96.

82. Sakaba T, Neher E. Calmodulin mediates rapid recruitment of fast-releasing synaptic vesicles at a calyx-type synapse. *Neuron.* 2001;32(6):1119–1131.

83. Stevens CF, Wesseling JF. Activity-dependent modulation of the rate at which synaptic vesicles become available to undergo exocytosis. *Neuron.* 1998;21(2):415–424.

84. Wang LY, Kaczmarek LK. High-frequency firing helps replenish the readily releasable pool of synaptic vesicles. *Nature.* 1998;394(6691):384–388.

85. Fuhrmann G, Segev I, Markram M, Tsodyks H. Coding of temporal information by activity-dependent synapses. *J Neurophysiol.* 2002;87(1):140–148.

86. Varela JA, Sen K, Gibson J, Fost J, Abbott LF, Nelson SB. A quantitative description of short-term plasticity at excitatory synapses in layer 2/3 of rat primary visual cortex. *J Neurosci.* 1997;17(20):7926–7940.

87. Tsodyks M, Pawelzik K, Markram H. Neural networks with dynamic synapses. *Neural Comput.* 1998;10:821–835.

88. Hennig MH. Theoretical models of synaptic short term plasticity. *Front Comput Neurosci.* 2013;7:45.

89. Wu LG, Borst JG. The reduced release probability of releasable vesicles during recovery from short-term synaptic depression. *Neuron.* 1999;23(4):821–832.

90. Nieus T, Sola E, Mapelli J, Saftenku E, Rossi P, D'Angelo E. LTP regulates burst initiation and frequency at mossy fiber-granule cell synapses of rat cerebellum: experimental observations and theoretical predictions. *J Neurophysiol.* 2006;95(2):686–699.

91. Wadiche JI, Jahr CE. Multivesicular release at climbing fiber-Purkinje cell synapses. *Neuron.* 2001;32(2):301–313.

92. Kuno M. Quantal components of excitatory synaptic potentials in spinal motoneurons. *J Physiol.* 1964;175(1):81–99.

93. Clements JD, Silver RA. Unveiling synaptic plasticity: a new graphical and analytical approach. *Trends Neurosci.* 2000;23(3):105–113.

94. Silver RA. Estimation of nonuniform quantal parameters with multiple-probability fluctuation analysis: theory, application and limitations. *J Neurosci Methods.* 2003;130(2):127–141.

95. Minneci F, Kanichay RT, Silver RA. Estimation of the time course of neurotransmitter release at central synapses from the first latency of postsynaptic currents. *J Neurosci Methods.* 2012;205(1):49–64.

96. Maass W, Zador AM. Dynamic stochastic synapses as computational units. *Neural Comput.* 1999;11(4):903–917.

97. Fogelson AL, Zucker RS. Presynaptic calcium diffusion from various arrays of single channels. Implications for transmitter release and synaptic facilitation. *Biophys J.* 1985;48(6):1003–1017.

98. Roberts WM. Localization of calcium signals by a mobile calcium buffer in frog saccular hair cells. *J Neurosci.* 1994;14(5 Pt. 2):3246–3262.

99. Bucurenciu I, Kulik A, Schwaller B, Frotscher M, Jonas P. Nanodomain coupling between Ca^{2+} channels and Ca^{2+} sensors promotes fast and efficient transmitter release at a cortical GABAergic synapse. *Neuron.* 2008;57(4):536–545.

100. Franks KM, Stevens CF, Sejnowski TJ. Independent sources of quantal variability at single glutamatergic synapses. *J Neurosci.* 2003;23(8):3186–3195.

101. Carnevale NT, Hines ML. *The NEURON Book.* Cambridge, UK: Cambridge University Press; 2006.

102. Bower JM, Beeman D. *The Book of GENESIS: Exploring Realistic Neural Models with the GEneral NEural SImulation SYstem.* New York: Springer-Verlag; 1998.

103. Gleeson P, Steuber V, Silver RA. neuroConstruct: a tool for modeling networks of neurons in 3D space. *Neuron.* 2007;54(2):219–235.

CHAPTER FOURTEEN

Multiscale Modeling and Synaptic Plasticity

Upinder S. Bhalla
National Centre for Biological Sciences, Bangalore, Karnataka, India

Contents

Progress in Molecular Biology and Translational Science, Volume 123
ISSN 1877-1173
http://dx.doi.org/10.1016/B978-0-12-397897-4.00012-7

Abstract

Synaptic plasticity is a major convergence point for theory and computation, and the process of plasticity engages physiology, cell, and molecular biology. In its many manifestations, plasticity is at the hub of basic neuroscience questions about memory and development, as well as more medically themed questions of neural damage and recovery. As an important cellular locus of memory, synaptic plasticity has received a huge amount of experimental and theoretical attention. If computational models have tended to pick specific aspects of plasticity, such as STDP, and reduce them to an equation, some experimental studies are equally guilty of oversimplification each time they identify a new molecule and declare it to be the last word in plasticity and learning. Multiscale modeling begins with the acknowledgment that synaptic function spans many levels of signaling, and these are so tightly coupled that we risk losing essential features of plasticity if we focus exclusively on any one level. Despite the technical challenges and gaps in data for model specification, an increasing number of multiscale modeling studies have taken on key questions in plasticity. These have provided new insights, but importantly, they have opened new avenues for questioning. This review discusses a wide range of multiscale models in plasticity, including their technical landscape and their implications.

1. INTRODUCTION

The concept of multiscale modeling is familiar in many fields, including materials science, meteorology, cardiac physiology, and neuroscience. In all these cases, phenomena emerge from interactions across scales. For example, in materials science, the interaction is between bulk material properties and microscopic physics such as lattices and polymers. In meteorology, the interaction is between short space and timescales for local weather, and much larger global weather models. Cardiac modeling addresses many of the same interacting physiological phenomena as synaptic plasticity, but at a much larger length scale and with mechanics and fluid dynamics thrown in.

Philosophically, multiscale modeling runs squarely into one of the key dilemmas of all modeling studies: how much detail to incorporate? This problem is especially acute for multiscale models because by definition they incorporate events across scales and have to justify their choices at all these levels. The topic of detail has been elegantly dealt with elsewhere,[1] but the added burden of proof for a multiscale model could be summarized as: do the cross-scale interactions form an essential part of the modeled system, and do you lose important properties if you ignore these interactions?

Multiscale models span different physical/chemical processes, different lengths, and different timescales. There is also the implication that these scales interact bidirectionally and continuously within the simulation. For example, a model where an electrical component computes calcium influx, and then just passes this into a chemical model as a boundary condition, would not be truly multiscale. One could just as well make a table of values of calcium influx and feed it into the chemical model. Likewise, a model that simulates movement of molecules simply by placing them in a different chemical compartment would only be a surrogate for true spatial chemical modeling with explicit representations of space. This is not to say that these halfway multiscale models do not provide insights, just that in making these abstractions they lose some relevant and potentially crucial contacts with the different scales.

Synaptic plasticity is a particularly fertile arena for cross–scale interactions. There is much activity on technical and scientific questions at these interfaces between scales. In Section 2, I follow the sequence of events in synaptic plasticity to highlight the key role of multiscale signaling in these processes. In subsequent sections, I discuss some of the key questions and recent insights from multiscale models and then consider technical issues in multiscale computations and model specification.

2. EVENTS IN SYNAPTIC PLASTICITY

As a road map to the kinds of multiscale signaling considered in this review, I will first walk through a sequence of major events in plasticity, indicating where multiscale modeling has a key role to play.

1. I begin with the arrival of an action potential at a presynaptic terminal. An intricate series of cell-biological events then takes place to transduce the action potential into neurotransmitter release. Remarkably, this typically happens within a millisecond. The key synaptic strength variables controlled by presynaptic events include the probability that a given action potential will elicit vesicle release, the number of vesicles released, and the neurotransmitter content of each vesicle. The process of conversion of action potential to calcium influx signals to vesicle fusion and reuptake, combines electrical and chemical signaling with the mechanical events of vesicle–membrane interactions. There is also a significant component of signaling through presynaptic receptors. Together, this is a challenging multiscale system to simulate, especially given the difficulties of getting good experimental parameters for each step. It is clear that

these presynaptic processes are involved in both short- and long-term forms of synaptic plasticity.

2. The released neurotransmitter diffuses to the postsynaptic membrane. Along the way, the neurotransmitter may be degraded, diffuse in other directions, and be taken up by the synaptic terminal or by astrocytes. The length scale of diffusion here ranges from a few nanometers for the synaptic cleft to many microns into the surrounding tissue.

3. The neurotransmitter opens ionotropic receptors and activates metabotropic receptors on the postsynaptic junction; many electrical and chemical signaling events follow. Electrical signals propagate into the dendrite relatively easily, but in synapses localized on dendritic spines, there is at least some degree of spatial isolation of the chemical signaling.[2,3] Key postsynaptic signaling events include the induction of plasticity due to associations between pre- and postsynaptic activity and a multistage process of establishment of long-lasting changes in synaptic conductance. The signaling in the postsynaptic terminal is itself multiscale. Ionotropic receptors not only mediate electrical signaling but also let calcium to initiate chemical cascades. The receptors are targets for signaling in the form of chemical modifications, trafficking, and structural changes. There is an immense range of timescales at work, from millisecond receptor opening and spike-timing-dependent plasticity (STDP) signaling to hour-long processes for protein-synthesis-dependent consolidation of potentiation. The functional outcomes of this signaling include an exquisite selectivity for input patterns, and the decision between a range of possible forms of plasticity. Once synaptic strength has been changed, the maintenance of the changed synaptic state is itself an interesting multiscale problem with structural, electrical, and chemical aspects. Overall, postsynaptic signaling is a multiscale problem in many respects.

4. Signals continue into the dendrite, where they combine with parallel and often very different inputs from other synapses. Electrical integration of signals occurs, both with other synapses[4] and with signals propagating back up from the soma. Some interesting multiscale signaling logic occurs to decide how the spatial and temporal organization of inputs triggers plasticity and in which synapses. One manifestation of this is synaptic tagging, where a strong input at one synapse may help to potentiate a nearby synapse that only gets a weak input.[5] Another major topic is cellular homeostasis, which addresses how the overall excitability of the cell is maintained despite changes in synaptic weight and modulation

of voltage-gated ion-channel conductances. As with postsynaptic signaling, electrical, chemical, and structural events interact in the dendrite at many length- and timescales.

5. Input signals converge to the soma, where transcriptional control brings synaptic plasticity signaling into contact with genetic networks. It is clear that many genes are specifically turned on by synaptic activity, and that these genes are required for subsequent plasticity. Genetic networks have their own modeling formalisms and timescales; thus, their interaction with electrical and biochemical signaling is yet another domain for multiscale modeling.

6. Proteins and mRNA, whose synthesis is triggered by all this activity, are now exported out to the dendrites and axonal branches. These cargoes are directed by a transport and tagging system whose details are far from understood. Some seem to be selectively taken up by recently active synapses. Many of the newly made mRNA molecules are stationed in dendrites or even synaptic spines, awaiting further activity to trigger translation. It is likely that mechanical and chemical steps interact in the cell transport decisions about which branch to take or when to unload the cargo.

7. Structural and morphological changes frequently accompany the electrical and signaling events of synaptic and cellular plasticity.[6] These range from modest morphological changes in dendritic spine shape (e.g., from filopodia to mushroom spines) to large-scale cellular reorganization including growth and retraction of spines, dendrites, and axons. These events are mediated by a range of chemical processes with mechanical and structural outcomes, including modification of the cytoskeleton, transport processes, and molecular motors. Technically, morphological changes pose interesting challenges for most existing forms of electrical and chemical models, which are based on static spatial discretization of the cell.

8. In the aftermath of synaptic input, both at somatic and dendritic levels, there is another whole layer of molecular signaling that goes on outside the neuron. The spines and dendrites themselves generate rapidly diffusing messengers, such as NO and arachidonic acid, that do not respect cellular boundaries but have a relatively short range. Growth factors such as BDNF are also secreted following activity. Outside these cellular-scale events, there are many tissue-signaling and extracellular diffusion events that impinge upon and modulate plasticity. Some of these events include the effects of extracellular ion gradients on intracellular potential, and the

diffusion of growth factors, neurotransmitters, and messengers between cells. This may lead to substantial network-scale reorganization and rewiring. In the midst of all this, many existing synapses undergo plasticity due to activity and these long-range molecular signals. These are all cases where subcellular-length-scale events lead to supracellular signaling consequences that affect many neurons.

This brief road map indicates the main territories of multiscale signaling and modeling that relate to synaptic plasticity. I now turn to some specific questions and studies in these domains.

3. MULTISCALE PROCESSES IN PLASTICITY

Synaptic plasticity is central to many questions in neuroscience, and it is probably safe to say that most of these questions span many scales of biological function. Memory, with its many ramifications, is obviously a major topic. Concise mathematical expressions for activity-dependent weight change are much beloved by abstract network modelers. The biological reality is considerably messier and sprawls over several kinds of electrical, chemical, and structural mechanisms, not to mention thousands of signaling molecules. Mechanisms to selectively trigger weight change are inextricably linked with those that maintain the weight change. Temporal input patterning is also tied tightly with spatial patterns of input, and with modulation through a wide range of signals.

Beyond memory, there is a wide range of multiscale processes that engage synaptic plasticity. Broadly, these include building up, keeping up, and the breaking down of neurons and networks.

Building up networks—neural development—includes studies on the specification and growth of neurons, their migration and axonal pathfinding, and how genes and activity combine to specify network connectivity. In addition to a strong genetic component, tissue-level modeling is essential for understanding gradients and mechanical events in neural development.

The process of "keeping up" networks starts to edge into medical questions: homeostasis of cells and networks, recovery and rewiring following injury, aging, or stroke, even the process of fitting newly born neurons into an existing network. Here again, tissue gradients of morphogens and their mapping with activity are crucial, as are processes of cellular structural change. One needs extensive data about signaling pathways for regulation and pharmacological intervention.

The breakdown of networks is squarely medical. We are still struggling with single-neuron mechanisms of neurodegeneration—the tissue-level interactions are harder still. More subtly, what goes wrong with the balance of activity and signals that throws network wiring off kilter in psychiatric disorders? Is there a way to map high-level theories of network (dys)function and behavior to underlying molecular events?

Because of the considerable overlap between these scientific questions, this section is subdivided into a discussion of processes rather than questions. I somewhat artificially separate many closely coupled processes to examine their multiscale implications and prior work. Some of the biases of computational neuroscience show up in an overemphasis on memory. Nevertheless, the range of processes considered here is quite broad. Arguably, if it were easier to incorporate all these processes into data-driven, predictive models, there would be far more engagement of the developmental and medical research efforts with such modeling.

3.1. Temporal pattern decoding for synaptic change

Donald Hebb's classic specification of the rule for association between pre- and postsynaptic activity was one of the oldest in the large and still growing family of abstract rules governing synaptic weight change.[7] In its early interpretations, it seemed that the NMDA channel was a single-molecule implementation of the Hebb rule.[8] As the field of synaptic plasticity has become increasingly well populated, a more complete specification has emerged of the kinds of input patterns that can elicit long-lasting synaptic change.[9,10] The long duration and temporal complexity of some of these patterns have posed a challenge for mechanisms that could accomplish this kind of filtering. Another challenge was that quite different plasticity outcomes can arise depending on input pattern.[11] This was most compactly stated by another venerable mathematically defined rule—the BCM rule—which simply suggested that varying just one parameter, the intensity of an input, could generate three different synaptic learning outcomes: stability, depression, or facilitation.[12] This input parameter was equated to postsynaptic calcium levels, thus bringing the mathematical question into the language of cellular signaling. Many models have addressed how calcium levels might bidirectionally control plasticity.[13–16] An old but still well-supported model is that a high-affinity but lower-concentration phosphatase, such as calcineurin, might give rise to LTD, and a low-affinity but high-concentration kinase, such as CaMKII, might be implicated in LTP.[17,18] A number of chemical kinetic models have shown that this form of decoding is plausible.[19–22] Relatively few studies have embedded this rather simple chemical network into an electrical model including synaptic calcium influx through the NMDA channel, thus closing the loop and tackling this as a multiscale problem.[23] Numerous studies of much greater chemical and electrical sophistication have followed. It is now clear that many signaling pathways are good at decoding temporal input patterns, including more complicated ones such as theta–burst and massed or spaced tetanic stimuli.[24,25]

Coming full circle, Hebb's rule has been given a much more precise mechanistic meaning through the discovery of STDP.[26,27] In STDP, the precise difference between input and output spike timings determines the amount and direction of synaptic plasticity. This form of plasticity poses particularly keen multiscale questions. It seems essential that fast electrical signaling must play a role in setting up the precise timing relationships, but the requirement that such pairings must be repeated many times suggests that there is a cumulative, chemical signaling stage as well.[28] One suggestion is that allosteric modifications of NMDA channels due to ligand, coupled with the rapid magnesium block, achieve the fast-time-course selectivity,[29] and the resultant modulated influx of calcium feeds into a slow integrating signaling pathway. Very few studies have actually implemented both biophysical and chemical portions of this model.

3.2. Spatial patterning of inputs

There are thousands of synapses on a typical dendrite. Plasticity depends on the placement as well as timing of their activity. Electrical signals flow quite readily from synapses into dendrites; chemical signals also propagate but much more slowly.[3,30–32] This introduces cross talk between synapses, which enriches their repertoire of pattern detection with a spatial dimension.[33] The phenomenon of synaptic tagging is one of the better studied manifestations of how synapses incorporate spatial patterns into their decision making.[5,34] Recent simulation studies have begun to address spatial pattern decoding, using discrete chemical compartment models[35] or full stochastic reaction–diffusion calculations.[36–38] This is a topic that seems tailor-made for the application of multiscale methods, particularly the incorporation of electrical signaling. Even a preliminary analysis of coupled electrical–chemical signaling of multiple spines along a dendrite shows interesting cross-scale feedback effects.[23] This is clearly an area that seems likely to be catalyzed by the availability of modeling tools with multiscale capabilities and of signaling models in exchangeable formats.

3.3. Synaptic weight change maintenance

Many studies leave the problem of plasticity at the input selectivity level, ignoring the fact that any selected form of plasticity must somehow be maintained if it is to be part of a long-term change. Another whole family of models has arisen which asks how state changes might be preserved. There are serious physiological challenges that stand in the way of achieving long-lasting state changes. First, synaptic molecules turn over with a lifetime of a few hours to a few days.[39] The synapse itself would decay away if there were not some process at work to preserve synapses and their states despite turn-over. Second, synaptic chemistry is noisy. Synapses are very small and have small integer numbers of key signaling molecules, so synaptic signaling reactions, including those that might store information, are unreliable and fluctuating.[40] Other conceptual difficulties with long-lasting state changes include the problem of how chemical changes map to structural ones and whether synaptic changes are analog or digital.

The traditional solution to the problem of turnover is bistability, implemented through chemical feedback reactions. If the switch is in its basal state, then few receptor molecules are phosphorylated (or present at the postsynaptic density), and hence the synapse is weak. When the switch is "on," then the converse is true and the synapse is strong. The role of the plasticity rule or decoding pathway is to flip the switch to and from the "on" state, and the role of the switch is to "remember" the state.

If the system has settled into any of its stable states, the addition or removal of a few reactant molecules is a small perturbation, and the system soon returns to the nearest stable state. The original synaptic bistable circuit involved CaMKII,[41] and since then many candidate synaptic feedback mechanisms have been proposed.[19,42–48] Bistable switches can also be devised that address the problem of chemical noise.[49] There are two related solutions to the problem of synaptic "forgetting" due to spontaneous stochastic state transitions. First, the combination of molecule numbers and rate constants in some signaling networks does seem to allow for long-term state retention over many years.[21,50] Second, the coupling of fast and slow switches can give the best of both worlds without particularly difficult constraints on kinetics. The fast switch responds quickly to plasticity stimuli, and this holds its state long enough to trigger much more stable changes in a slow-moving switch.[21,51–53]

An apparent problem with bistable switches as a mechanism for synaptic plasticity is the familiar observation that synapses are not, according to most experiments, two-state systems. Leaving aside the debate over the experimental evidence for and against this, it turns out that chemical systems can, in principle, form multistable systems. One approach is the line-attractor mechanism to achieve continuous, stable states.[20,54] Another possibility is simply to couple a number of loosely interacting bistable switches within the synapse so that the total synaptic weight is a function of the number of switches in the "on" state.

The variety of possible memory mechanisms, such as bistable switches, multiplies hugely with multiscale interactions. Just as chemical feedback switches can store information through bistability, electrical signaling can store information at the network level through bistable reverberatory activity in circuit loops.[7,55] These feedback mechanisms have many potential points of coupling: essentially all receptors and voltage-gated ion channels are susceptible to signaling-mediated modulation, and control of expression levels and insertion into the membrane. Electrical signaling couples to the chemical side through numerous ligand-gated ion channels and voltage-gated calcium channels. Structural changes triggered by chemical signaling have implications both for other chemical events and for the electrical properties of neurons. In the limit, one can regard the formation and removal of synaptic connections, and even dendrites and axonal branches, as a fundamental form of plasticity under the control of signaling. Few of these interactions have been modeled in a truly multiscale sense: usually, the electrical or chemical signal has been reduced to an external event or lookup table. Many interesting possibilities of multiscale feedback dynamics in memory storage remain to be explored.

3.4. Homeostasis and synaptic normalization

Cellular activity homeostasis has been the subject of intense theoretical and experimental activity.[56,57] An emerging conceptual framework is that neurons appear to have a "set point" of activity, including complex attributes such as bursting behavior as well as firing rate. Deviations from this set point lead to changes in channel expression that tend to restore the cell activity toward the set point. The details of these electrical changes have been worked on in a few cases, but their signaling mechanisms remain elusive. A minimal but remarkably effective suggestion is that pathways responding to different timescales of intracellular calcium are sufficient to account for many of the observed homeostasis properties.[58]

A related form of multiscale homeostasis is synaptic normalization. Here, synaptic plasticity modifies some synaptic weights, and the role of homeostasis is to restore the overall excitability of the cell. An exploratory study incorporating both chemical and electrical signaling in a dendritic segment decorated with a few spines has proposed a simple calcium-dependent mechanism for this.[23] In this model, the increased excitability raises basal calcium levels to a point where they induce LTD in susceptible synapses through a calcium–calcineurin-dependent mechanism.

Homeostasis at the network level has recently been studied with abstracted representations of neurons and signaling.[59] Using a higher-level learning rule for homeostatic intrinsic plasticity, this study shows that the system as a whole can achieve a robust set point of network dynamics. Thus even abstracted representations of synapse and cell-level plasticity rules give insights into network-scale homeostasis.

Neuronal homeostasis clearly deserves a much more complete signaling account. It has the canonical form of a multiscale problem: an interplay between electrical activity and chemical signaling, and a vast range of timescales, from millisecond spiking activity to days and weeks of homeostatic recovery.

3.5. Synaptic and neuronal restructuring

Neurons and synapses are highly dynamic, and synaptic plasticity is known to be closely correlated with structural changes.[60] Several models have explored this class of multiscaling. One of the simplest variants on this was to treat insertion of receptors as an event driven by the location of already-inserted receptors.[54] In this model, the structural component was a lattice of putative receptor sites, and the rates of insertion were a function of occupancy. Another instance of the convergence of plasticity and structural change was a recent study illustrating bistability in synapse size arising from volume changes, and the reciprocal enhancement of calcium influx by larger volumes.[61] This study also spanned electrical and structural domains but relied on abstractions for the mechanisms for inducing the structural change. A family of models involving receptor insertion and bistability[21,62] is also rather short of true structural change, as these models are entirely in the chemical domain with implicit structural effects rather than any mechanical calculations. Various other models have considered structural changes arising from plasticity stimuli with simplified chemical kinetics and abstractions of the mechanical events.[63]

In sum, structural change is clearly a feature whose multiscale implications are recognized, but explicit and mechanistic models relating this to electrical and chemical signaling are yet to be fully realized.

3.6. The role of genes and protein synthesis

Gene-regulatory and mRNA-regulatory networks are a largely unexplored aspect of synaptic plasticity. It is well established that synaptic activity induces local protein synthesis,[64] and this process has even been modeled in some biochemical detail.[65] There are also several plasticity models which invoke individual, specific gene activation and subsequent protein synthesis.[45,46] However, these models have barely scratched the surface of the pattern-selection, state-dependence and modulatory capabilities well known in gene regulation.[66,67] Gene-regulatory networks have been extensively modeled, both with digital[68] and analog[69] formalisms (as described below). Developmental pathways and cell-fate decision making are among the major applications for gene-network models (e.g., Ref. 70). Thus, the technical aspects of making such multiscale models seem well within reach. As often the case, precise quantitative data are limiting.

A further level of possible regulation of protein synthesis may reside in the transport of proteins and mRNA to remote dendritic and axonal sites. There are increasingly precise observations about the decisions that govern whether transport complexes choose one branch or another, or decide to unload. These could form the substrate for some very interesting, multiscale cellular decisions that control plasticity.

3.7. Tissue-level effects

Every neuron is embedded in a riotously tangled fabric of axons, dendrites, astrocytes, and blood vessels. All of these are emitters and decoders of signals, and many of these signals affect plasticity. Very few simulations have yet addressed this surrounding tissue environment. An exquisitely detailed series of studies follows individual neurotransmitter molecules as they diffuse from the pre- to the postsynaptic receptors.[71–74] This has important implications for the distribution of neurotransmitters at the target receptors, at perisynaptic receptors, and, indeed, at receptors on other dendrites and neurons. The challenge is to scale such models up to incorporate electrical and chemical signaling throughout the tissue. Numerous experimental observations point to the importance of short[75] and long-range[76] gradients in directing the outgrowth of spines, axons, and dendrites. Similarly, there are numerous experimental suggestions of extensive bidirectional cross talk across the synapse and dendrites. Messengers such as NO and arachidonic acid do not respect cell-membrane boundaries and seem likely to play a role in such bidirectional signaling.[77,78] At a larger, tissue-level scale, diffusible signals, including NO, play a key role in regulating blood flow. Apart from the obvious implications in relating neuronal activity to intrinsic signal and fMRI-based imaging techniques, this kind of signaling plays a core part in metabolic regulation whose implications for plasticity are yet to be explored.

In brief, relating tissue-level signaling to cellular and subcellular events is by definition a multiscale problem. There are diverse questions in this area, and these have barely been touched.

3.8. Multiscale questions at the edge

If there is one common theme from the above exploration of plasticity-related questions, it is that these studies push the limits of what is possible. In every domain, there are glaringly obvious questions, yet we have only begun to nibble around the edges of these possibilities. As in almost all biological modeling, it is easy to point to lack of data as a major constraint. However, this does not account for all of the difficulty in going from point to spatial models, or in tying together well-known models from electrical and chemical domains. Instead, I suggest that there are critical technical difficulties that hold us back: numerical and computational frameworks, and the ability to exchange and incorporate each other's models. Sections 4 and 5 examine the technical limits of where we are and ask how we can cross these barriers.

4. NUMERICAL FRAMEWORKS FOR MULTISCALE MODELING IN PLASTICITY

Just as biology demands different formalisms to account for multiscale processes, diverse numerical techniques are required to model them. There is an interesting debate about whether one should do this in the context of a monolithic, jack-of-all-trades simulator, or whether one should have specialized, separate packages for each technique, and somehow glue together the whole. Here, I go through each problem domain in relative isolation. The thorny problem of performing calculations in a truly multiscale manner, with tightly coupled numerical engines, is an evolving topic. At the end of this section, I consider what has been done, and some of what is planned for multiscale model integration, based on recent work in the field.

4.1. Chemical signaling calculations: Deterministic

Synaptic plasticity suffers from an embarrassment of riches when it comes to biochemical signaling. There are over 1400 known proteins in the postsynaptic density alone,[79] and presumably similar numbers of molecular species regulate events presynaptically and in the dendrite. A profusion of proteins does not in itself make a system multiscale, but the ways they interact do. A significant fraction of Table 14.1 deals with different scales of time and space within chemical signaling.

Chemical kinetic equations are readily solved using simple Ordinary Differential Equation (ODE) solution methods. They have some higher-order terms and nonlinearities, but not much by way of numerical difficulty. The computational cost of ODE calculations increases linearly with model complexity (number of species or reactions) but does not depend on the volume. The classic explicit Runge–Kutta method and its variable timestep derivatives are generally effective at solving such systems. In a few pathways, especially those involving rapid calcium binding, some reactions may take place at a much shorter timescale than the rest. In these *stiff* systems, it may be preferable to use an implicit numerical method such as backward Euler.

Even today, the majority of chemical signaling calculations are done using an ODE formulation. Many models incorporate a small number of discrete spatial compartments to represent different cellular organelles and subdivisions. This is a simple and computationally very effective way to approximate the reaction–diffusion calculations discussed below.

4.2. Chemical signaling calculations: Stochastic

ODE models are efficient and easy to model, but unfortunately not well suited to synaptic plasticity calculations. A simple calculation demonstrates that a half-micron diameter dendritic spine will have about five or six free calcium ions in the resting state. Most other important signaling molecules are also present in the tens to low hundreds at the synapse. This is clearly a domain for stochastic calculations of signaling. In contrast, the dendrite and soma, and certainly the bulk extracellular domain, occupy much larger volumes. In principle, it is correct to apply stochastic signaling numerical methods throughout a model: deterministic calculations are only an approximation, though a very good one at large volumes. This is impractical

Table 14.1 Functional levels and physical scales for stages in synaptic plasticity

Stage	Electrical	Chemical	Mechanical	Genetic	Length scale			Timescale		
					≪1 u	1 u	≫1 u	≪1 s	1 s	≫1 s
Presynaptic	X	X	X			X		X	X	
Transmitter diffusion		X			X	X		X	X	
Postsynaptic	X	X	X			X		X	X	X
Dendrite	X	X	X			X	X	X	X	X
Soma	X	X		X		X		X	X	X
Protein synthesis/transport		X	X			X	X			X
Morphological change		X	X		X	X	X		X	X
Tissue		X	X			X	X		X	X

Crosses indicate where the specified scale applies to each stage of plasticity.

because, as detailed below, stochastic calculations get slower with larger volumes and molecule counts. Given the expensive scaling of simulation time of stochastic methods with numbers of molecules, this seems to be a good place to apply multiscale methods. In other words, combine stochastic signaling methods at the synapse with deterministic methods in the rest of the model. Surprisingly, only a few studies have gone the multiscale numerical route. Instead, various ingenious algorithmic approaches to stochastic calculations have been coupled with brute-force computer power so that a single method will suffice. This approach has been used to develop models that span the sub-femtoliter scales around the synapse, to the picoliter volumes of the dendrite and surrounds.

The most commonly used family of stochastic numerical methods for chemical systems take their origin from a classic paper by Gillespie.[80] Gillespie devised an algorithm (the Gillespie Stochastic Systems Algorithm, or GSSA) for finding (stochastic) trajectories through the possible evolution of states of the system such that these trajectories followed the distribution required by the chemical master equation. All the GSSA-derived methods get slower with greater numbers of distinct chemical species (complexity of the reaction system) as well as with larger volumes (number of molecules of each species). "Stiff" reaction systems, with very rapid reactions, also slow down the Gillespie method. This happens because the waiting time between transitions for these few reactions becomes small, so most of the simulation time is spent updating just these reactions. The classic Gillespie method scales linearly with the number of reactions, and also with the volume (number of reactant molecules). Gibson and Bruck[81] introduced a number of optimizations to this that gives a logarithmic scaling with the number of reactions. Still more recent optimizations are independent of the number of reactions.[82] All the above methods are "exact" methods: they generate exact trajectories sampled from the distribution predicted from the chemical master equation. One can go much faster with inexact methods. Gillespie and colleagues have adapted his original method to scale well with larger numbers of molecules and volumes, through the use of approximations in computing the number of molecules making reaction (or diffusional) transitions.[83] This is called the tau-leap family of methods. This is nearly independent of reaction volume.

An important point is that if a chemical system is stochastic, a single Gillespie-derived run is just a single trajectory through the possible evolution of states of the system. To properly characterize the system, one needs to run the calculations enough times to generate a proper distribution.

This considerably adds to the numerical cost of running stochastic calculations.

A potentially scalable numerical approach is to combine stochastic and deterministic calculations. One class of methods partitions the reaction system between stochastic and deterministic portions and computes each sub-division accordingly.[84] This can be done at the start, in which case it runs the risk of classifications that should change as the reaction proceeds. Dynamic partitioning methods adaptively decide which reactions should be computed stochastically or deterministically.[85] Yet another plausible approach is to subdivide the system into different spatial zones or compartments, having different numerical treatments. As indicated above, one would typically compute reactions stochastically in the dendritic spine, and deterministically in the soma and dendrite.

4.3. Reaction–diffusion computations

Where does space come into the picture? The above numerical methods have been framed in terms of coupled well-stirred spines or dendritic compartments, and possibly a few coupled compartments. Two main approaches are used to incorporate space. In the first, the volume of interest is explicitly subdivided into voxels, each of which carries the whole complement of chemical reactions, and each of which permits diffusion of molecules to its neighbors. In the second, the algorithm is designed around individual molecular interactions and movements, which are constrained by compartmental boundaries rather than voxels. Each molecule can diffuse and react throughout this permitted volume.

4.3.1 Reaction–diffusion methods using spatial discretization

The general approach of this first category of methods is to partition the reaction–diffusion geometry of interest into spatial volume elements (voxels). In principle, these could be planar (on the membrane) or even linear (along a dendrite or axon). For the subdivision to be numerically valid, these voxels have to be smaller than the smallest geometrical features of interest, and also smaller than the diffusion length of any of the reactants in the timescales of interest. The numerical properties of such systems lead to rather unfortunate scaling effects if one is interested in fine geometrical features (like dendritic spines), since the diffusion timescales as the square of the length scale. Thus, the required timestep for accuracy (let alone stability) becomes a hundred times smaller if one goes from a 1-µm to a 0.1-µm voxel. In a three-dimensional system, the computations take *one hundred thousand* times longer, since the cubic mesh now has a thousand times as many voxels. Implicit methods are less sensitive to time-step instabilities, but scale just as poorly with mesh size.

With this caveat in mind, the typical approach to handling reaction–diffusion calculations is to replicate the chemical reaction system in each voxel and to interleave calculations for chemistry with calculations for diffusion. There is an enormous literature on numerical methods for solution of diffusive partial differential equations, and reaction–diffusion systems, which is out of the scope of this review.[86]

Modulo the atrocious scaling problem, the voxelization approach also works well for stochastic models. The Gillespie algorithm lends itself to spatial discretization to handle reaction–diffusion systems.[87–89] In fact, it scales better than deterministic methods because the GSSA runs faster for smaller molecule counts. As discussed below, variants on this have also been used to fold in electrical signaling.

4.3.2 Single-particle reaction–diffusion methods in continuous space

A completely different numerical approach is to do calculations at the level of single molecules and avoid spatial discretization altogether. This is, of course, an approach which only works for stochastic calculations. Here, each molecule moves around and reacts in a continuous space with cellular or organelle boundaries. MCell and Smoldyn are two simulators that use quite different reaction mathematical treatments to achieve this. MCell computes the likelihood of molecular collisions following random-walk jumps and then computes reaction probabilities from mass-action kinetic rates. Smoldyn instead computes effective collision cross-sections for reactants, in which the cross-section area scales with reaction rate. In both cases, the diffusing molecules traverse continuous space. The simulators use a range of geometrical constructs including finite-element surface meshes to find out when the molecular path intersects with a compartmental boundary. Despite these fundamental differences about chemical assumptions, both simulators converge very closely to analytic distributions for test reaction–diffusion situations and give very similar computational results. Both simulators scale almost linearly in the number of reactant molecules, with a penalty for complicated surface geometries.

One nonintuitive outcome of the computational scaling properties of these methods is that when voxel sizes become very small, the continuous space methods win over voxel methods even if the voxel methods are deterministic. This may seem surprising: why is it more efficient to track individual molecules? The reason is the previously mentioned steep scaling of computation time with spatial resolution. This makes the diffusion calculations, whether stochastic or deterministic, extremely slow for small length-scales. In effect, if one wants to model reaction–diffusion on submicron length scales, it may well be more efficient to do this using a single-particle Monte Carlo method.

4.4. Genetic and transcription networks

Gene-regulatory networks have been studied since the early days of the lac operon and bacteriophage lambda. Arguably, many of the core concepts date back to Mendelian genetics: the preponderance of switch-like actions, suppression and expression of phenotypes, and interactions between genes. These are reflected in the current mathematical and computational approaches to their analysis.

The simplest set of approaches is to take switch-like behavior to an extreme and treat the network like a collection of Boolean logic gates. There are variants on this: synchronous versus asynchronous updates, incorporation of time delays, or addition of noise and jitter.[68] These elaborations have important implications for network behavior.

A somewhat more biologically motivated approach is to go analog and describe the activation of each gene through a power-law transform.[69] The sum of input activations is transformed through this nonlinear function into an output state, in a manner reminiscent of certain classes of neural networks.

Overall, it is remarkable that these very abstracted views of an immensely complex chemical and physical process do such a good job in encapsulating the behavior of gene-regulatory networks. From the viewpoint of multiscale modeling, the inputs and outputs of such networks can rather easily be mapped to protein concentrations, thus tying them to chemical signaling formalisms.

4.5. Single-neuron compartmental models

The most familiar form of multiscale model involving synaptic plasticity is where physiologically and morphologically detailed single-neuron models are connected up into a network. Such models map closely to a wide range of experiments, ranging from cell physiology to network activity recordings, and also to a variety of ways to measure connectivity. Most single-neuron calculations express the compartmental structure of the neuron in an ordered matrix form such that it can be solved in a single pass through Gaussian elimination and back substitution. This was independently devised by Hines.[90]

Hines devised a staggered time-step approach to couple this to the nonlinear evaluation of the channel state equations. These matrix calculations scale essentially linearly for large numbers of neurons. Parallel computer implementations typically keep the matrices intact and decompose neurons between nodes, but there are also methods for subdividing individual neurons between processors.[91] Synaptic transmission has both sparse and computationally demanding aspects. It is sparse because even a very active neuron only emits a spike every 10 ms or so, and most systems have an average firing rate in the order of 10 Hz. This is a much longer time frame than the time steps needed for neuronal biophysics. On the other, the dispersion of synaptic events to very large numbers of synapses on neurons over the network can be expensive,[92] and if each synapse performs a complex weighting calculation on each event, this can multiply. Efficient synapse implementations are therefore important, especially for plasticity models.

4.6. Network models

The technical requirements for multiscale network modeling are first to scale the single-neuron calculations many-fold and second to handle transmission of action potentials to synapses.

For multiscale modeling, the key requirement of these methods is mostly their efficiency and scalability to parallel computer architectures. The NEURON[93] and NEST[94] simulators are highly advanced in this regard, and several groups are working on GPU implementations to further increase speed.[95] On the modeling front, large multiscale network models have been around for a long time.[96] Recent incarnations of such models have many thousands of neurons.[97,98] In most such models, a relatively high-level synaptic learning rule is applied to the synapses. The multiscale aspect of these models therefore resides not in their synapses, but in the way that network dynamics emerge from and contribute to single-cell physiology.

A closely related derivative of multiscale network models is where extracellular fields are computed from the detailed ion current distribution in the networked cells in the model. One phenomenon of interest is the interaction of these fields with various kinds of recording electrodes.[99] Here, the field computation lends another dimension to the multiscale phenomenology. The technical requirements for extracellular field calculations are rather straightforward extensions of the capability to devise multiscale network models. The field computation requires geometry and membrane current information for each membrane compartment of each neuron in the network. It is worth noting that these calculations could not be done using integrate-and-fire or other reduced detail network models which lack the essential spatial and physiological detail.

A related problem arises in simulating neuromodulators and how they modulate plasticity. Neuromodulators may be released at multiple points in the network by different neurons, so the influence of the modulator on any given synapse depends on the spatial configuration of release and receptor sites. The general approach to solving this problem is to carry out reaction–diffusion calculations in a geometrically detailed network. However, an efficient abstraction of this calculation has been implemented in the simulator NEST that is considerably less computationally expensive.[100]

4.7. Mechanical and morphological modeling

In contrast to some of the other rather tightly defined technical domains discussed above, mechanical modeling of cells and tissues is wildly diverse.[101,102] While it is not quite the case that every model comes with its own numerical approaches, it is quite difficult to imagine a common computational infrastructure that would fit them all. Very few studies have explicitly modeled forces in the context of synaptic plasticity,[103,104] so we do not have a clear idea of the terrain of capabilities that will be needed. At one extreme, there are elaborate finite-element computations that derive their technology from structural engineering. This would probably suffice for many of the currently envisioned plasticity models involving neurite outgrowth or growth cone movement. However, it seems likely that it would be far more computationally tractable to build problem-specific calculations. This raises interesting problems for integrating with other numerical approaches in multiscale models.

4.8. Bridging numerical engines for multiscale calculations

In the sections above, we have discussed modeling techniques for chemical reaction–diffusion systems, neuronal electrical and network computations, gene networks, and mechanical models. Many of the interesting questions in plasticity draw upon all of these. How does one numerically tie these together?

Based on current research in the area of synaptic plasticity, the big two computational methods are chemical and electrical. Gene-regulatory networks rather easily link to chemical calculations, and mechanical/

morphological changes look to be quite hard to integrate with anything. We now discuss the approaches to linking all these methods together.

4.8.1 Chemical–electrical computations

The multiscale models with the most direct relevance to synaptic plasticity seek to analyze combined electrical and chemical signaling. This is a classic multiscale problem. The physical processes are different and involve different kinds of equations and numerical methods. The length- and timescales are different. Nevertheless, the two kinds of signaling are so closely intertwined in synaptic plasticity that they really must be modeled together. The most direct points of interface are:

a. (electrical model) influx of calcium ions leading to (chemical model) downstream signaling events;

b. (chemical model) diffusion, buffers, pumps, and compartmental uptake and release of calcium ions leading to (electrical model) gating effects on calcium-dependent ion channels;

c. (chemical model) presynaptic dynamics of calcium and vesicles leading to modulation of neurotransmitter release and thus (electrical model) synaptic weight;

d. (chemical model) phosphorylation and other modifications of channel proteins leading to (electrical model) conductance changes;

e. (chemical model) transport and trafficking of receptors to and from the synapse, leading to (electrical model) conductance changes;

f. (electrical model) synaptic activity leading to release of neurotransmitters which (chemical model) activate metabotropic receptors.

Indeed, items (b) and (c) are so ubiquitous that most electrical compartmental modeling programs have built in capabilities for simple versions of calcium dynamics and synaptic facilitation (Fig. 14.1).

The large majority of current models with pretensions to multiscaling still use the table-lookup approach: make a table of calcium concentrations from an electrical model and feed it into a chemical model. True multiscaling has the electrical and chemical models running simultaneously and updating each other during the course of the simulation. As simulators have inched toward this capability, a variety of methods have been deployed to couple distinct solution engines. In one approach, the electrical and chemical models ran on separate simulation threads and communicated with each other through the filesystem.[31] Another variant used the language Python as a common scripting interface between two simulators, each running a portion of the multiscale model in its own domain.[105] The end-point of this

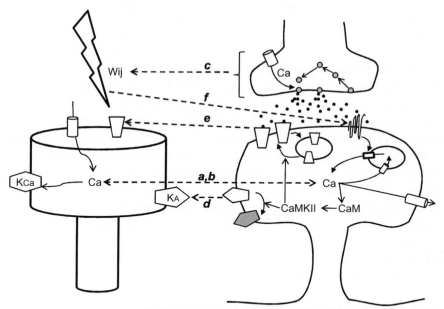

Figure 14.1 Coupling between electrical and chemical signaling in a multiscale model of the dendritic spine. Left: Electrical model cartoon, right: chemical model cartoon. Jagged arrow on the electrical model represents synaptic input with a weight Wij. The dashed arrows between components of the models indicate aspects of coupling. Bidirectional arrow *a,b* between the calcium icons represents mutual effect of models on each other's calcium levels. Arrow *c* represents presynaptic events influencing synaptic weight Wij in electrical model. Many chemical events are represented here by CaM and CaMKII signaling. The diagram shows CaMKII-mediated phosphorylation of channel proteins to alter their kinetics or conductance (arrow *d*) or altering receptor insertion (arrow *e*). Arrow *f* represents activation of metabotropic receptors by synaptic activity and long-range neurotransmitters.

technical evolution, which many simulators are approaching, is to have a single simulator with systematic coupling mechanisms between the numerical engines.

As an example of a tightly coupled approach, the STEPS simulator explicitly models channel state transitions as well as chemical events through a kinetic formalism.[89] Ion fluxes are thus modeled as an explicit movement of charged particles, which can undergo subsequent chemical interactions. These events are spatially mapped onto a finite-element grid, thus supporting arbitrary geometries and reaction–diffusion calculations. This approach has the merit that the numerical methods are completely consistent and uniform. The domain of applicability of this approach is for highly detailed spatial models of subcellular regions, such as a segment of dendrite with a few spines.

The simulator MOOSE uses a more loosely coupled approach in order to run larger-scale cell and tissue-level models.[23] Chemical calculations can use any of a range of numerical engines (deterministic, stochastic, and reaction–diffusion with branching or cubic meshes) and likewise the electrical calculations. This permits mix-and-match between high detail and high efficiency calculations even in subparts of the same cell. All numerical engines map transparently to biologically defined simulation entities such as molecules and ion channels. The coupling between the integration methods is accomplished through adaptor objects which transform quantities back and forth between these simulation entities. For example, the calcium current is converted into a flux of the chemical species for calcium ions. The two numerical engines have to update their respective cross-scale mappings at periodic checkpoints, typically between 1 and 5 ms.

The TimeScales framework[106] also uses loose coupling between numerical engines, with an event-driven framework for efficient synchronization between the chemical and electrical formalisms.

4.8.2 Chemical and gene-network computations

The interface to gene network models, as indicated above, is a rather trivial extension of chemical models. Since transcription factors and genes themselves are also chemical entities, the gene-network formalism maps transparently to the chemical rate equation form. Some existing simulators already handle such combination models.[107] Based on this body of experience with such combined models, it seems likely that the addition of rather simple formalisms to reaction systems should suffice to extend their domain to gene networks.

4.8.3 Interfacing to mechanical computations

The signaling inputs to mechanical force calculations are mostly chemical in nature. There are many examples of such models, as discussed above.[101,102] The reverse transform, from mechanical changes to chemical and electrical consequences, is likely to be more difficult. There is every reason to expect that these multiscale calculations will need substantial reworking of existing methods. The fundamental problem is that the partial differential equations in both reaction–diffusion and cable-equation calculations require spatial discretization, and any nontrivial morphological changes will alter these numerical subdivisions. Current simulator designs do the spatial discretization at setup time, as this tends to be quite expensive. The brute-force approach to incorporating morphological change is to rebuild the spatial mesh every so often, as the simulation progresses. This is computationally costly. The saving grace here is that morphological changes at the spine are likely to be rather slow compared to the time-course of chemical and electrical signaling. This should allow long intervals between the checkpoints at which the neuron model is rebuilt. In the longer term, there are some possible approaches to be adopted from other fields. For example, variable mesh methods[108] have been used in cardiac models.

A less numerically stressful approach is to keep the whole process *ad hoc*. This is very much in line with the rather opportunistic nature of the algorithms for mechanical change in many models. In practice, this would entail defining certain scaling laws for chemical and electrical properties as a function of morphological changes. One could envision doing this through a set of objects that carry out arbitrary mathematical transforms and map into the conventional electrical/chemical solution engines. At this point, the feasibility of such mappings is mostly a matter of informed guesswork, since there have not been enough use cases to generalize.

4.9. Simulators and hardware

The general simulator has been essential for the development of computational neuroscience, and more recently for systems biology. The essence of such tools is that they take model specifications, for example, in an XML-based form and are able to instantiate and efficiently run a very wide range of models. This contrasts with special purpose simulators (or the ubiquitous C, Matlab or Python scripts) which run just one specific model. Examples of the latter class include cardiac simulators like the University of Tokyo heart simulator.[109]

At the time of writing, I am aware of a handful of simulators that have some multiscale chemical–electrical modeling capability. Two simulators have been built from the ground up with multiscaling (MOOSE[105] and STEPS[89]), and two are retrofitting (NEURON[93] and VCELL[110]). A recent publication refers to the TimeScales framework for multiscale modeling.[106] GENESIS[111] is a legacy simulator that has been used to run multiscale simulations[31] but is no longer being actively developed. It is extremely likely that there are other simulation tools in this domain. Rather than enumerate individual tool features, I will here discuss the general capabilities and indicate directions.

The widely used NEURON simulator already contains the capability to implement arbitrary kinetic calculations through the compiled "hoc" language. To make chemical calculations more accessible, a NumPy component for carrying out deterministic reaction–diffusion calculations is under development. This transparently builds the branching diffusion equations from the electrical model specification. The electrical computations remain in compiled C code, and it would appear that some level of "hoc" interfacing will be required to map entities between the numerical schemes.

In a complementary development, the team for the Virtual Cell simulator (which is one of the established simulators capable of running spatial chemical models) has been developing a compartmental neuron simulation capability.[112]

All the simulators support deterministic reaction modeling capabilities, and several also support stochastic and reaction–diffusion calculations. There is also general support for electrical modeling, though the numerical approaches and size scales of STEPS are distinctive and much finer-grained than the others. Most of the simulators have parallel versions, or are developing them, and most are working on developing GPU capabilities.

There are several clear directions that these simulators are likely to evolve toward. To the best of my knowledge, none at this point explicitly support structural changes coupled to the chemistry, though there have been proposals for some creative hacks to confer the semblance of such features. None of them at this point talk to the major single-particle reaction–diffusion simulation engines MCELL[74] and SMOLDYN,[113] though again there are moves to achieve this. Perhaps most importantly, as mentioned above, these simulators have yet to converge to a standard XML-based format for reading and writing multiscale models, not least because no such standard format exists.

To summarize, the tools for multiscale modeling are evolving rapidly, much like the models themselves. It would hugely benefit the field to have some standards in place for model specification.

5. MULTISCALE MODEL SPECIFICATION

Which comes first: the model format or the simulator to run the model? Historically, most neuronal simulators have just made it up as they went along. The outcome is that there are several divergent and completely incompatible model definition formats, a regrettable situation that groups have been trying to resolve for many years.[114] There is now general agreement among almost all simulator developers that simulator-independent standards are valuable. This section discusses the shape that a multiscale model standard might take.

5.1. Model formats

"The most general simulator," a colleague of mine once said, "is C." Arguably, the same is true for model specifications. The point was that if you wanted to be able to specify any arbitrary model, in a form that could run on any computer, you could do it in C. There has, nevertheless, been a steady march of simulator design, from hand-crafting code for each model, through scripting languages that drive simulators which do much of the tedious work, toward the goal of simulator-independent specifications.

Model specification formats should be mathematically unambiguous, portable, concise, and readable. It has been cogently argued that declarative specifications, such as XML-based languages, are preferable to procedural ones such as script-language specifications.[115] This applies even to high-level languages such as Python, which is rapidly becoming the common language of many simulators. The reason, in brief, is that a declarative specification tells the simulator *what* is in the model, rather than a script, which tells it *how* to set up the model. C is even worse because it also tells the system exactly how to do every calculation. Each simulator has its own semantics, and if you want portability, you cannot rely on simulator-specific commands. There are high-level model scripting extensions (PyNN[116] comes to mind), which define a set of model setup functions and lay down what they are supposed to do within the framework of a common scripting language such as Python. This does achieve a significant level of portability, though the process of defining the model becomes interleaved with the process of running it. Standalone, simulator-independent model specification files encapsulate the model definition and separate it from other operations. XML is the base language for most such specification standards, in large part because XML comes with an array of tools that facilitate specification of the language syntax, conversion of specifications, and checks on internal consistency. It should be noted, however, that XML fares rather poorly on readability and conciseness.

Enormous strides have been made on domain-specific model specification. Biochemical models, in particular, benefit from a very widely accepted standard, SBML[117] (Systems Biology Markup Language), which has spawned a lively ecosystem of tools. SBML specializes in chemical rate equation models. However, the current set of spatial extensions in SBML does not readily map to multiscale models embedded in neurons.

Neuronal models have two somewhat newer standards. NeuroML[114] is widely used for single-neuron and network models, as well as by numerous tools including simulators, rendering software, and model setup software. The NineML specification is a work in progress and is currently designed for network models. It has a special place for plasticity rules, but typically of a higher order of abstraction. Of these neuronal standards, NeuroML is most relevant for this discussion because of its emphasis on geometrically detailed models. There are already a number of geometrically detailed multiscale simulations specified in NeuroML, but none of these emphasizes plasticity.

Yet another layer of specification resides in the determination of input, runtime setup, and output for the model. At least some of this has now been formalized through the Simulation Experiment Description Markup Language (SED-ML)[118] definition. One way of thinking of the specification is to ask what it would take to define all the steps to take a model and generate a specific figure in a paper. This, in principle, is completely independent of any of the other aspects of model specification and is useful for any model specification, provided the variables and parameters can be unambiguously accessed.

At this point, there are no standards for structural/mechanical biological model specification that I am aware of. NeuroML specifies neuronal geometry, but not its evolution in time.

In summary, all of these formats specify necessary components needed to define a full multiscale model, but they are not sufficient.

5.2. Bridging formats

The logical next level of multiscale model specification would be to take electrical and chemical models defined in these respective formats, and combine them. It is clear that setting up such a multiscale model is considerably more involved than the sum of the parts. To first order, the bridge between electrical and chemical signaling models requires the definition of the mapping of entities between them. This is usually mathematically simple. For example, the electrically computed net flux of calcium current into a compartment is readily converted to concentration using a scale factor with an offset to account for basal calcium. However, this mapping typically takes place over compartments with different spatial discretizations, and different mappings may apply in different locations. So, even this first-order operation requires significant extra specification.

A more subtle issue concerns the specification of numerical properties of the simulation system. It is a matter of debate whether this is the realm of the model definition or of the simulation setup. For example, should reactions be computed deterministically or stochastically, or in a mixed manner depending on volume? How should the spatial and temporal discretization of the models be carried out? How often should chemical and electrical models update each other? In an ideal world, these decisions would be derived by the numerical engines themselves based on a simple specification of numerical precision, but this is not currently possible. Furthermore, if you want to replicate a specific model or figure in a paper, you quite likely need to know something about the numerical choices made by the original author. It would not do to try to replicate a stochastic finding with a deterministic calculation. SED-ML is not a good format for this glue, either, since many numerical details are very specific to this problem domain.

It may be argued that these model and simulation specification issues are moot because there is no extant simulator that can read all these formats. Nevertheless, such a format, even if it is a work in progress and designed to support capabilities that do not yet exist, has enormous value. Even a mature format—SBML—which is very widely used, is not handled in its entirety by any simulator.[117] This does not lessen its impact as a medium for model exchange. Different tools have sprung up which address different aspects of models defined in SBML.

As discussed in the Section 4.9, developers are only just beginning to endow simulators with the multiscale capabilities that could even in principle simulate such models. The presence of a model definition standard sets targets for simulator capabilities, and even as the standard evolves, it keeps the models from becoming unusable as simulators change.

Overall, it seems clear that a standard for multiscale model specification would be highly desirable. It is far from clear at this point whether this would be a wrapper standard that could refer to distinct electrical (e.g., NeuroML) and chemical (e.g., SBML) model components, and then stitch them together, or an extension to an existing standard, or whether it would be something monolithic.

5.3. Model verification and validation

How does one know that a multiscale model is correct? Even the Hodgkin–Huxley equations defeat analytic solutions. The traditional approach, used, for example, by the Rallpacks,[119] is to pit two simulators with quite independent code-bases against each other, and confirm that their solutions converge. Among the suggestions at a recent meeting on multiscale modeling was the definition of a number of use cases to build upon for reference models. For multiscale models, one would like to establish a substantial library of such use cases and their reference solutions, so as to explore the wide range of scales, physiological processes, and numerical approaches that we have discussed. This depends hugely on the ecosystem of tools that we have discussed: portable model standards, simulators that can read the model specifications, and not least, the availability of experimental data to reassure us that our models track reality.

6. CONCLUSION

Models should bring clarity, not confusion. Multiscale plasticity models are at the limit of so many of our current capabilities—data, software, tools, and hardware—that the science they illuminate is at enormous risk of being hidden by their complexity and technical virtuosity. This review seeks to outline an ecosystem of methodology and tools where the science can flourish. The field is at an inflection, where developments in tools and standards will allow researchers to spend more of their time on developing the models rather than on the software. Fascinating and fundamental scientific questions are at stake, and equally critical implications for analysis and medicine. We do not know if there will be new one-line learning rules to account for long-range synapse formation, nor if understanding the activity-driven orchestration of molecular gradients in the brain will help us better treat stroke patients. We can be confident, however, that multiscale synaptic modeling will be needed as the theoretical counterpart of experiments to answer these questions.

REFERENCES

1. Abbott LF. Theoretical neuroscience rising. *Neuron*. 2008;60(3):489–495. http://dx.doi.org/10.1016/j.neuron.2008.10.019.
2. Sabatini BL, Oertner TG, Svoboda K. The life cycle of Ca(2+) ions in dendritic spines. *Neuron*. 2002;33(3):439–452.
3. Woolf TB, Greer CA. Local communication within dendritic spines: models of second messenger diffusion in granule cell spines of the mammalian olfactory bulb. *Synapse*. 1994;17(4):247–267. http://dx.doi.org/10.1002/syn.890170406.
4. Branco T, Clark BA, Häusser M. Dendritic discrimination of temporal input sequences in cortical neurons. *Science*. 2010;329(5999):1671–1675. http://dx.doi.org/10.1126/science.1189664.
5. Frey U, Morris RG. Synaptic tagging and long-term potentiation. *Nature*. 1997;385(6616):533–536. http://dx.doi.org/10.1038/385533a0.

6. Bhatt DH, Zhang S, Gan W-B. Dendritic spine dynamics. *Annu Rev Physiol.* 2009;71:261–282. http://dx.doi.org/10.1146/annurev.physiol.010908.163140.
7. Hebb DO. *The Organization of Behavior.* New York: Wiley; 1949.
8. Zador A, Koch C, Brown TH. Biophysical model of a Hebbian synapse. *Proc Natl Acad Sci USA.* 1990;87(17):6718–6722.
9. Bliss TV, Collingridge GL. A synaptic model of memory: long-term potentiation in the hippocampus. *Nature.* 1993;361(6407):31–39. http://dx.doi.org/10.1038/361031a0.
10. Caporale N, Dan Y. Spike timing-dependent plasticity: a Hebbian learning rule. *Annu Rev Neurosci.* 2008;31:25–46. http://dx.doi.org/10.1146/annurev.neuro.31.060407. 125639.
11. Dudek SM, Bear MF. Bidirectional long-term modification of synaptic effectiveness in the adult and immature hippocampus. *J Neurosci.* 1993;13(7):2910–2918.
12. Bienenstock EL, Cooper LN, Munro PW. Theory for the development of neuron selectivity: orientation specificity and binocular interaction in visual cortex. *J Neurosci.* 1982;2(1):32–48.
13. Keller DX, Franks KM, Bartol TM, Sejnowski TJ. Calmodulin activation by calcium transients in the postsynaptic density of dendritic spines. *PLoS One.* 2008;3(4):e2045. http://dx.doi.org/10.1371/journal.pone.0002045.
14. Mihalas S. Calcium messenger heterogeneity: a possible signal for spike timing-dependent plasticity. *Front Comput Neurosci.* 2011;4:158. http://dx.doi.org/10.3389/fncom.2010.00158.
15. Naoki H, Sakumura Y, Ishii S. Local signaling with molecular diffusion as a decoder of $Ca2+$ signals in synaptic plasticity. *Mol Syst Biol.* 2005;1:2005.0027. http://dx.doi.org/10.1038/msb4100035.
16. Pepke S, Kinzer-Ursem T, Mihalas S, Kennedy MB. A dynamic model of interactions of $Ca2+$, calmodulin, and catalytic subunits of $Ca2+$/calmodulin-dependent protein kinase II. *PLoS Comput Biol.* 2010;6(2):e1000675. http://dx.doi.org/10.1371/journal.pcbi.1000675.
17. Lisman J. A mechanism for the Hebb and the anti-Hebb processes underlying learning and memory. *Proc Natl Acad Sci USA.* 1989;86(23):9574–9578.
18. Bush D, Jin Y. Calcium control of triphasic hippocampal STDP. *J Comput Neurosci.* 2012;33(3):495–514. http://dx.doi.org/10.1007/s10827-012-0397-5.
19. Bhalla US, Iyengar R. Emergent properties of networks of biological signaling pathways. *Science.* 1999;283(5400):381–387.
20. D'Alcantara P, Schiffmann SN, Swillens S. Bidirectional synaptic plasticity as a consequence of interdependent $Ca2+$−controlled phosphorylation and dephosphorylation pathways. *Eur J Neurosci.* 2003;17(12):2521–2528.
21. Hayer A, Bhalla US. Molecular switches at the synapse emerge from receptor and kinase traffic. *PLoS Comput Biol.* 2005;1(2):137–154.
22. Stefan MI, Edelstein SJ, Le Novère N. An allosteric model of calmodulin explains differential activation of PP2B and CaMKII. *Proc Natl Acad Sci USA.* 2008;105(31):10768–10773. http://dx.doi.org/10.1073/pnas.0804672105.
23. Bhalla US. Multiscale interactions between chemical and electric signaling in LTP induction, LTP reversal and dendritic excitability. *Neural Netw.* 2011;24(9):943–949. http://dx.doi.org/10.1016/j.neunet.2011.05.001.
24. Bhalla US. Biochemical signaling networks decode temporal patterns of synaptic input. *J Comput Neurosci.* 2002;13(1):49–62.
25. Bhalla US. Mechanisms for temporal tuning and filtering by postsynaptic signaling pathways. *Biophys J.* 2002;83(2):740–752. http://dx.doi.org/10.1016/S0006-3495(02)75205-3.
26. Markram H, Lübke J, Frotscher M, Sakmann B. Regulation of synaptic efficacy by coincidence of postsynaptic APs and EPSPs. *Science.* 1997;275(5297):213–215.

27. Bi GQ, Poo MM. Synaptic modifications in cultured hippocampal neurons: dependence on spike timing, synaptic strength, and postsynaptic cell type. *J Neurosci.* 1998;18(24):10464–10472.

28. Hunzinger JF, Chan VH, Froemke RC. Learning complex temporal patterns with resource-dependent spike timing-dependent plasticity. *J Neurophysiol.* 2012;108(2):551–566. http://dx.doi.org/10.1152/jn.01150.2011.

29. Urakubo H, Honda M, Froemke RC, Kuroda S. Requirement of an allosteric kinetics of NMDA receptors for spike timing-dependent plasticity. *J Neurosci.* 2008;28(13):3310–3323. http://dx.doi.org/10.1523/JNEUROSCI.0303-08.2008.

30. Harvey CD, Yasuda R, Zhong H, Svoboda K. The spread of Ras activity triggered by activation of a single dendritic spine. *Science.* 2008;321(5885):136–140. http://dx.doi.org/10.1126/science.1159675.

31. Ajay SM, Bhalla US. A propagating ERKII switch forms zones of elevated dendritic activation correlated with plasticity. *HFSP J.* 2007;1(1):49–66. http://dx.doi.org/10.2976/1.2721383.

32. Earnshaw BA, Bressloff PC. A diffusion-activation model of CaMKII translocation waves in dendrites. *J Comput Neurosci.* 2010;28(1):77–89. http://dx.doi.org/10.1007/s10827-009-0188-9.

33. Aihara T, Kobayashi Y, Tsukada M. Spatiotemporal visualization of long-term potentiation and depression in the hippocampal CA1 area. *Hippocampus.* 2005;15(1):68–78. http://dx.doi.org/10.1002/hipo.20031.

34. Frey S, Frey JU. "Synaptic tagging" and "cross-tagging" and related associative reinforcement processes of functional plasticity as the cellular basis for memory formation. *Prog Brain Res.* 2008;169:117–143. http://dx.doi.org/10.1016/S0079-6123(07)00007-6.

35. Smolen P, Baxter DA, Byrne JH. Molecular constraints on synaptic tagging and maintenance of long-term potentiation: a predictive model. *PLoS Comput Biol.* 2012;8(8):e1002620. http://dx.doi.org/10.1371/journal.pcbi.1002620.

36. Khan S, Reese TS, Rajpoot N, Shabbir A. Spatiotemporal maps of CaMKII in dendritic spines. *J Comput Neurosci.* 2012;33(1):123–139. http://dx.doi.org/10.1007/s10827-011-0377-1.

37. Khan S, Zou Y, Amjad A, et al. Sequestration of CaMKII in dendritic spines in silico. *J Comput Neurosci.* 2011;31(3):581–594. http://dx.doi.org/10.1007/s10827-011-0323-2.

38. Kim B, Hawes SL, Gillani F, Wallace LJ, Blackwell KT. Signaling pathways involved in striatal synaptic plasticity are sensitive to temporal pattern and exhibit spatial specificity. *PLoS Comput Biol.* 2013;9(3):e1002953. http://dx.doi.org/10.1371/journal.pcbi.1002953.

39. Ehlers MD. Activity level controls postsynaptic composition and signaling via the ubiquitin-proteasome system. *Nat Neurosci.* 2003;6(3):231–242. http://dx.doi.org/10.1038/nn1013.

40. Bhalla US. Signaling in small subcellular volumes. II. Stochastic and diffusion effects on synaptic network properties. *Biophys J.* 2004;87(2):745–753. http://dx.doi.org/10.1529/biophysj.104.040501.

41. Lisman JE. A mechanism for memory storage insensitive to molecular turnover: a bistable autophosphorylating kinase. *Proc Natl Acad Sci USA.* 1985;82(9):3055–3057.

42. Lisman JE, Zhabotinsky AM. A model of synaptic memory: a CaMKII/PP1 switch that potentiates transmission by organizing an AMPA receptor anchoring assembly. *Neuron.* 2001;31(2):191–201.

43. Kuroda S, Schweighofer N, Kawato M. Exploration of signal transduction pathways in cerebellar long-term depression by kinetic simulation. *J Neurosci.* 2001;21(15):5693–5702.

44. Graupner M, Brunel N. STDP in a bistable synapse model based on CaMKII and associated signaling pathways. *PLoS Comput Biol*. 2007;3(11):e221. http://dx.doi.org/10.1371/journal.pcbi.0030221.

45. Pettigrew DB, Smolen P, Baxter DA, Byrne JH. Dynamic properties of regulatory motifs associated with induction of three temporal domains of memory in aplysia. *J Comput Neurosci*. 2005;18(2):163–181. http://dx.doi.org/10.1007/s10827-005-6557-0.

46. Aslam N, Kubota Y, Wells D, Shouval HZ. Translational switch for long-term maintenance of synaptic plasticity. *Mol Syst Biol*. 2009;5:284. http://dx.doi.org/10.1038/msb.2009.38.

47. Nakano T, Doi T, Yoshimoto J, Doya K. A kinetic model of dopamine- and calcium-dependent striatal synaptic plasticity. *PLoS Comput Biol*. 2010;6(2):e1000670. http://dx.doi.org/10.1371/journal.pcbi.1000670.

48. Smolen P, Baxter DA, Byrne JH. Bistable MAP kinase activity: a plausible mechanism contributing to maintenance of late long-term potentiation. *Am J Physiol Cell Physiol*. 2008;294(2):C503–C515. http://dx.doi.org/10.1152/ajpcell.00447.2007.

49. Bialek W. Stability and noise in biochemical switches. *Adv Neural Inf Process Syst*. 2001;13:103–109.

50. Miller P, Zhabotinsky AM, Lisman JE, Wang X-J. The stability of a stochastic CaMKII switch: dependence on the number of enzyme molecules and protein turnover. *PLoS Biol*. 2005;3(4):e107. http://dx.doi.org/10.1371/journal.pbio.0030107.

51. Brandman O, Ferrell JE, Li R, Meyer T. Interlinked fast and slow positive feedback loops drive reliable cell decisions. *Science*. 2005;310(5747):496–498. http://dx.doi.org/10.1126/science.1113834.

52. Fusi S, Drew PJ, Abbott LF. Cascade models of synaptically stored memories. *Neuron*. 2005;45(4):599–611, S0896-6273(05)00117-0.

53. Smolen P, Baxter DA, Byrne JH. Interlinked dual-time feedback loops can enhance robustness to stochasticity and persistence of memory. *Phys Rev E Stat Nonlin Soft Matter Phys*. 2009;79(3 Pt. 1):031902.

54. Shouval HZ. Clusters of interacting receptors can stabilize synaptic efficacies. *Proc Natl Acad Sci USA*. 2005;102(40):14440–14445. http://dx.doi.org/10.1073/pnas.0506934102.

55. Lorente de No R. Analysis of the activity of the chains of internuncial neurons. *J Neurophysiol*. 1938;1:195–206.

56. Marder E, Goaillard J-M. Variability, compensation and homeostasis in neuron and network function. *Nat Rev Neurosci*. 2006;7(7):563–574. http://dx.doi.org/10.1038/nrn1949.

57. Turrigiano G. Too many cooks? Intrinsic and synaptic homeostatic mechanisms in cortical circuit refinement. *Annu Rev Neurosci*. 2011;34:89–103. http://dx.doi.org/10.1146/annurev-neuro-060909-153238.

58. Liu Z, Golowasch J, Marder E, Abbott LF. A model neuron with activity-dependent conductances regulated by multiple calcium sensors. *J Neurosci*. 1998;18(7):2309–2320.

59. Naudé J, Cessac B, Berry H, Delord B. Effects of cellular homeostatic intrinsic plasticity on dynamical and computational properties of biological recurrent neural networks. *J Neurosci*. 2013;33(38):15032–15043. http://dx.doi.org/10.1523/JNEUROSCI.0870-13.2013.

60. Park M, Salgado JM, Ostroff L, et al. Plasticity-induced growth of dendritic spines by exocytic trafficking from recycling endosomes. *Neuron*. 2006;52(5):817–830. http://dx.doi.org/10.1016/j.neuron.2006.09.040.

61. O'Donnell C, Nolan MF, van Rossum MCW. Dendritic spine dynamics regulate the long-term stability of synaptic plasticity. *J Neurosci*. 2011;31(45):16142–16156. http://dx.doi.org/10.1523/JNEUROSCI.2520-11.2011.

62. Bhalla US. Trafficking motifs as the basis for two-compartment signaling systems to form multiple stable states. *Biophys J.* 2011;101(1):21–32. http://dx.doi.org/10.1016/j.bpj.2011.05.037.

63. Crook SM, Dur-E-Ahmad M, Baer SM. A model of activity-dependent changes in dendritic spine density and spine structure. *Math Biosci Eng.* 2007;4(4):617–631.

64. Sutton MA, Schuman EM. Dendritic protein synthesis, synaptic plasticity, and memory. *Cell.* 2006;127(1):49–58.

65. Jain P, Bhalla US. Signaling logic of activity-triggered dendritic protein synthesis: an mTOR gate but not a feedback switch. *PLoS Comput Biol.* 2009;5(2):e1000287. http://dx.doi.org/10.1371/journal.pcbi.1000287.

66. Tyson JJ, Chen KC, Novak B. Sniffers, buzzers, toggles and blinkers: dynamics of regulatory and signaling pathways in the cell. *Curr Opin Cell Biol.* 2003;15(2):221–231.

67. Alon U. Network motifs: theory and experimental approaches. *Nat Rev Genet.* 2007;8(6):450–461. http://dx.doi.org/10.1038/nrg2102.

68. Naldi A, Berenguier D, Fauré A, Lopez F, Thieffry D, Chaouiya C. Logical modelling of regulatory networks with GINsim 2.3. *Biosystems.* 2009;97(2):134–139.

69. Savageau MA. Design principles for elementary gene circuits: elements, methods, and examples. *Chaos.* 2001;11(1):142–159. http://dx.doi.org/10.1063/1.1349892.

70. Von Dassow G, Meir E, Munro EM, Odell GM. The segment polarity network is a robust developmental module. *Nature.* 2000;406(6792):188–192. http://dx.doi.org/10.1038/35018085.

71. Anglister L, Stiles JR, Salpeter MM. Acetylcholinesterase density and turnover number at frog neuromuscular junctions, with modeling of their role in synaptic function. *Neuron.* 1994;12(4):783–794.

72. Coggan JS, Bartol TM, Esquenazi E, et al. Evidence for ectopic neurotransmission at a neuronal synapse. *Science.* 2005;309(5733):446–451. http://dx.doi.org/10.1126/science.1108239.

73. Nadkarni S, Bartol TM, Sejnowski TJ, Levine H. Modelling vesicular release at hippocampal synapses. *PLoS Comput Biol.* 2010;6(11):e1000983. http://dx.doi.org/10.1371/journal.pcbi.1000983.

74. Stiles JR, Bartol TM. Monte Carlo methods for simulating realistic synaptic microphysiology using MCell. In: De Schutter Erik, ed. *Computational Neuroscience: Realistic Modeling for Experimentalists.* Boca Raton: CRC Press; 2001:87–127.

75. Kwon H-B, Sabatini BL. Glutamate induces de novo growth of functional spines in developing cortex. *Nature.* 2011;474(7349):100–104. http://dx.doi.org/10.1038/nature09986.

76. Liston C, Gan W-B. Glucocorticoids are critical regulators of dendritic spine development and plasticity in vivo. *Proc Natl Acad Sci USA.* 2011;108(38):16074–16079. http://dx.doi.org/10.1073/pnas.1110444108.

77. Feil R, Kleppisch T. NO/cGMP-dependent modulation of synaptic transmission. *Handb Exp Pharmacol.* 2008;184:529–560. http://dx.doi.org/10.1007/978-3-540-74805-2_16.

78. Leu BH, Schmidt JT. Arachidonic acid as a retrograde signal controlling growth and dynamics of retinotectal arbors. *Dev Neurobiol.* 2008;68(1):18–30. http://dx.doi.org/10.1002/dneu.20561.

79. Bayés A, van de Lagemaat LN, Collins MO, et al. Characterization of the proteome, diseases and evolution of the human postsynaptic density. *Nat Neurosci.* 2011;14(1):19–21. http://dx.doi.org/10.1038/nn.2719.

80. Gillespie DT. Exact stochastic simulation of coupled chemical reactions. *J Phys Chem.* 1977;81(25):2340–2361. http://dx.doi.org/10.1021/j100540a008.

81. Gibson MA, Bruck J. Efficient exact stochastic simulation of chemical systems with many species and many channels. *J Phys Chem A.* 2000;104(9):1876–1889. http://dx.doi.org/10.1021/jp993732q.

82. Slepoy A, Thompson AP, Plimpton SJ. A constant-time kinetic Monte Carlo algorithm for simulation of large biochemical reaction networks. *J Chem Phys*. 2008;128(20):205101. http://dx.doi.org/10.1063/1.2919546.

83. Cao Y, Gillespie DT, Petzold LR. Adaptive explicit-implicit tau-leaping method with automatic tau selection. *J Chem Phys*. 2007;126(22):224101. http://dx.doi.org/10.1063/1.2745299.

84. Haseltine EL, Rawlings JB. Approximate simulation of coupled fast and slow reactions for stochastic chemical kinetics. *J Chem Phys*. 2002;117(15):6959. http://dx.doi.org/10.1063/1.1505860.

85. Vasudeva K, Bhalla US. Adaptive stochastic-deterministic chemical kinetic simulations. *Bioinformatics*. 2004;20(1):78–84.

86. Press WH, Teukolsky SA, Vetterling WT, Flannery BP. *Numerical Recipes in C: The Art of Scientific Computing*. Cambridge: Cambridge University Press; 1992.

87. Fange D, Mahmutovic A, Elf J. MesoRD 1.0: stochastic reaction-diffusion simulations in the microscopic limit. *Bioinformatics*. 2012;28(23):3155–3157.

88. Oliveira RF, Terrin A, Di Benedetto G, et al. The role of type 4 phosphodiesterases in generating microdomains of cAMP: large scale stochastic simulations. *PLoS One*. 2010;5(7):e11725. http://dx.doi.org/10.1371/journal.pone.0011725.

89. Wils S, De Schutter E. STEPS: modeling and simulating complex reaction-diffusion systems with Python. *Front Neuroinform*. 2009;3:15. http://dx.doi.org/10.3389/neuro.11.015.2009.

90. Hines M. Efficient computation of branched nerve equations. *Int J Biomed Comput*. 1984;15(1):69–76.

91. Hines ML, Markram H, Schürmann F. Fully implicit parallel simulation of single neurons. *J Comput Neurosci*. 2008;25(3):439–448. http://dx.doi.org/10.1007/s10827-008-0087-5.

92. Lytton WW, Omurtag A, Neymotin SA, Hines ML. Just-in-time connectivity for large spiking networks. *Neural Comput*. 2008;20(11):2745–2756. http://dx.doi.org/10.1162/neco.2008.10-07-622.

93. Carnevale NT, Hines ML. *The NEURON Book*. Cambridge, UK: Cambridge University Press; 2006.

94. Gewaltig M-O, Diesmann M. NEST (NEural Simulation Tool). *Scholarpedia*. 2007;2(4):1430. http://dx.doi.org/10.4249/scholarpedia.1430.

95. De Camargo RY, Rozante L, Song SW. A multi-GPU algorithm for large-scale neuronal networks. *Concurr Comput*. 2011;23(6):556–572. http://dx.doi.org/10.1002/cpe.1665.

96. Wilson M, Bower JM. Cortical oscillations and temporal interactions in a computer simulation of piriform cortex. *J Neurophysiol*. 1992;67(4):981–995.

97. Traub RD, Contreras D, Cunningham MO, et al. Single-column thalamocortical network model exhibiting gamma oscillations, sleep spindles, and epileptogenic bursts. *J Neurophysiol*. 2005;93(4):2194–2232. http://dx.doi.org/10.1152/jn.00983.2004.

98. Reimann MW, Anastassiou CA, Perin R, Hill SL, Markram H, Koch C. A biophysically detailed model of neocortical local field potentials predicts the critical role of active membrane currents. *Neuron*. 2013;79(2):375–390. http://dx.doi.org/10.1016/j.neuron.2013.05.023.

99. Łęski S, Lindén H, Tetzlaff T, Pettersen KH, Einevoll GT. Frequency dependence of signal power and spatial reach of the local field potential. *PLoS Comput Biol*. 2013;9(7):e1003137. http://dx.doi.org/10.1371/journal.pcbi.1003137.

100. Potjans W, Morrison A, Diesmann M. Enabling functional neural circuit simulations with distributed computing of neuromodulated plasticity. *Front Comput Neurosci*. 2010;4:141. http://dx.doi.org/10.3389/fncom.2010.00141.

101. Mogilner A. Mathematics of cell motility: have we got its number? *J Math Biol*. 2009;58(1–2):105–134. http://dx.doi.org/10.1007/s00285-008-0182-2.

102. Holmes WR, Edelstein-Keshet L. A comparison of computational models for eukaryotic cell shape and motility. *PLoS Comput Biol*. 2012;8(12):e1002793. http://dx.doi.org/10.1371/journal.pcbi.1002793.

103. Holcman D, Schuss Z, Korkotian E. Calcium dynamics in dendritic spines and spine motility. *Biophys J*. 2004;87(1):81–91. http://dx.doi.org/10.1529/biophysj.103.035972.

104. Rvachev MM. Neuron as a reward-modulated combinatorial switch and a model of learning behavior. *Neural Netw Off J Int Neural Netw Soc*. 2013;46:62–74. http://dx.doi.org/10.1016/j.neunet.2013.04.010.

105. Ray S, Bhalla US. PyMOOSE: interoperable scripting in Python for MOOSE. *Front Neuroinform*. 2008;2:6. http://dx.doi.org/10.3389/neuro.11.006.2008.

106. Mattioni M, Le Novère N. Integration of biochemical and electrical signaling-multiscale model of the medium spiny neuron of the striatum. *PLoS One*. 2013;8(7): e66811. http://dx.doi.org/10.1371/journal.pone.0066811.

107. Dräger A, Hassis N, Supper J, Schröder A, Zell A. SBMLsqueezer: a Cell Designer plug-in to generate kinetic rate equations for biochemical networks. *BMC Syst Biol*. 2008;2:39. http://dx.doi.org/10.1186/1752-0509-2-39.

108. Espino DM, Shepherd DET, Hukins DWL. Evaluation of a transient, simultaneous, arbitrary Lagrange-Euler based multi-physics method for simulating the mitral heart valve. *Comput Methods Biomech Biomed Engin*. 2012;17(4):450–458. http://dx.doi.org/10.1080/10255842.2012.688818.

109. Sugiura S, Washio T, Hatano A, Okada J, Watanabe H, Hisada T. Multi-scale simulations of cardiac electrophysiology and mechanics using the University of Tokyo heart simulator. *Prog Biophys Mol Biol*. 2012;110(2–3):380–389. http://dx.doi.org/10.1016/j.pbiomolbio.2012.07.001.

110. Schaff JC, Slepchenko BM, Loew LM. Physiological modeling with virtual cell framework. *Methods Enzymol*. 2000;321:1–23.

111. Bower JM, Beeman D. *The book of GENESIS*. Santa Clara: TELOS, Springer-Verlag; 1995.

112. Brown S-A, Moraru II, Schaff JC, Loew LM. Virtual NEURON: a strategy for merged biochemical and electrophysiological modeling. *J Comput Neurosci*. 2011;31(2): 385–400. http://dx.doi.org/10.1007/s10827-011-0317-0.

113. Andrews SS. Spatial and stochastic cellular modeling with the Smoldyn simulator. *Methods Mol Biol*. 2012;804:519–542. http://dx.doi.org/10.1007/978-1-61779-361-5_26.

114. Gleeson P, Crook S, Cannon RC, et al. NeuroML: a language for describing data driven models of neurons and networks with a high degree of biological detail. *PLoS Comput Biol*. 2010;6(6):e1000815. http://dx.doi.org/10.1371/journal.pcbi.1000815.

115. Cannon RC, Gewaltig M-O, Gleeson P, et al. Interoperability of neuroscience modeling software: current status and future directions. *Neuroinformatics*. 2007;5(2):127–138.

116. Davison AP, Bruderle D, Eppler J, et al. PyNN: a common interface for neuronal network simulators. *Front Neuroinform*. 2009;2:11. http://dx.doi.org/10.3389/neuro.11.011.2008.

117. Hucka M, Finney A, Sauro HM, et al. The systems biology markup language (SBML): a medium for representation and exchange of biochemical network models. *Bioinformatics*. 2003;19(4):524–531.

118. Waltemath D, Adams R, Bergmann FT, et al. Reproducible computational biology experiments with SED-ML–the Simulation Experiment Description Markup Language. *BMC Syst Biol*. 2011;5:198. http://dx.doi.org/10.1186/1752-0509-5-198.

119. Bhalla US, Bilitch DH, Bower JM. Rallpacks: a set of benchmarks for neuronal simulators. *Trends Neurosci*. 1992;15(11):453–458.

INDEX

Note: Page numbers followed by "*f*" indicate figures and "*t*" indicate tables.

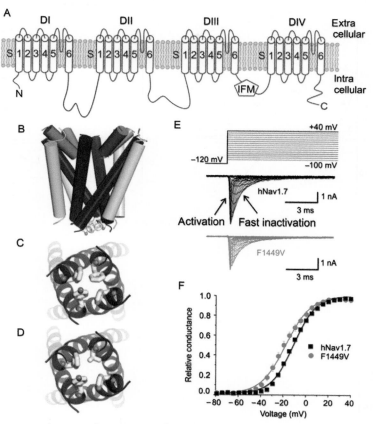

A

DI   DII   DIII   DIV   Extra
cellular

S 1 2 3 4 5 6   S 1 2 3 4 5 6   S 1 2 3 4 5 6   S 1 2 3 4 5 6

N   IFM   Intra
cellular

                                       C

B

C

D

E

+40 mV

−120 mV   −100 mV

hNav1.7   1 nA

3 ms

Activation   Fast inactivation

F1449V   1 nA

3 ms

F

Relative conductance

1.0
0.8
0.6
0.4
0.2
0.0

−80 −60 −40 −20 0 20 40
Voltage (mV)

■ hNav1.7
● F1449V

ANGELIKA LAMPERT AND ALON KORNGREEN, FIGURE 1.1

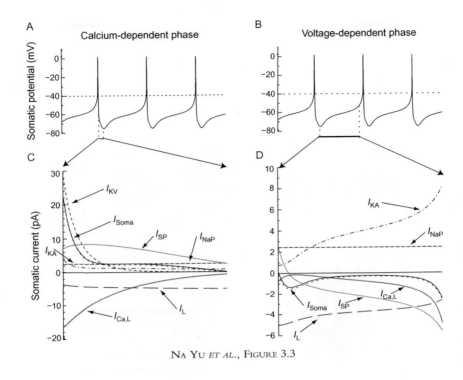

NA YU ET AL., FIGURE 3.3

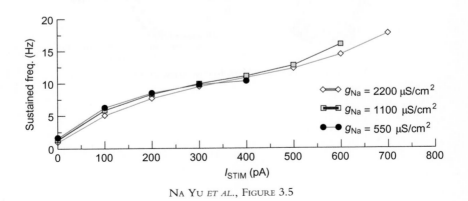

NA YU ET AL., FIGURE 3.5

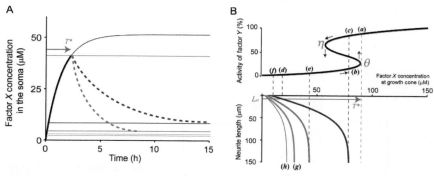

Honda Naoki and Shin Ishii, Figure 6.4

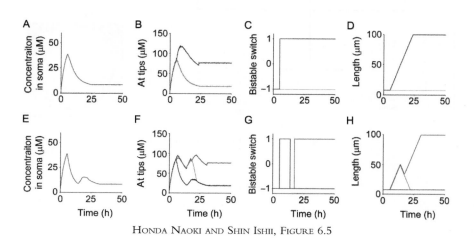

Honda Naoki and Shin Ishii, Figure 6.5

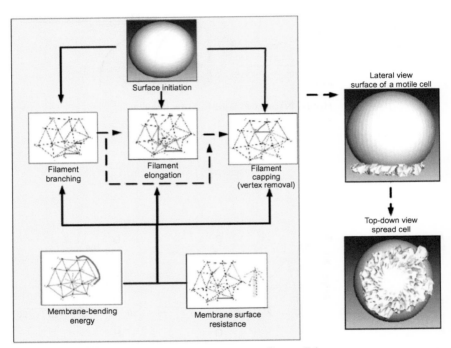

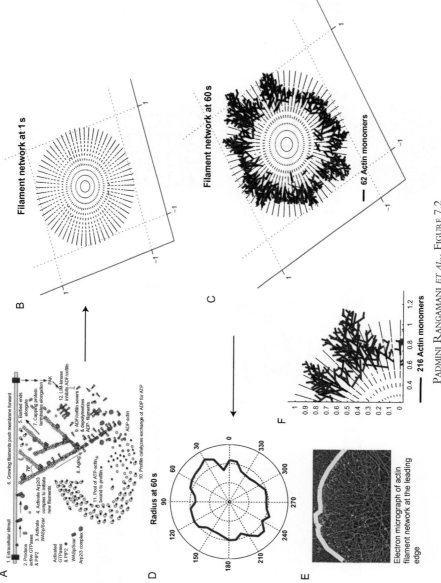

B Filament network at 1 s

C Filament network at 60 s

— 62 Actin monomers

A

1. Extracellular stimuli

6. Growing filaments push membrane forward

5. Barbed ends elongate

7. Capping protein terminates elongation

12. LIM-kinase inhibits ADF/cofilin

PAK

2. Produce active GTPases & PIP2

3. Activate WASp/Scar

4. Activate Arp2/3 complex to initiate new filaments

9. ADF/cofilin severs & depolymerizes ADP-filaments

8. Aging

ADP-actin

Activated GTPases & PIP2

WASp/Scar

Arp2/3 complex

10. Profilin catalyzes exchange of ADP for ATP

11. Pool of ATP-actin bound to profilin

D Radius at 60 s

0
30
60
90
120
150
180
210
240
270
300
330

E

Electron micrograph of actin filament network at the leading edge

F

1
0.9
0.8
0.7
0.6
0.5
0.4
0.3
0.2
0.1
0

0.4 0.6 0.8 1 1.2

— 216 Actin monomers

PADMINI RANGAMANI *ET AL.*, FIGURE 7.2

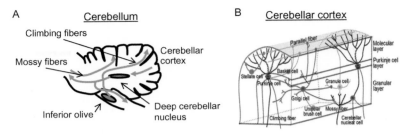

TOMA M. MARINOV AND FIDEL SANTAMARIA, FIGURE 8.1

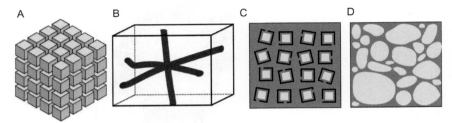

TOMA M. MARINOV AND FIDEL SANTAMARIA, FIGURE 8.9

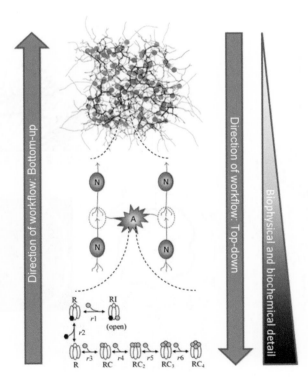

MARJA-LEENA LINNE AND TUULA O. JALONEN, FIGURE 9.1

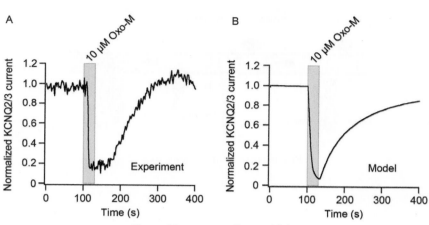

BERTIL HILLE *ET AL.*, FIGURE 10.1

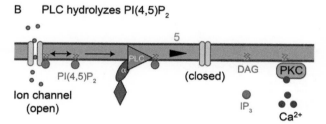

BERTIL HILLE *ET AL.*, FIGURE 10.2

Measuring Gα_q dissociation, PLC activation

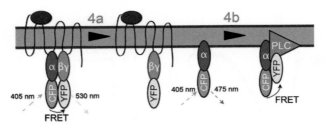

Bertil Hille *et al.*, Figure 10.3

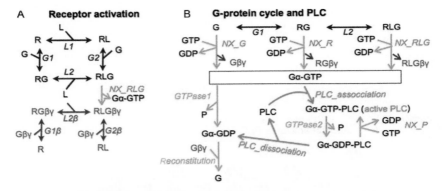

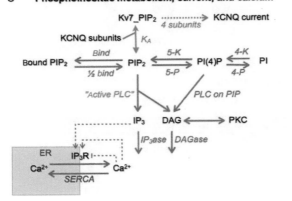

Bertil Hille *et al.*, Figure 10.4

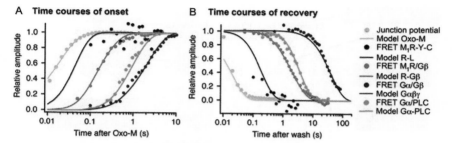

A Time courses of onset

B Time courses of recovery

Junction potential
Model Oxo-M
FRET M₁R-Y-C
Model R-L
FRET M₁R/Gβ
Model R-Gβ
FRET Gα/Gβ
Model Gαβγ
FRET Gα/PLC
Model Gα-PLC

C Concentration–response curves for high receptor density

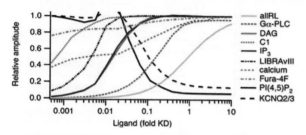

allRL
Gα-PLC
DAG
C1
IP₃
LIBRAvIII
calcium
Fura-4F
PI(4,5)P₂
KCNQ2/3

D Concentration–response curves for low receptor density

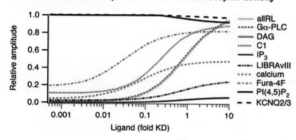

allRL
Gα-PLC
DAG
C1
IP₃
LIBRAvIII
calcium
Fura-4F
PI(4,5)P₂
KCNQ2/3

BERTIL HILLE *ET AL.*, FIGURE 10.5

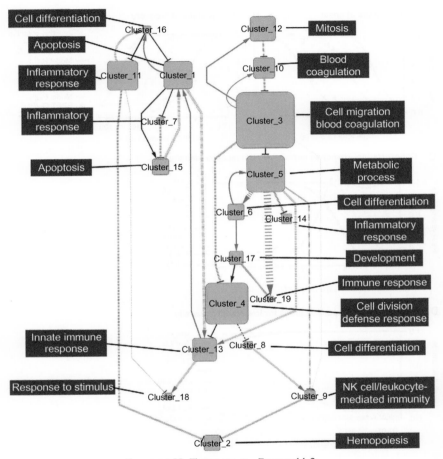

ZACHARY H. TAXIN *ET AL.*, FIGURE 11.2

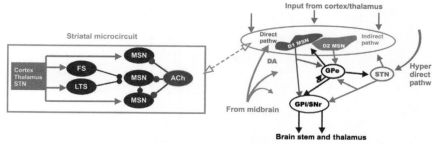

Anu G. Nair *et al.*, Figure 12.1

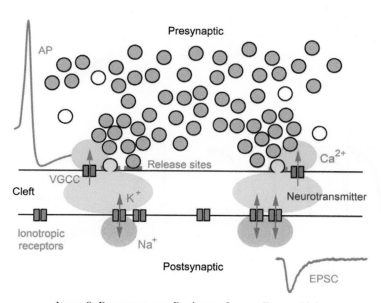

Jason S. Rothman and R. Angus Silver, Figure 13.1

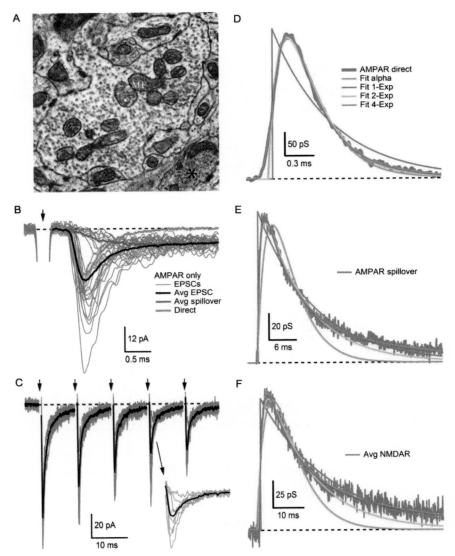

JASON S. ROTHMAN AND R. ANGUS SILVER, FIGURE 13.2

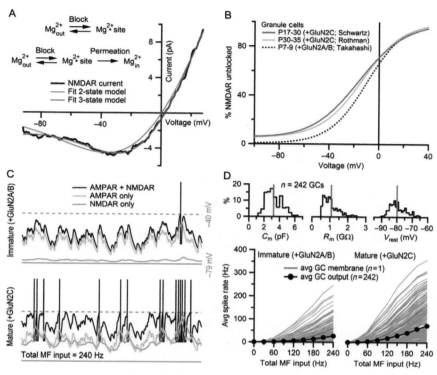

JASON S. ROTHMAN AND R. ANGUS SILVER, FIGURE 13.3

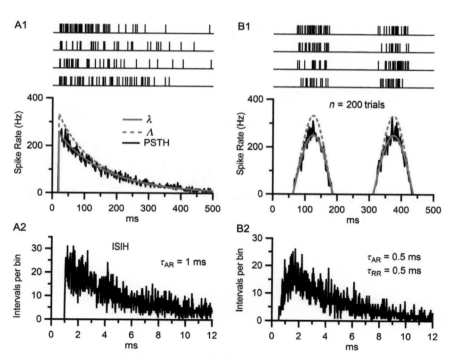

JASON S. ROTHMAN AND R. ANGUS SILVER, FIGURE 13.4

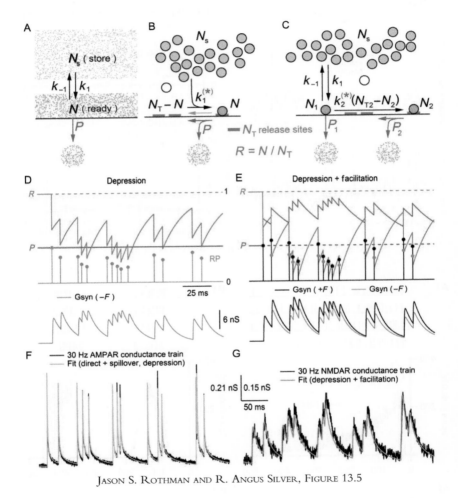

JASON S. ROTHMAN AND R. ANGUS SILVER, FIGURE 13.5

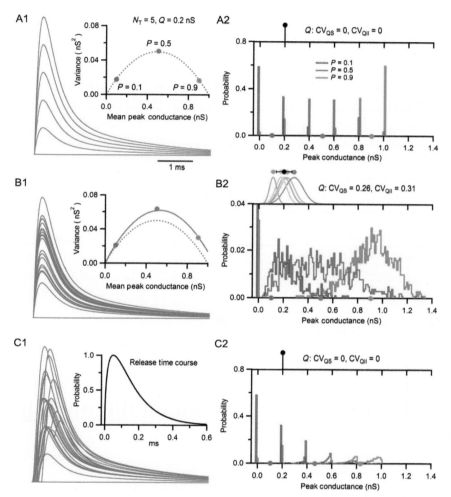

JASON S. ROTHMAN AND R. ANGUS SILVER, FIGURE 13.6

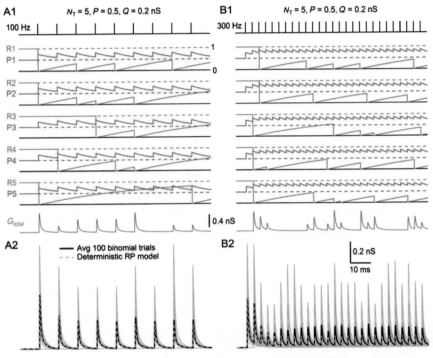

A1 $N_T = 5, P = 0.5, Q = 0.2$ nS

100 Hz

R1
P1

R2
P2

R3
P3

R4
P4

R5
P5

G_{total} 0.4 nS

A2
— Avg 100 binomial trials
--- Deterministic RP model

B1 $N_T = 5, P = 0.5, Q = 0.2$ nS

300 Hz

B2
0.2 nS
10 ms

JASON S. ROTHMAN AND R. ANGUS SILVER, FIGURE 13.7

Printed and bound by CPI Group (UK) Ltd, Croydon, CR0 4YY

08/05/2025

01864957-0003